D1163158

PESTICIDES IN AGRICULTURE AND THE ENVIRONMENT

BOOKS IN SOILS, PLANTS, AND THE ENVIRONMENT

Soil Biochemistry, Volume 1, edited by A. D. McLaren and G. H. Peterson
Soil Biochemistry, Volume 2, edited by A. D. McLaren and J. Skujiņš
Soil Biochemistry, Volume 3, edited by E. A. Paul and A. D. McLaren
Soil Biochemistry, Volume 4, edited by E. A. Paul and A. D. McLaren
Soil Biochemistry, Volume 5, edited by E. A. Paul and J. N. Ladd
Soil Biochemistry, Volume 6, edited by Jean-Marc Bollag and G. Stotzky
Soil Biochemistry, Volume 7, edited by G. Stotzky and Jean-Marc Bollag
Soil Biochemistry, Volume 8, edited by Jean-Marc Bollag and G. Stotzky
Soil Biochemistry, Volume 9, edited by G. Stotzky and Jean-Marc Bollag
Soil Biochemistry, Volume 10, edited by Jean-Marc Bollag and G. Stotzky

Organic Chemicals in the Soil Environment, Volumes 1 and 2, edited by C. A. I. Goring and J. W. Hamaker
Humic Substances in the Environment, M. Schnitzer and S. U. Khan
Microbial Life in the Soil: An Introduction, T. Hattori
Principles of Soil Chemistry, Kim H. Tan
Soil Analysis: Instrumental Techniques and Related Procedures, edited by Keith A. Smith
Soil Reclamation Processes: Microbiological Analyses and Applications, edited by Robert L. Tate III and Donald A. Klein
Symbiotic Nitrogen Fixation Technology, edited by Gerald H. Elkan
Soil–Water Interactions: Mechanisms and Applications, Shingo Iwata and Toshio Tabuchi with Benno P. Warkentin

Soil Analysis: Modern Instrumental Techniques, Second Edition, edited by Keith A. Smith
Soil Analysis: Physical Methods, edited by Keith A. Smith and Chris E. Mullins
Growth and Mineral Nutrition of Field Crops, N. K. Fageria, V. C. Baligar, and Charles Allan Jones
Semiarid Lands and Deserts: Soil Resource and Reclamation, edited by J. Skujiņš
Plant Roots: The Hidden Half, edited by Yoav Waisel, Amram Eshel, and Uzi Kafkafi
Plant Biochemical Regulators, edited by Harold W. Gausman
Maximizing Crop Yields, N. K. Fageria
Transgenic Plants: Fundamentals and Applications, edited by Andrew Hiatt
Soil Microbial Ecology: Applications in Agricultural and Environmental Management, edited by F. Blaine Metting, Jr.
Principles of Soil Chemistry: Second Edition, Kim H. Tan
Water Flow in Soils, edited by Tsuyoshi Miyazaki
Handbook of Plant and Crop Stress, edited by Mohammad Pessarakli
Genetic Improvement of Field Crops, edited by Gustavo A. Slafer
Agricultural Field Experiments: Design and Analysis, Roger G. Petersen
Environmental Soil Science, Kim H. Tan
Mechanisms of Plant Growth and Improved Productivity: Modern Approaches, edited by Amarjit S. Basra
Selenium in the Environment, edited by W. T. Frankenberger, Jr., and Sally Benson
Plant–Environment Interactions, edited by Robert E. Wilkinson
Handbook of Plant and Crop Physiology, edited by Mohammad Pessarakli
Handbook of Phytoalexin Metabolism and Action, edited by M. Daniel and R. P. Purkayastha
Soil–Water Interactions: Mechanisms and Applications, Second Edition, Revised and Expanded, Shingo Iwata, Toshio Tabuchi, and Benno P. Warkentin
Stored-Grain Ecosystems, edited by Digvir S. Jayas, Noel D. G. White, and William E. Muir
Agrochemicals from Natural Products, edited by C. R. A. Godfrey
Seed Development and Germination, edited by Jaime Kigel and Gad Galili
Nitrogen Fertilization in the Environment, edited by Peter Edward Bacon
Phytohormones in Soils: Microbial Production and Function, William T. Frankenberger, Jr., and Muhammad Arshad
Handbook of Weed Management Systems, edited by Albert E. Smith
Soil Sampling, Preparation, and Analysis, Kim H. Tan
Soil Erosion, Conservation, and Rehabilitation, edited by Menachem Agassi
Plant Roots: The Hidden Half, Second Edition, Revised and Expanded, edited by Yoav Waisel, Amram Eshel, and Uzi Kafkafi
Photoassimilate Distribution in Plants and Crops: Source–Sink Relationships, edited by Eli Zamski and Arthur A. Schaffer
Mass Spectrometry of Soils, edited by Thomas W. Boutton and Shinichi Yamasaki

Handbook of Photosynthesis, edited by Mohammad Pessarakli
Chemical and Isotopic Groundwater Hydrology: The Applied Approach, Second Edition, Revised and Expanded, Emanuel Mazor
Fauna in Soil Ecosystems: Recycling Processes, Nutrient Fluxes, and Agricultural Production, edited by Gero Benckiser
Soil and Plant Analysis in Sustainable Agriculture and Environment, edited by Teresa Hood and J. Benton Jones, Jr.
Seeds Handbook: Biology, Production, Processing, and Storage, B. B. Desai, P. M. Kotecha, and D. K. Salunkhe
Modern Soil Microbiology, edited by J. D. van Elsas, J. T. Trevors, and E. M. H. Wellington
Growth and Mineral Nutrition of Field Crops: Second Edition, N. K. Fageria, V. C. Baligar, and Charles Allan Jones
Fungal Pathogenesis in Plants and Crops: Molecular Biology and Host Defense Mechanisms, P. Vidhyasekaran
Plant Pathogen Detection and Disease Diagnosis, P. Narayanasamy
Agricultural Systems Modeling and Simulation, edited by Robert M. Peart and R. Bruce Curry
Agricultural Biotechnology, edited by Arie Altman
Plant–Microbe Interactions and Biological Control, edited by Greg J. Boland and L. David Kuykendall
Handbook of Soil Conditioners: Substances That Enhance the Physical Properties of Soil, edited by Arthur Wallace and Richard E. Terry
Environmental Chemistry of Selenium, edited by William T. Frankenberger, Jr., and Richard A. Engberg
Principles of Soil Chemistry: Third Edition, Revised and Expanded, Kim H. Tan
Sulfur in the Environment, edited by Douglas G. Maynard
Soil–Machine Interactions: A Finite Element Perspective, edited by Jie Shen and Radhey Lal Kushwaha
Mycotoxins in Agriculture and Food Safety, edited by Kaushal K. Sinha and Deepak Bhatnagar
Plant Amino Acids: Biochemistry and Biotechnology, edited by Bijay K. Singh
Handbook of Functional Plant Ecology, edited by Francisco I. Pugnaire and Fernando Valladares
Handbook of Plant and Crop Stress: Second Edition, Revised and Expanded, edited by Mohammad Pessarakli
Plant Responses to Environmental Stresses: From Phytohormones to Genome Reorganization, edited by H. R. Lerner
Handbook of Pest Management, edited by John R. Ruberson
Environmental Soil Science: Second Edition, Revised and Expanded, Kim H. Tan
Microbial Endophytes, edited by Charles W. Bacon and James F. White, Jr.
Plant–Environment Interactions: Second Edition, edited by Robert E. Wilkinson
Microbial Pest Control, Sushil K. Khetan
Soil and Environmental Analysis: Physical Methods, Second Edition, Revised and Expanded, edited by Keith A. Smith and Chris E. Mullins

The Rhizosphere: Biochemistry and Organic Substances at the Soil–Plant Interface, Roberto Pinton, Zeno Varanini, and Paolo Nannipieri
Woody Plants and Woody Plant Management: Ecology, Safety, and Environmental Impact, Rodney W. Bovey
Metals in the Environment: Analysis by Biodiversity, M. N. V. Prasad
Plant Pathogen Detection and Disease Diagnosis: Second Edition, Revised and Expanded, P. Narayanasamy
Handbook of Plant and Crop Physiology: Second Edition, Revised and Expanded, edited by Mohammad Pessarakli
Environmental Chemistry of Arsenic, edited by William T. Frankenberger, Jr.
Enzymes in the Environment: Activity, Ecology, and Applications, edited by Richard G. Burns and Richard P. Dick
Plant Roots: The Hidden Half, Third Edition, Revised and Expanded, edited by Yoav Waisel, Amram Eshel, and Uzi Kafkafi
Handbook of Plant Growth: pH as the Master Variable, edited by Zdenko Rengel
Biological Control of Crop Diseases, edited by Samuel S. Gnanamanickam
Pesticides in Agriculture and the Environment, edited by Willis B. Wheeler
Mathematical Models of Crop Growth and Yield, Allen R. Overman and Richard V. Scholtz III
Plant Biotechnology and Transgenic Plants, edited by Kirsi-Marja Oksman-Caldentey and Wolfgang H. Barz

Additional Volumes in Preparation

Handbook of Postharvest Technology, edited by A. Chakraverty, Arun S. Mujumdar, G. S. V. Raghavan, and H. S. Ramaswamy
Handbook of Soil Acidity, edited by Zdenko Rengel

PESTICIDES IN AGRICULTURE AND THE ENVIRONMENT

EDITED BY

WILLIS B. WHEELER
University of Florida
Gainesville, Florida, U.S.A.

MARCEL DEKKER, INC. NEW YORK • BASEL

ISBN: 0-8247-0809-1

This book is printed on acid-free paper.

Headquarters
Marcel Dekker, Inc.
270 Madison Avenue, New York, NY 10016
tel: 212-696-9000; fax: 212-685-4540

Eastern Hemisphere Distribution
Marcel Dekker AG
Hutgasse 4, Postfach 812, CH-4001 Basel, Switzerland
tel: 41-61-260-6300; fax: 41-61-260-6333

World Wide Web
http://www.dekker.com

The publisher offers discounts on this book when ordered in bulk quantities. For more information, write to Special Sales/Professional Marketing at the headquarters address above.

Current printing (last digit):
10 9 8 7 6 5 4 3 2 1

PRINTED IN THE UNITED STATES OF AMERICA

Preface

This volume is designed to fill the niche established in the early 1970s by *Pesticides in the Environment*, edited by Robert White-Stephens, at the time a member of the Rutgers University faculty. The three-volume work represented a state-of-the-art description of the field of pesticides in a different time and different place.

The arena of pest management has changed dramatically in the past 30-plus years. *Pesticides in Agriculture and the Environment* is designed to summarize the state of the various aspects of pest management, some of which did not exist a generation ago and all of which have changed dramatically. It does not focus on the chemistry of the various pest management tactics as did White-Stephens's book. The present volume describes the current status of pesticide issues and those related to the broader topic of pest management. It discusses integrated pest management (IPM) and how it came to be, the current state of risk assessment, biological control techniques, the economics of pest management and pest management legislation, and the current state of analytical methods used by international regulators and offers a state-of-the-art description of the science of environmental fate. It also presents specific issues for pest management on "minor crops," the current approach and issues related to chemical application technology, the important issues of resistance of pests to pesticides and management of that resistance, and, finally, a look to the future for both pest management chemistry and the state of the pest management industry. The authors of these chapters represent the best expertise in the field.

The enactment of the Food Quality Protection Act (FQPA) of 1996 has had a major impact on contemporary pest management regulation. Its far-reaching consequences are discussed in essentially every chapter. Owing to its importance, I summarize a number of its provisions in the following paragraphs.

The FQPA of 1996 amended the Federal Insecticide, Fungicide, and Rodenticide Act (FIFRA) and the Federal Food, Drug, and Cosmetic Act (FFDCA). These amendments fundamentally changed the way the U.S. Environmental Protection Agency (USEPA) regulates pesticides. The requirements include a new safety standard—reasonable certainty of no harm—that must be applied to all pesticides used on foods. The FQPA was designed to resolve the Delaney Paradox, protect children from pesticides, and address endocrine disruption. To accomplish these goals, the law provides that:

- The USEPA is to reregister pesticides every 15 years using the best available data.
- There is a specific definition of minor (use) crops: The definition includes crops grown on fewer than 300,000 acres *or* a minor use may be defined on an economic basis if the pesticide use on a crop is very limited. It may also be defined as minor if the pesticide use is the only alternative, or if it is safer than other alternatives, or if it is needed for IPM and resistance management. The FQPA also provided incentives to develop and maintain minor uses, and to implement a faster approval of reduced-risk pesticides and those used on minor crops.
- The zero-tolerance standard for certain pesticides in processed foods be eliminated (the old Delaney Clause) and that we establish new standards for setting tolerances in both fresh and processed foods.
- Tolerances (maximum residue value) must be safe, i.e., "provide a reasonable certainty that no harm will result from aggregate exposure." All tolerances must be reviewed by 2006, and the most toxic materials must be reviewed first.
- Risks from pesticides must be based on exposures to all chemicals that have a common mode of toxicity. In the past, exposure was based on pesticides in food only. Now all exposures must be considered: dietary, water, and household.
- Safety factors formerly included intra- and interspecies variation (ranging from 100- to 1000-fold); now safety factors must also include factors for infants and children. Thus additional safety factors can give a

1000–10,000-fold safety factor. To implement evaluation of the safety factor for infants and children, the USEPA has looked at the foods that make up large percentages of the diets of infants and children, including apples, peaches, soybeans, pears, and carrots.

- Endocrine disruptors are compounds that mimic or block the effects of hormones, such as estrogen, or act on the endocrine system and may cause developmental or reproductive problems. These must be considered when registering a pesticide.

Pesticides in Agriculture and the Environment discusses issues that are essential components of the contemporary pest management arena. The chapter topics include:

Chapter 1: A description of the major policy considerations that have shaped federal IPM programs over the past three decades.

Chapter 2: A description of the approaches to nonchemical pest management; discussions of definitions of biological control, benefits and limitations, and its ecological basis.

Chapter 3: An in-depth discussion of major pesticide use trends in the United States; the effects of such factors as pesticide productivity, farm programs, and pesticide regulations on use; and changing law and policy.

Chapter 4: An introduction to pesticide safety and the framework of health risk assessment, specifically pesticide risk assessment and ecological risk assessment.

Chapter 5: A description of the processes of transport and fate of pesticides in the environment. It examines dissipation, leaching, and degradation and models for predicting these processes.

Chapter 6: A discussion of the analytical process as it is practiced in the regulatory arena, including approaches to monitoring the food supply in many countries around the world.

Chapter 7: The issues of pest management related specifically to low-acreage, high-value crops. There are economic and other issues for pesticide manufacturers and producers of minor crops.

Chapter 8: A discussion of the importance of pesticide resistance for pest management in agriculture and human health protection and description of a publicly available resistance database.

Chapter 9: A review of efforts to increase pesticide applicator safety and to improve the efficacy and effectiveness of the application techniques.

Chapter 10: An analysis of the current state of the crop protection industry and a projection of the future. The discussion includes company mergers and acquisitions, generic pesticide producers, seed companies, new chemistries of pesticides, plant biotechnology, and major trends in the industry.

It is my hope that readers will find this book an informative reference on pest management in the modern world.

Willis B. Wheeler

Contents

Contributors

Jerry J. Baron, Ph.D. IR-4 Project, The Technology Centre of New Jersey, Rutgers University, North Brunswick, New Jersey, U.S.A.

Patrick S. Bills, B.S. Department of Entomology and Center for Integrated Plant Systems, Michigan State University, East Lansing, Michigan, U.S.A.

S. Chandramohan, Ph.D. Department of Plant Pathology, University of Florida, Gainesville, Florida, U.S.A.

Raghavan Charudattan, Ph.D. Department of Plant Pathology, University of Florida, Gainesville, Florida, U.S.A.

Michael S. Fitzner, Ph.D. Cooperative State Research, Education, and Extension Service, U.S. Department of Agriculture, Washington, D.C., U.S.A.

Richard T. Guest, Ph.D.* IR-4 Project, The Technology Centre of New Jersey, Rutgers University, North Brunswick, New Jersey, U.S.A.

Robert E. Holm, Ph.D. IR-4 Project, The Technology Centre of New Jersey, Rutgers University, North Brunswick, New Jersey, U.S.A.

S. Mark Lee, Ph.D. Center for Analytical Chemistry, California Department of Food and Agriculture, Sacramento, California, U.S.A.

David Mota-Sanchez, M.S. Department of Entomology and Center for Integrated Plant Systems, Michigan State University, East Lansing, Michigan, U.S.A.

Craig D. Osteen, Ph.D. Resource Economics Division, Economic Research Service, U.S. Department of Agriculture, Washington, D.C., U.S.A.

Merritt Padgitt, Ph.D. Resource Economics Division, Economic Research Service, U.S. Department of Agriculture, Washington, D.C., U.S.A.

Nu-may Ruby Reed, Ph.D. Department of Pesticide Regulation, California Environmental Protection Agency, Sacramento, California, U.S.A.

Sylvia J. Richman, Ph.D. Center for Analytical Chemistry, California Department of Food and Agriculture, Sacramento, California, U.S.A.

Paul H. Schwartz, Ph.D. Agricultural Research Service, U.S. Department of Agriculture, Beltsville, Maryland, U.S.A.

James N. Seiber, Ph.D. Western Regional Research Center, Agricultural Research Service, U.S. Department of Agriculture, Albany, California, U.S.A.

Mark E. Whalon, Ph.D. Department of Entomology and Center for Integrated Plant Systems, Michigan State University, East Lansing, Michigan, U.S.A.

Robert E. Wolf, Ph.D. Department of Biological and Agricultural Engineering, Kansas State University, Manhattan, Kansas, U.S.A.

Gabriela S. Wyss, Ph.D. Division of Plant Protection, Research Institute of Organic Agriculture, Frick, Switzerland

* Retired.

PESTICIDES IN AGRICULTURE AND THE ENVIRONMENT

1

Three Decades of Federal Integrated Pest Management Policy

Michael S. Fitzner
Cooperative State Research, Education, and Extension Service
U.S. Department of Agriculture
Washington, D.C., U.S.A.

1 INTRODUCTION

The scientific and technical development of integrated pest management* (IPM) methods during the twentieth century were covered by several publications of the 1990s [1–4], but the policy aspects of IPM have received less attention. This is unfortunate, because policy and politics have been as much a part of the history of IPM in the United States as the science. This chapter provides a summary of the major policy considerations that shaped federal IPM programs over the last three decades of the century.

A great deal of discussion preceded the first major allocation of federal funds for IPM programs in 1972. A review of policy documents from this period provides a fascinating look at a national debate regarding the hazards of pesticide use. Then, as today, policy makers and technical experts struggled over the trade-offs between agricultural productivity and environmental impacts. Perhaps Dr. Gordon Guyer, a professor of entomology at Michigan State University, best

* The terminology has evolved over time, but the basic concepts have remained fairly constant. For consistency, this chapter uses the term "IPM" in most cases.

summed up the dilemma during his testimony before Congress in 1971: "Whereas chemicals have allowed for the greatest agricultural production in history and made major contributions to world health programs, they have also contaminated the environment" [5]. Policy discussions during the early 1970s conveyed a sense of urgency in dealing with serious environmental impacts of the use of pesticides but never lost sight of the importance of maintaining agricultural productivity and profitability.

In recent years, the sense of purpose that underlays the policy discussions of the early 1970s appears to have been replaced by debates on whether IPM programs have been true to concept or to the goals established in the early 1970s [1–3,5–7]. This chapter traces the evolution of federal IPM policy over the past three decades in an attempt to understand the goals established for federal IPM programs. Considerable attention is given to the early 1970s, when policy objectives for federal IPM efforts were first articulated. Perhaps by better understanding the evolution of federal IPM policy we will be better prepared to guide IPM programs in the decades to come.

2 A CALL TO ACTION

The late 1960s and early 1970s were pivotal in the evolution of the policies that still serve as the basis for federal IPM programs. There were several reasons for the attention given to pest management issues at this time. Public concerns about the environmental effects of pesticides were at a high level, heightened by the publication of *Silent Spring* [8] in 1962 and other emerging evidence concerning the environmental impacts of highly persistent chlorinated hydrocarbons such as DDT, Dieldrin, Aldrin, and Mirex. President Nixon reflected public concerns about pesticides in his Environmental Message of 1971 [9]:

> Pesticides have provided important benefits by protecting man from disease and increasing his ability to produce food and fiber. However, the use and misuse of pesticides has become one of the major concerns of all who are interested in a better environment. The decline in numbers of several of our bird species is a signal of the potential hazards of pesticides to the environment. We are continuing a major research effort to develop nonchemical methods of pest control, but we must continue to rely on pesticides for the foreseeable future. The challenge is to institute the necessary mechanisms to prevent pesticides from harming human health and the environment.

Concerns about pesticides were at least as strong on America's farms and ranches as they were in other communities. After all, the environmental damage and health effects attributed to pesticides were more likely to affect those who worked and lived on farms than the rest of the population. But of even greater importance to this discussion, in the early 1970s agricultural producers were struggling with the loss or increased cost of their "old standby" pesticides as a result of pest resistance and greater scrutiny of their persistence, biomagnification, and toxicity to nontarget organisms [10].

In the late 1960s, concerns about pesticides took center stage in the federal policy arena. The federal government's first step in addressing the pesticide problem came in the form of the National Environmental Policy Act (NEPA) of 1969. NEPA established [11]

> a national policy which will encourage productive and enjoyable harmony between man and his environment; to promote efforts which will prevent or eliminate damage to the environment and biosphere and stimulate the health and welfare of man; to enrich the understanding of the ecological systems and natural resources to the Nation; and to establish a Council on Environmental Quality.

Congress established the Council on Environmental Quality (CEQ) to promote "the advancement of scientific knowledge of the effects of actions and technology on the environment and [to encourage] the development of the means to prevent or reduce adverse effects that endanger the health and well-being of man."

The establishment of the CEQ proved to be a key event in the development of federal IPM policy. Soon after its formation, CEQ used the legal authority and rationale provided by NEPA to recommend that the federal government support the development and promotion of IPM programs. These recommendations were backed by a variety of governmental agencies, university researchers, industry representatives, and public interest groups who called for a concerted effort to develop and implement IPM methods nationwide. The basis for the CEQ's recommendation was solid: Published research results provided strong evidence that IPM methods worked. Scientists and research administrators had been advocating for federal funding to develop interdisciplinary systems approaches to pest management for several years, and a group of scientists from 18 universities had developed a proposal for a large interdisciplinary project; this project, later known as the Huffaker Project, proved to be a major stimulus for the development of federal IPM policy.

The case for increased federal support for IPM was further strengthened by the results from IPM "pilot" projects designed to refine, test, and evaluate available technology on crops where "intensive chemical pest control is presently practiced" [12]. The objective of the pilot projects was to "limit the use of pesticides to situations in which they are needed to prevent economic damage to a crop. This will not only result in savings in cost of production, but will also reduce the overall amount of pesticide being added to the environment" [12].

The development of federal IPM policy took a major step forward on September 20, 1971, when a subcommittee of the U.S. Senate's Committee on Agriculture and Forestry began a two-day hearing on Senate Bill 1794, "A bill to authorize pilot field-research programs for the control of agricultural and forest pests by integrated biological-cultural methods" [5]. The legislation proposed to direct the Secretary of Agriculture to conduct pilot field-research programs to (1) develop and test biological-cultural methods for the control of agricultural and forest pests, (2) determine the economic and environmental consequences of implementing multidisciplinary and integrated biological-cultural methods, and (3) develop methods for collecting and interpreting data obtained from the pilot research programs. The legislation also proposed to authorize the Secretary of Agriculture to reimburse farmers and ranchers for any losses resulting from their participation in the pilot research program. The bill authorized the appropriation of $2 million per year for up to five years to the U.S. Department of Agriculture (USDA) for this effort, plus $2 million per year for up to five years to the National Science Foundation (NSF) to expand its fundamental research on integrated biological-cultural principles and techniques to control agricultural and forest pests.

The Congressional hearings on Senate Bill 1794, which were titled "Pest Control Research," were a crucial step in the development of federal IPM policy. A total of 35 witnesses—senators, government officials, farmer representatives, environmentalists, and university researchers—provided 174 pages of testimony during the hearings. Together, the witnesses represented one of the most experienced and knowledgeable panels ever assembled to discuss what was then referred to as "integrated control" (the term "integrated pest management" would become the predominant term after President Nixon used it in his 1972 Environmental Message) [3]. The hearings were introduced by the author of Senate Bill 1794, Senator Gaylord Nelson of Wisconsin, who remarked that the bill had strong bipartisan support from its 24 cosponsors. Senator Nelson said the bill would provide for the establishment of demonstration projects and expanded basic research in the principles of integrated pest control. He remarked that leaders in agriculture and the environmental movement were in agreement on the need to provide "food and fiber for a growing society without depending on broad-spectrum, persistent chemicals to control insect pests" and further stated that "with the single strategy of chemical pest control we have not only saturated the environment with deadly poisons that endanger a wide spectrum of living organisms, including man himself, but we have begun to seriously disrupt the economic stability of the farming community." Nelson then articulated the goal of the proposed effort:

> There is a compelling and urgent need to reconsider our approach to pest control by recognizing a very basic ecological principle. That is, each integral part of the natural system survives in balance with—not at the expense of—the other parts. I believe that integrated control offers

> the alternative that recognizes this principle. Integrated control offers the use of the best-suited combination of alternate pest control methods to suppress pest insects in a given crop situation below the economically disruptive threshold. We are not talking about a unilateral, one-method approach that we have become accustomed to in the application of broad-spectrum chemicals. And we aren't ruling out the use of chemicals in an integrated control program, because some situations may call for selective chemical applications during a particular phase of the overall program. But the use of chemicals—particularly broad-spectrum chemicals—necessarily is very limited in integrated pest control so as not to interfere with other aspects of the program, most notably the use of beneficial insects.

Neslon concluded his statement by saying,

> The idea of a pilot program has been under discussion for several years and it has not happened. I think the real import and the real importance of this matter is that it directs the establishment of a pilot project which would involve various crops in the South, Southwest, Midwest, East, and Far West, so that we can have a genuine, scientific demonstration program to discover what successes we can have and to educate farmers and the country on the effectiveness of a rational use of a scientific integrated program.

The hearings on pest control research represent a guidebook on integrated control that remains relevant to this day and should be considered required reading for anyone involved in pest management policy, research, or implementation. In spite of the large number of witnesses and diversity of organizations represented, all were in agreement on two points: (1) The problems associated with the use of broad-spectrum pesticides had to be addressed, and (2) the programs authorized by S 1794 were greatly needed, but needed to be authorized at a higher level of funding than was provided in the bill. Selected statements* made by several witnesses are provided below to illustrate the thoughts and concerns that helped shape federal IPM policy at this early stage in its development.

2.1 Selected Statements Made by Senators

Senator John Tunney of California, one of the sponsors of S 1794, stated

> Pesticides are most valuable tools when used properly and in the context of the entire ecological system of an environment, but they are not ulti-

* Quotes are true to the published transcript. However, in some cases sentences from separate portions of the testimony are merged together to enhance readability.

> mate solutions to pest control. Their widespread use has brought a number of pressing problems, including pollution caused by toxic chemical residues and the development of insect resistance to such chemicals. We must develop methods that integrate not only chemical and biological control techniques but all other control procedures and agricultural production practices that man has developed through the ages into single systems approaches aimed at profitable production of high-quality products in a manner not inimical to our environment. . . . We must continue to recognize that there will continue to be a role for pesticides in agriculture, but we must also develop integrated pest control techniques that make use of chemicals as only one of a number of tools without disruption of the ecological systems in our agricultural production areas.

Senator Lawton Chiles of Florida, one of the sponsors of S 1794, indicated that pilot field projects would demonstrate integrated biological-cultural methods to facilitate a

> change in our present method of attack of agricultural insects—a change in strategy—a change directed toward helping the farmer, who is bearing the burden of increasing costs of pesticides and yet what he receives for his product seems to be the same; a change aimed at reestablishing the natural ecological balance now being damaged on an appalling scale and rate; a change that would provide the much needed funds and leadership to substantially reduce our single prolonged reliance on pesticides. It is about time we face the fact that pest control practices have been fraught with many grave problems. Ecological disaster can and must be prevented. The farmer can and must be helped to produce a reasonably pest-free crop efficiently and ecologically. This legislation offers the framework and incentive to prevent that disaster, and offers assistance in an expending of energies and funds in a positive direction: research for ways and means of controlling our agricultural pests, using natural predators and parasites of harmful insects in a correct balance. We must seek practical, economically and ecologically feasible alternatives to pesticides. This bill would aid that search.

Senator Chiles also emphasized his belief that the federal government has a responsibility to replace pesticides lost as a result of regulatory action with the following rhetorical question:

> Do you feel Government has a responsibility then, to at a time we say to the farmer, you cannot use DDT or you cannot use one of these pesticides that you have used, that Government owes the responsibility to him to try to give him an alternate method of trying to control the pests, if Government is going to take away his right, restrict his right? For

> Congress to say to the farmer to unilaterally stop using pesticides, without making a sincere effort to help the farmer find alternate means of pest control, is grossly unfair. I feel we owe a responsibility to the farmer by giving him an acceptable viable alternative to the use of pesticides. We had to use more and more pesticides in attempting to control the pests, and therefore to the farmer we tremendously raised the cost and the frequency with which he then had to apply the pesticides, and also increased greatly the resulting harm that happened to the environment because of the tremendous usage.

Senator Allen of Alabama, one of the sponsors of S 1794 and chairman of the subcommittee, stated that

> encouraging the use of the integrated control methods to control insects and pests probably offers the best mechanism for a reduction in the use of pesticides and insecticides that could cause damage to our environment. So rather than having overkill with insecticides and pesticides, under this system, you could use a small amount of the chemical and integrate that with the biological control and in that way get at the problem better than resort to only one method.

Senator Allen further stated, "I am sure the ultimate purpose of it is to provide for the gradual withdrawal of the use of pesticides."

2.2 Selected Statements by Farmer Representatives

Mr. Harry Bell of the National Cotton Council said, "The cotton industry has for years recognized that there should be better alternatives to wide-scale poisoning with broad-spectrum insecticides." But he also was careful to include pesticides in describing the focus of the research effort when he stated that cotton producers support "biological-cultural-chemical approaches." In describing the goals for the effort, Bell said, "In our opinion, the development and use of practical integrated control techniques would reduce our production costs, cut environmental contamination, lessen pesticide residues in cottonseed, avoid or delay the onset of pest resistance to chemicals, and reduce toxicity hazards to people and other animal life."

Mr. B. F. Smith of the Delta Council indicated that his organization supports the development of effective control methods that have a reasonable cost and make limited use of insecticides. He said, "Farmers certainly are willing to give up the use of insecticides and pesticides if acceptable and effective methods can be discovered or developed to combat the pests and insects" and that he believed that this would result in less pollution, less resistance, and better biological balance between insects on different crops so minor pests do not become major pests.

Dr. Weldon Barton of the National Farmers Union said,

> Farmers Union urges effective regulation—in combination with governmental research and educational programs—aimed at proper use of pesticides, herbicides, fertilizer, and other chemicals if they constitute a source of pollution. The development and usage of integrated methods, whereby we attempt to eliminate agricultural pests by working primarily with biological means from within their cultures rather that by applying chemical insecticides "from the outside," is increasingly being recognized as essential . . . for at least two reasons: (1) because the continuous usage of chemical insecticides and pesticides pollutes our water, land, and other natural resources: and (2) because from a strictly economic standpoint, farmers can be the real losers from the continued reliance on nonintegrated chemical applications. For the benefit of farmers as well as the protection of our natural environment, we must develop integrated pest control methods that can help us to move away from this spiral of chemical pesticide usage.

Mr. Charles Frazier of the National Farmers Union urged that

> we not undertake to resolve the future of all chemicals in agricultural production by sweeping, widespread actions based on emotional reactions . . . but rather it would be preferable to approach each of the major insect problems in some realistic and dispassionate manner that would move the control of insects of such economic importance to the nation from chemicals to biological means or to combinations of the methods that may be available.

2.3 Selected Statements by Environmentalists

Mr. William Butler of the Environmental Defense Fund supported S 1794 and emphasized the "intense need for more research on integrated biological-cultural methods of insect and pest control to complement, reduce, and in some instances entirely supplant current overreliance upon chemical controls for pests." He also said that overreliance on pesticides has resulted in resistance, destruction of natural enemies, and environmental harm to nontarget species.

Ms. Linda Billings of the Sierra Club indicated that conservationists support the legislation and reminded the subcommittee that they have long protested the "exorbitant" use of chemical pesticides. She stated that the "reckless use of chemical pesticides has wrought ecological havoc and . . . threatens production of vital food and fiber crops, forest products, and endangers human health." She further stated, "I hope the lessons of the past will not be ignored by those developing new pest control methods and that care will be taken to note and guard against adverse environmental effects."

2.4 Selected Statements by University Researchers

Dr. Carl Huffaker of the University of California stated that

> the long-term interests of the grower and of the environment are both served by a balanced biological and multidisciplinary approach to pest control. The goal of this program is to place pest control on a more scientific basis, wherein the grower can manage his crop pests in a more reliable and predictable manner without need for the extensive use of broadly disturbing toxic chemicals.

Dr. Gordon Guyer of Michigan State University acknowledged the enormous benefits resulting from the use of synthetic pesticides but indicated that undesirable side effects have not been fully assessed:

> Whereas, chemicals have allowed for the greatest agricultural production in history and made major contributions to world health programs, they have also contaminated the environment. It is generally agreed that the use of pesticides should be reduced and only used when and where necessary. However, few effective alternatives have been developed which compare with insecticides as being quick acting, consistently effective, economically feasible, technologically adaptable to grower implementations and applicable to a broad range of crops under diverse environmental conditions. One alternative approach to the unilateral use of pesticides is integrated pest control, which envisions maximum use of nonchemical—biological, cultural, genetic, et cetera—control methods and the minimization of chemical control tactics. This philosophy is advanced . . . as the most practical and realistic alternative for reorienting plant and animal protection practices away from the excessive use of chemicals.

Dr. Charles Lincoln of the University of Arkansas said that

> primary dependence on broad spectrum insecticides, which makes that cheap program possible, is no longer a tenable approach to insect control, however. Resistance of insect pests to insecticides, pollution, and disruption of populations of nontarget species have reached critical levels. We must, therefore, place much more emphasis on biological and cultural methods. In a pilot program, all available methods of cultural and biological control will be brought together to obtain acceptable yields of crops and forest products. Insecticide use will be kept to a minimum, with emphasis on the use of safe, selective insecticides. In a pilot test, it will be necessary to monitor populations of many of these species. . . . A pilot test must, therefore, include several hundred to a

few thousand acres as a minimum, and require a great deal of manpower and instrumentation.

Dr. Perry Adkisson of Texas A&M University spoke of the pesticide resistance problem, saying,

> As insects become resistant to pesticides, the common reaction is to apply more toxic pesticides in greater dosages at shorter intervals. The result is increased production costs, increased hazards to applicators and farm laborers, and increased contamination of the environment. Many of these hazards may be averted by a system of pest management known as integrated control. This system, which brings all known suppression measures to bear . . . offers the greatest promise for keeping our agricultural production viable and environmental contamination by agricultural chemicals at a minimum.

Dr. Robert Van Den Bosch of the University of California at Berkeley said, "Perhaps the greatest attribute of integrated control is . . . it automatically assures a high level of environmental quality. . . . A second major advantage of integrated control is economy, which again derives from its heavy reliance on natural controls and minimal dependence on costly artificial measures." He also spoke of the "ever greater use of pesticides" resulting from pest resistance and cautioned that "there simply is no measure, method, or material which in itself will prove to be a panacea."

Dr. J. Lawrence Apple of North Carolina State University indicated that he was concerned about the focus on nonchemical approaches:

> I am concerned about some comments made in these hearings relative to the use of chemicals. We look upon a pest management system as one that will involve the use of all of the tools at our command in controlling pests. Undoubtedly, this shall continue to involve—and in some cases very heavily—the use of chemicals. We want to minimize the chemical load to the extent possible for several reasons. But for many of the major crops, we cannot foresee the day when we will no longer need chemicals.

He also stated that the tendency of farmers to overuse pesticides is understandable "in that they do not have the guidance that is necessary to make a rational decision as to when to use and when not to use pesticides. That is the type of information we need to supply the farmer." Dr. Apple also encouraged the subcommittee to broaden the scope of the bill to include all pests, not just insects.

Dr. Theo Watson of the University of Arizona cautioned that

> ecological disruptions have taken place which may take several years to correct. The gradual changeover to a truly integrated control program

> will encompass continued use of insecticides in the conventional sense, but with greater care exercised in their selection and use. It will also require greater emphasis on augmentation and conservation of natural enemies and the use of biotic insecticides. Cultural practices which are beneficial to crop production and which adversely affect the pest complex will need to be incorporated in the overall integrated system. The problem remains of how to obtain grower acceptance of this approach which necessarily requires more time and consideration in management decisions but on the other hand ultimately improves production efficiency as well as environmental quality. The reward . . . will be an agriculture aimed not at maximum immediate profit, but rather at optimum sustained production, year after year, with minimum detriment or hazard to nearby food-or-feed-producing enterprises, to agricultural workers, to wildlife, and to the general consumer. Integrated pest control will utilize all available tools, including the discriminate use of pesticides.

Dr. Ernest Bay of the University of Maryland spoke of the likelihood of yield reductions as biological and integrated control methods are developed. He also cautioned against "the widely held but seldom spoken skepticism that the term 'integrated control' is an ecological platitude, and that our only practical reliance can continue to be on strict chemical schedules."

Dr. H. T. Reynolds of the University of California at Riverside stated his hope that pesticide use could eventually be reduced by at least 50% on those crops that rely heavily on pesticides, such as cotton.

An important exchange regarding the use of pesticides in IPM systems occurred between Senator Allen and Dr. Bay:

Sen. Allen: "Even an integrated system would not necessarily eliminate, certainly at the outset, chemical methods of control?"

Dr. Bay: "No. Your chemical methods are entwined with this. The chemical method is absolutely a part of it."

Sen. Allen: "You think with the gains the insects are making even under the integrated method of control, we are going to have to continue using pretty nearly the same amount of chemicals?"

Dr. Bay: "I would like to think not, but maybe we would be at least able to develop a system where we could hold our own. But without the use of integrated control, the use of chemicals will have to be increased with the population increase."

3 FROM POLICY TO PROGRAMS

By the end of the congressional hearings on pest control research, the nature of the pesticide problem and the need for federal support for IPM research and extension had been well established. The momentum created by the hearings had an effect on budget priorities at the USDA and the Office of Management and Budget. On January 1, 1972, Agriculture Secretary Butz announced that funding would be provided for a new pest management action program and an expanded research program [13]. The news release stated that the programs will "help farmers control pests more economically and effectively. At the same time it will reduce the amount of DDT and other chemical pesticides currently being used." The new pest management effort was conducted jointly with the USDA, the NSF, and the USEPA and in cooperation with state departments of agriculture, state agricultural experiment stations, and state extension services.

The foundation for federal IPM policy was prepared by NEPA, the CEQ, the congressional hearings, and agricultural scientists, but if one specific point in time were picked to mark the "ribbon cutting" for federal involvement in IPM, it would be February 8, 1972, when President Nixon transmitted the Environmental Message to Congress [14]. The President's Environmental Message represented the final piece in the IPM puzzle because it signified that the executive branch and Congress were in agreement on the need for a concerted federal IPM effort. In a section of his Environmental Message titled "Making Technology an Environmental Ally," President Nixon announced a comprehensive IPM initiative, including funding for research and development, field testing and demonstrations of new techniques, and the development of training programs for crop consultants. Nixon reflected the heightened environmental awareness of the time when he said, "Our destiny is one: This environmental awakening has taught America in the first years of the seventies. Let us never forget, though, that it is not a destiny of fear, but of promise." Referring to pesticides as an "example of a technological innovation which has provided important benefits to man but which has also produced unintended and unanticipated harm," he declared that "new technologies of integrated pest management must be developed so that agricultural and forest productivity can be maintained together with, rather than at the expense of, environmental quality." He went on to state, "Integrated pest management means judicious use of selective chemical pesticides in combination with nonchemical agents and methods. It seeks to maximize reliance on such natural pest population controls as predators, sterilization, and pest diseases." He announced a plan to

1. Launch a large-scale IPM research and development effort to develop integrated pest management techniques. (USDA, NSF, and the USEPA with leading universities)
2. Increase field testing of promising new methods of pest detection and control and the incorporation of new pest management techniques into existing federal pesticide application programs. (USDA)
3. Develop training and certification programs for crop consultants at ap-

propriate academic institutions. (USDA and the Health, Education, and Welfare Department)

4. Expand the field scout demonstration program to cover 4 million acres by the upcoming growing season. (USDA)

President Nixon's Environmental Message was followed nine months later by a CEQ report on IPM that provided the policy analysis and recommendations that shaped federal IPM policy for the following three decades [10]. The report acknowledged the "dilemma of increasing food production on the one hand and maintaining environmental quality on the other" but cautioned against being "complacent about environmental damages and health threats that can occur from pesticide use, especially when pesticides are used improperly." The report concluded that "the accumulation of pesticides in the food chain, the possible reduction in the populations of some fish and wildlife, and the potential threat to man's health posed by some pesticides have shown the need to seek new methods of pest control to supplement current practices."

Based on an analysis of published research findings and the preliminary results from pilot pest management projects, the CEQ report concluded, "In general, the use of integrated pest management should lead to greatly reduced environmental contamination from pesticide use and to many fewer problems with pest resistance and secondary outbreaks while maintaining or improving our current ability to prevent pest damage." The report went on to state that "pest control can be improved, with reduced environmental impact and often at lower costs to the user." The report also stated that IPM represents an improved method of pest control but does not accomplish this through the elimination of pesticides, which are an important component of IPM programs when used properly and only when needed. Finally, the report indicated that although the evidence of the "overall economic advantage of integrated pest management is still incomplete, it seems reasonably well established for crops such as cotton, apples, and citrus" and predicted that the economic incentive would be smaller for crops using less pesticide.

The CEQ report defined IPM as "an approach that employs a combination of techniques to control the wide variety of potential pests that may threaten crops" and went on to say,

> It involves maximum reliance on natural pest population controls, along with a combination of techniques that may contribute to suppression . . . to affect the potential pests adversely and to aid natural enemies of the pests. Once these preventive measures are taken, the fields are monitored to determine the levels of pests, their natural enemies, and important environmental factors. Only when the threshold level at which significant crop damage from the pest is likely to be exceeded should suppressive measures be taken. If these measures are required, then the

most suitable technique or combination of techniques, such as biological control, use of pest-specific diseases, and even selective use of pesticides, must be chosen to control a pest while causing minimum disruption of its natural enemies.

The report anticipated the multitude of debates and discussions that would ensue in the following decades by stating "the purpose of integrated pest management is not to avoid the use of chemicals but to use the most effective and environmentally sound pest control technique or combination of techniques for long-range pest control. A pest management system is not simply biological control or the use of any single technique."

Finally, after several years of consideration, the federal government was ready to begin implementing its IPM policy. The first significant federal support for IPM programs resulting from the new federal IPM policy came in fiscal 1972, when funding was provided for a project that was titled "The Principles, Strategies, and Tactics of Pest Population Regulation and Control in Major Crop Ecosystems" but was better known as simply "the Huffaker Project." The IPM effort was conducted in partnership with a number of the nation's leading universities and included extensive field tests of promising new methods of pest detection and control. Six major cropping systems were included in the project: alfalfa, citrus, cotton, pine, pome and stone fruits, and soybeans. The project was jointly funded by the NSF, the USDA, and the USEPA. The federal agencies coordinated their efforts, with NSF supporting basic research and the USDA supporting applied research, development, and testing. Over the course of this seven-year project, the federal government provided $13 million for research conducted by 18 universities on six crops (representing approximately 70% of pesticide use). By the end of the project, advances had been made in methods for timely application of insecticides, the development of insect-resistant crops, new appreciation for biocontrol tactics, and the design of methods for the evaluation of the economic and environmental impacts of IPM programs [15].

In addition to providing funds for the Huffaker Project, the USDA expanded the pilot pest management program in fiscal years 1972 and 1973 to include cotton in all major cotton-producing states and initiated demonstration projects for alfalfa, apple, citrus, corn, grain sorghum, peanut, potato, sweet corn, tobacco, and some vegetable crops in 17 states. These demonstration projects were structured so that participating farmers would help pay the cost of scouts during the first three years of the demonstration project, then assume the full cost. There were three goals for these projects: (1) Ensure maximum production of food and fiber; (2) reduce farm operating costs; and (3) enhance the quality of the environment [16]. From 1971 to 1974, a total of 39 three-year pilot pest management projects were conducted. The USDA provided funding for additional "pilot application" projects in fiscal 1975 and established the following objectives [17]:

1. Develop and implement effective integrated approaches to "prevent or mitigate losses caused by pests through use of biological, cultural, chemical, and varietal methods of control."
2. Field test a combination of suppression tactics.
3. Provide "grower exposure" (information and training) to gain their support and the adoption of IPM practices.

A fourth objective was added in fiscal 1976: "To monitor field population levels of pests" [18]. An excellent summary of the organization and accomplishments of the pilot pest management projects was prepared by Dr. Joe Good, the Extension Service's Director of Pest Management Programs [19].

Federal IPM policy took another step in its evolution in 1977, when President Carter stated that "environmental protection is no longer just a legislative job, but one that requires—and will now receive—firm and unsparing support from the Executive Branch" [20]. He then announced a "coordinated attack on toxic chemicals in the environment" and instructed the CEQ to "recommend actions which the federal government can take to encourage the development and application of pest management techniques which emphasize the use of natural biological controls like predators, pest-specific diseases, pest-resistant plant varieties, and hormones, relying on chemical agents only as needed." In response, the Secretary of Agriculture issued a 1977 memorandum that declared, "It is the policy of the U.S. Department of Agriculture to develop, practice, and encourage the use of integrated pest management methods, systems, and strategies that are practical, effective, and energy efficient" and "to seek adequate protection against significant pests with the least hazard to man, his possessions, wildlife, and the natural environment" [21]. The Secretary's memorandum was followed by the establishment of the Work Group on Pest Management to provide leadership and information exchange among the 11 USDA agencies actively engaged in pest management programs and to coordinate USDA pest management activities with those of the USEPA and other federal and state agencies [22].

4 THE REALITIES SET IN

The 1970s were a decade of great optimism and creativity as IPM research and extension programs responded to a call to action issued by a country concerned about the effects of pesticides. By the end of the decade, however, the euphoria of a new effort had faded and the practical realities of altering pest management practices on millions of farms became apparent. At congressional hearings in late 1977, environmentalists complained that IPM was moving at a snail's pace [23].

Witnesses at the hearing provided estimates on what it would take to speed up implementation of IPM methods on farms across the country. Dr. Fowden Maxwell of the University of Florida indicated that one Extension pest management specialist would be needed in each of Florida's 67 counties to fully implement IPM methods. Dr. Joe Good of the USDA Extension Service estimated that 300–400 Extension IPM specialists, 3,000 private consultants, and 63,000 scouts would be needed to adequately handle about one-third of the nation's agricultural lands. Dr. Good later indicated that it would take 10 years and 500–600 additional Extension IPM agents and specialists (including 53 state and federal IPM coordinators and 330 area Extension IPM agents) to implement a well-planned IPM effort to increase IPM implementation nationwide; he estimated that this level of effort would cost $20.4 million per year [19].

Confusion over the meaning and goals of IPM programs was already apparent in the late 1970s. One of the architects of the USDA IPM effort cautioned against portraying IPM in an abstract way "as a technological fix, a placebo, a mystical cure for environmental and agricultural problems attendant to pest control" and stressed the need to "start developing specific IPM practices and programs with prescribed methodologies and technologies" [24]. He further stated, "Frequently, we hear the 'use of IPM' will protect the environment. However, it will be possible to develop specific pest management practices or regulatory procedures to protect the environment when *and only when* specific pesticide related environmental problems are identified and understood. The specific problem or need must be identified before a corrective program can be launched."

In 1979 the CEQ published a second report on IPM [25] that cautioned,

> The recent accomplishments of integrated pest management and continued public interest in alternatives to conventional pesticide programs have resulted in some uncritical endorsement of IPM programs without regard to their feasibility and in some confusion about the concept. IPM is not a panacea; nor is it a term which embraces all programs that employ more than one control technique.

The report concluded that "the lack of understanding and support for interdisciplinary research projects and companion educational and demonstration programs at public institutions is a major impediment to IPM" as is the fact that "public agricultural research and extension institutions are frequently required to produce quick, simple answers to complex problems that are not well understood because of pressure from commodity groups or from elected federal and state officials."

The incorporation of biological control methods into IPM strategies had already become a point of contention among IPM supporters and critics by the late 1970s. The IPM leadership at the USDA were concerned about the tendency to think about IPM as being synonymous with biological control. "Too often the term IPM is equated with biological control or nonchemical control. In most instances, farmers do not accept this because their experience proves otherwise" [24].

5 A SHIFT IN EMPHASIS

In the 1980s, economics and the farm financial crisis became the dominant force in agricultural policy and in the evolution of IPM policy. The shift toward greater emphasis on economics was seen in the research effort that followed the Huffaker Project: the Consortium for Integrated Pest Management (CIPM), funded by the USEPA from 1979 to 1981 and the USDA's Cooperative State Research Service (CSRS) from 1981 to 1985. CIPM built on the foundation laid by the Huffaker Project, using systems science and modeling to organize and quantify biological systems in alfalfa, apple, cotton, and soybean. Although addressing environmental concerns continued to be a goal of CIPM, the research conducted by the 17 participating universities was focused on the economics and profitability of crop production [15].

The 1980s also saw a decline in support for IPM when President Reagan's fiscal 1986 budget proposed elimination of funding for the Extension Service's IPM implementation program. However, congressional supporters were able to restore funding for the IPM program. The House Agriculture Committee included the following statement in its report on the appropriations bill [26]:

> The Committee strongly urges the Department to maintain and strengthen its efforts to assist producers in the development, understanding and implementation of integrated pest management practices. It has been demonstrated that existing IPM efforts have greatly aided producers in the responsible and effective control of pests, and additional efforts are needed as the problem of pest management continues to grow.

Federal IPM programs had a number of champions in Congress, including Representative de la Garza of Texas, who said, "Money alone will not solve our problems in pest management, just as money by itself is not the full answer to problems in other agricultural policy fields. But if we believe in the effectiveness of programs like IPM, we must be prepared to support the government's ability to cooperate with farmers and state extension programs in pest management" [27].

6 A PUSH TO INCREASE IPM ADOPTION

By the 1990s the benefits of IPM methods had been well established, but adoption rates remained lower than hoped. The USEPA and the USDA sponsored a series of workshops in 1992 and 1993 to identify factors constraining adoption of IPM systems [28,29]. Workshop participants identified a large number of impediments to greater adoption of IPM methods, including inadequate knowledge of currently available IPM tactics, a shortage of consultants and other pest management professionals to provide IPM services, the high level of management input required for implementation of some IPM systems, and the lack of alternative pest control tactics for some pests. Another major constraint identified at the workshops—the lack of a national commitment to IPM—was addressed in 1993, when the Clinton Administration announced at a congressional hearing that "implementing IPM practices on 75% of the nation's crop acres by the year 2000" was a national goal [30]. Benbrook [2] reported that the IPM adoption goal apparently was a compromise worked out during negotiations between the USEPA and the USDA policy staff; the USEPA staff had wanted the administration to set "tangible pesticide use reduction goals, patterned after successful European programs."

On December 14, 1994, the USDA announced a plan to provide agricultural producers with the tools they needed to deal with the environmental and economic problems of pest control and to help them implement IPM methods on 75% of U.S. crop acreage by the year 2000 [31]. The plan called for the development of the knowledge and technologies that would make it possible for the majority of U.S. farmers to reduce production costs by implementing biologically based pest management systems and substantially reduce their reliance on broadly toxic chemical pesticides. The plan also called for the establishment of a process for setting priorities at the local or regional level, linking research and education efforts to meet those priorities, and coordinating USDA efforts across agencies.

After setting the 75% adoption goal, the USDA struggled to find a credible and practical way to obtain the data needed to measure progress. At the time, there was no established mechanism for obtaining the data, nor was there agreement on the survey questions that would be used to measure IPM use. There was, however, general consensus among government, industry, academia, and public interest groups that adoption of IPM systems should be measured along a "continuum" ranging from low to high levels of IPM adoption. A USDA report published in 1994 measured IPM adoption along the adoption continuum [32], and this approach was refined by the Consumers Union in its 1996 report *Pest Management at the Crossroads* [2]. These analyses estimated that more than half of U.S. crop acreage was being managed using IPM methods but that the majority of these acres were managed with basic IPM practices at the "low" end of the IPM continuum.

In 1998, the USDA developed a "rational working definition of what growers must do in order to be considered as IPM practitioners" [33]. The definition stated that:

> Adoption of IPM systems normally occurs along a continuum from largely reliant on prophylactic control measures and pesticides to multiple-strategy biologically intensive approaches, and is not usually an

"either/or" situation. It is important to note that the practice of IPM is site-specific in nature, and individual tactics are determined by the particular crop-pest-environment scenario. Where appropriate, each site should have in place a management strategy for Prevention, Avoidance, Monitoring, and Suppression of pest populations (the PAMS approach). In order to qualify as IPM practitioners, growers should be utilizing tactics in three or more of the PAMS components.

The USDA's National Agricultural Statistics Service used the working definition to design the survey instrument for "fall area" surveys conducted to measure farm-level adoption of IPM practices. A preliminary analysis of data from the 2000 growing season indicates that IPM practices had been implemented on approximately 70% of cropland, with the expected variations in adoption across crops and areas of the country.

7 THE ROLE OF PESTICIDE REGULATIONS

From the early days of federal involvement in IPM, the USDA and the USEPA were in agreement on the need to support the development and wide-scale implementation of IPM methods, and the leadership of the two agencies recognized the importance of developing a close working relationship on IPM and pesticide regulatory issues. In a 1977 speech, Dr. M. Rupert Cutler, USDA's Assistant Secretary for Conservation, Research, and Education, emphasized the USDA's obligation of working with the USEPA "as a team rather than as adversaries" [34]. However, in practice the degree of cooperation between the USDA and the USEPA on IPM programs has generally been poor, reflecting differences in the missions of the two agencies and the goals they have set for IPM programs.

Pesticide regulatory actions provide a powerful tool for reducing pesticide risks and have played an important role in the evolution of federal IPM policy. The demand for IPM programs increases when pesticide regulatory actions (or market forces or pest resistance) remove pesticides from the market. This fact was demonstrated most dramatically when the Indonesian government banned broad-spectrum insecticides [35]. Though the impacts may not be as dramatic, the passage of the Food Quality Protection Act (FQPA) of 1996 fundamentally changed the way the USEPA regulates pesticides.

However, regulatory activities can also divert attention away from research and extension programs that offer long-term solutions to pest management problems. This was certainly true with FQPA. Although FQPA increased the demand for IPM on farms, it also diverted the USDA's attention away from implementation of the IPM strategic plan. Pesticide regulatory issues took center stage as government agencies and the private sector struggled to implement the sweeping changes mandated by FQPA.

8 CONCLUSIONS

The policy landscape is unsettled as the federal government enters its fourth decade of involvement in IPM efforts. The potential for pesticides to cause huge environmental impacts has been greatly reduced since the early 1970s. Many of the dangerous pesticides in use at that time have been eliminated by regulatory action, voluntary withdrawal from the market by pesticide manufacturers, or loss of efficacy due to pest resistance. Pesticide residue concerns, especially for "at risk" populations such as infants and children, are being addressed with lower exposure levels mandated by FQPA. Health concerns related to farm workers and pesticide applicators are being addressed by new farm worker safety regulations and by education efforts such as the Pesticide Safety Education Program [36].

Few would disagree with a claim that progress has been made on the production efficiency and risk reduction goals set for IPM in the early 1970s. However, disagreements emerge when discussions turn to the success of IPM programs in reducing pesticide use. A review of policy documents indicates that pesticide use reduction was not the only goal of federal IPM programs initiated 30 years ago. The IPM policy established by the federal government recognized the need for flexibility in implementing IPM programs and consistently stated that the goal was to maximize production efficiency while minimizing environmental impacts (i.e., risk reduction). Clearly, federal IPM policy was designed to address health and environmental issues related to the use (and misuse) of pesticides available in the 1960s. However, federal IPM policy was just as clearly designed to address agricultural productivity concerns as they related to efficiency, cost of production, and the availability of effective pest control tactics. The goal was to develop pest management strategies that respected the right of growers to make a reasonable profit but that made it possible for them to do so without creating unreasonable risks to human health or the environment. Thus, it was a goal with two equal and inseparable objectives, and the "art" of successful IPM efforts was to find the proper balance between these two objectives. The authors of federal IPM policy considered pesticide use reduction to be a possible (even likely) primary objective for IPM programs once a serious environmental or health problem caused by use of a particular pesticide had been documented. Likewise, policy makers in the early 1970s acknowledged that the primary objective of an IPM program could be to increase the efficacy of control tactics when losses caused by pest damage became unacceptably high.

Federal IPM policy wisely sets a goal for federal IPM efforts that provides flexibility in focusing efforts on key environmental and health issues and problems yet requires that production efficiency and environmental objectives be balanced to the extent possible. It is this fact that makes stakeholder involvement in the development of IPM programs essential, because if key stakeholders are involved in finding the right balance between these two objectives they will understand the trade-offs involved and will continue to support the effort. The strength of the IPM concept lies in the flexibility to shift emphasis between the two objectives, depending on the specific situation. But the weakness of the IPM concept also lies in this flexibility, because it has caused confusion over what the goal of IPM actually is.

Progress has been made during the past 30 years, but many of the problems and issues that led to the pest control research hearings in 1971 still exist today: farm economic difficulties, the loss of effective pesticides due to pest resistance and regulatory action, and the emergence of new "invasive" pest species and new issues, such as those related to the use of genetically modified organisms. Concerns about the impacts of pesticides on human health and the environment still exist, though at a reduced and perhaps more localized level. And, for a variety of reasons, there is still a long way to go before integrated biological-cultural methods of pest control are the predominant way that pests are managed on America's farms and ranches [37]. In short, there are a variety of needs and issues that call for a renewed IPM effort. The USDA is working with IPM stakeholders in developing a plan to address these needs and to guide its IPM programs over the next decade.

The federal government recognized in 1972 that the successful development and implementation of effective IPM programs could be accomplished only through a partnership effort involving the private and public sectors. The legislative mandate that provided funding for the Huffaker Project acknowledged the importance of the partnership by stating, "The Federal Government can help, but the long-term success of IPM depends upon the states, the universities, the private IPM industry, and ultimately the farmer" (as quoted by Zalom et al. [4]). Hopefully, we will continue to see a strong federal commitment to research and extension programs conducted in partnership with universities and the private sector during its fourth decade of involvement in IPM efforts.

Future IPM efforts will be successful only if producers and their advisors, government, university researchers and extension specialists, and public interest groups come together to find science-based solutions for problems related to pest management and the use of pesticides. While some stakeholders grow impatient about the pace of transitions to new pest management approaches (some believe transitions are occurring too fast, others believe they are occurring too slowly), science serves as a safety mechanism that minimizes mistakes and protects livelihoods. IPM has successfully demonstrated that science provides a guidepost as concerns about pest management practices are addressed. Let us continue investing in IPM research and education efforts in the coming decades so that future crises can be prevented.

REFERENCES

1. JR Cate, MK Hinkle. Integrated Pest Management: The Path of a Paradigm. Washington, DC: Natl Audubon Soc, 1994.
2. CM Benbrook. Pest Management at the Crossroads. Yonkers, NY: Consumers Union, 1996.
3. M Kogan. Integrated pest management: Historical perspectives and contemporary developments. Annu Rev Entomol 43:243–270, 1998.
4. FG Zalom, RE Ford, RE Frisbie, CR Edwards, JP Tette. Integrated pest management: Addressing the economic and environmental issues of contemporary agriculture. In: FG Zalom, WE Fry, eds. Food, Crop Pests, and the Environment: The Need and Potential for Biologically Intensive Integrated Pest Management. St. Paul, MN: APS Press, 1992, pp 1–12.
5. US Congress. Pest Control Research. Hearings before the Subcommittee on Agricultural Research and General Legislation of the Committee on Agriculture and Forestry, Sept 30 and Oct 1, 1971. US Senate, 92nd Congress; S. 1794. Washington, DC: US Govt Printing Office, 1971.
6. National Research Council. Ecologically Based Pest Management: New Solutions for a New Century. Washington, DC: Natl Acad Press, 1996.
7. TA Royer, PG Mulder, GW Cuperus. Renaming (redefining) integrated pest management: Fumble, pass, or play? Am Entomol 45(3):136–139, 1999.
8. R Carson. Silent Spring. Boston: Houghton Mifflin, 1962.
9. R Nixon. Special Message to the Congress Proposing the 1971 Environmental Program. Transmitted Feb 8, 1971. In: Public Papers of the Presidents of the United States. Washington, DC: US Govt Printing Office.
10. Council on Environmental Quality. Integrated Pest Management. Washington, DC: Supt Documents, US Govt Printing Office, Stock No. 4111-0010, 1972.
11. US Congress. National Environmental Policy Act. Pub. L. 91-190, 42 U.S.C. 4321–4347, Jan 1, 1970.
12. US Department of Agriculture. Pest Management Program. A report submitted by the pest management committee to ND Bayley, Director of Science and Education, Office of the Secretary of Agriculture, Apr 14, 1971.
13. US Department of Agriculture. New pest management program to help farmers announced. USDA News Releases 137–172, 1972.
14. R Nixon. Special Message to the Congress Outlining the 1972 Environmental Program. Transmitted Feb 8, 1972. In: Public Papers of the Presidents of the United States. Washington, DC: US Govt Printing Office.
15. RE Frisbie. Consortium for Integrated Pest Management (CIPM)—Organization and administration. In: RE Frisbie, PL Adkisson, eds. Integrated Pest Management on Major Agricultural Systems. Texas Agric Exp Sta, 1985, MP-1616.
16. US Department of Agriculture. Pest Management Working Agreement Between the Extension Service and the Animal and Plant Health Inspection Service, USDA. Dated Nov 15, 1974. No. ANR-5-70. Transmitted to State Extension directors in a memo from Edwin Kirby, Extension Service Administrator, Dec 11, 1974.
17. US Department of Agriculture. Proposed Invitation for FY 1975 Pilot-Application

Pest Management Projects. Memo from Edwin Kirby, Extension Service Administrator, to State Extension Directors, dated Jan 10, 1975.
18. US Department of Agriculture. Request for Proposals for New FY 1976 Pilot Pest Management Projects. Memo from Edwin Kirby, Extension Service Administrator, to State Extension Directors, dated Dec 15, 1975.
19. JM Good. Integrated Pest Management—A Look to the Future. ESC 583. Washington, DC: Extension Service, USDA, 1977.
20. J Carter. Environmental Message to Congress. Transmitted May 23, 1977. In: Public Papers of the Presidents of the United States. Washington, DC: US Govt Printing Office.
21. US Department of Agriculture. USDA Policy on Management of Pest Problems. Secretary's Memorandum No. 1929, Dec 12, 1977.
22. US Department of Agriculture. Charter for the Work Group on Pest Management. Office of Environmental Quality Activities, Apr 8, 1977.
23. US Congress. Integrated Pest Management. Hearings before the Subcommittee on Agricultural Research and General Legislation of the Committee on Agriculture, Nutrition, and Forestry, Oct 31 and Nov 1, 1977. US Senate, 95th Congr; S. 1794. Washington, DC: US Govt Printing Office.
24. JM Good. Is integrated pest management succeeding? Presentation at the meeting of the Eastern Branch of the Entomological Society of America, Sept 26–28, 1979, Hershey, PA.
25. D Bottrell. Integrated Pest Management. Washington, DC: Supt Documents, US Govt Printing Office, 1979.
26. US Congress. Report of the Committee on Agriculture. US House of Representatives. Washington, DC: US Govt Printing Office, 1985.
27. E de la Garza. What IPM has meant to my congressional district. In: RE Frisbie, PL Adkisson, eds. Integrated Pest Management of Major Agricultural Systems. Texas Agric Exp Sta MP-1616, 1985.
28. AA Sorensen. Regional Producer Workshops: Constraints to the Adoption of Integrated Pest Management. Austin, TX: Natl Found for IPM Education, 1993.
29. AA Sorensen. Proc Natl Integrated Pest Management Forum, June 17–19, 1992, Arlington, VA. DeKalb, IL: Am Farmland Trust, Center for Agriculture and the Environment, 1994.
30. US Congress. Testimony of Carol M. Browner, Administrator EPA; Richard Rominger, Deputy Secretary of Agriculture; and David A. Kessler, Commissioner of FDA. Hearing before the Committee on Labor and Human Resources, US Senate, and Subcommittee on Health and the Environment, Committee on Energy and Commerce, US House of Representatives, Sept 21, 1993. Washington, DC.
31. US Department of Agriculture. USDA's Integrated Pest Management (IPM) Initiative. Washington, DC: Office of Communications, News Backgrounder, Release No. 0942.94, 1994.
32. US Department of Agriculture. Adoption of Integrated Pest Management in US Agriculture. Washington, DC: Economic Research Service, Agric Info Bull No. 707, 1994.
33. US Department of Agriculture. Determining the practice of integrated pest manage-

ment: A working definition for the year 2000 goal. USDA Integrated Pest Management Committee. October 1998.

34. MP Cutler. The role of USDA in integrated pest management. Speech presented at the Symposium on Pest Control Strategies—Understanding and Action, Cornell Univ, Ithaca, NY, June 22, 1977. In: EH Smith, D Pimentel, eds. Pest Control Strategies. New York: Academic Press, 1978.
35. PE Kenmore. Integrated pest management in rice. In: GJ Persley, ed. Biotechnology and Integrated Pest Management. Wellingford, UK: CAB Int, 1996.
36. US Department of Agriculture. Pesticide Safety Programs. Brochure developed by USDA and EPA. Washington, DC, 2001.
37. MS Fitzner. The role of education in the transfer of biological control technologies. In: RD Lumsden, JL Vaughn, eds. Pest Management: Biologically Based Technologies. Proc Beltsville Symp XVIII, May 2–6, 1993. Washington, DC: ACS, 1993, pp 382–387.

2

Biological Control

Raghavan Charudattan and S. Chandramohan
University of Florida
Gainesville, Florida, U.S.A.

Gabriela S. Wyss
Research Institute of Organic Agriculture
Frick, Switzerland

1 DEFINITION OF BIOLOGICAL CONTROL

It is difficult to define biological control in a manner that is universally acceptable to the diverse practitioners of this field. However, a clear definition is necessary to explain and delimit different biological control processes and methodologies. Definitions have evolved over the years to encompass different types of biologically based controls that are now considered under the umbrella of biological control. In this chapter, we follow the definition proposed by Charudattan et al. [1]:

> Biological control is the reduction or mitigation of pests and pest effects through the use of natural enemies. Biotechnologies dealing with the elucidation and use of natural enemy's genes and gene products for the enhancement of biological control agents are considered a relevant part of modern biological control.

There are several reasons for seeking biological control for pest and disease management. It is well known that chemical pesticides and chemically based controls have limitations, notwithstanding the fact that chemical pesticides and the chemical pesticide industry have been responsible to a great extent for enabling food production for the world's burgeoning population. Nonetheless, it must be remembered that chemical pesticides are, in essence, compounds that disrupt the normal metabolic functions of target organisms. They have side effects or nontarget effects that may lead to a series of changes that adversely affect organisms that constitute the ecological web. Some or all of these adverse changes may be passed along the food chain, ultimately affecting human and environmental health.

2 BENEFITS AND LIMITATIONS OF BIOLOGICAL CONTROL

Biological control has strengths as well as weaknesses. On the beneficial side, biocontrol agents are typically host-specific and therefore are less likely to inflict nontarget damage. As living organisms, biocontrol agents themselves are subject to mortality and hence are not likely to build up in nature and cause environmental problems. Some types of biological controls may provide benefits over a period of several years after an initial phase of establishment of the control agents. This is generally true with biocontrol agents that are self-sustaining and capable of multiplying in a density-dependent manner (i.e., when more food is available in the form of a host substrate, greater numbers of the biocontrol agent will build up through successful reproduction on the host, and when less food is available, lesser numbers). As a result, the cost of pest control may not be recurrent, and the cost is often limited to the initial research program, field release, and establishment of the biocontrol agent. As opposed to this example, in cases where annual or periodic applications of biocontrol agents are needed to ensure control, the costs will be higher. Typically, it is less costly to develop biological control agents than to develop chemical pesticides. Exact figures are hard to obtain owing to the proprietary nature of sales information, but it is claimed that it takes 8–10 years and $25–80 million to develop a new agrochemical product compared to 3 years and a cost of about $2 million for a biopesticide (see below for definition) [2]. Research and development costs of other types of biological control agents (e.g., inoculative agents) fall within the same range as those of biopesticides. Biological controls also have certain beneficial environmental advantages compared to chemical pesticides. Because biological control is slower acting than chemical pesticides, there is time for the ecosystem to readjust and restabilize. Hence, there is a gradual ecological change as the pest and disease problems are controlled. For this reason, biological control is less likely to create voids in ecosystems. Biological control, like many chemical pesticides, can be integrated with other pest management tactics. In nature, many different biological agents interact to cause pest suppression. Often a pest is a host to a number of natural enemies, and this natural association of interactive agents can be exploited to achieve integrated pest control (IPM). Finally, biocontrol has an overwhelming record of human and environmental safety compared to chemical pesticides.

Some of the disadvantages of biological control include the following.

1. As stated, biocontrol agents are generally host-specific. That is, typically, each agent is active against a single pest species or a disease. Therefore, the farmer or the user who is faced with several different pests must resort to many different biocontrol agents and must seek several supplementary control methods or use a broad-spectrum pesticide that will control all of the pests (e.g., methyl bromide as a soil fumigant) or certain categories of pests (e.g., broad-spectrum herbicides).
2. Because biological control agents, as living organisms, depend on multistep and multifactorial interactions to be effective, their success as biocontrol agents is notoriously unpredictable.
3. The slow rate of action of biological control may not satisfy the user's needs. Whereas the slower actions of biocontrol agents may have advantages (see above), the users may require quicker solutions to their pest problems. In some crops, there may be time constraints that preclude the use of biological control agents. For example, a crop may have a short period of pest attack during which a biological control agent must be effective to protect the crop. A biocontrol agent that requires a period of several weeks or months to be effective may not serve the purpose. However, the concept of "compound interest" may be applied to this scenario; a biocontrol agent may be introduced and allowed to build up over several years and provide gradual pest suppression. There are many examples in the literature attesting to the fact that this situation occurs. For example, fields that have been left untreated with chemical pesticides for several years tend to gradually build up a strong suite of beneficial agents that protect against deleterious organisms.
4. Performance of biocontrol is subject to environmental and ecological factors that are often site- and host-biotype-specific. Many biocontrol agents, because of their specific environmental and host adaptations, are not effective when used in sites removed from their original habitats or against host types that may have certain phenotypic or genotypic differences from the original type upon which the agents were found.
5. Biocontrol agents may suffer from short shelf life. The term "shelf life" is commonly used in the context of biocontrol agents that are

commercially produced, such as microbial biocontrol agents. It is the length of time that an agent can be left on the shelf under reasonable environmental conditions before use. A biocontrol agent should be viable and capable of remaining efficacious during its predicted shelf life.
6. Although biological control agents have a proven record of safety that outweighs their potential risks, some agents, such as certain microorganisms, can produce metabolites that are highly toxic to humans and other animals. Also, fungal biocontrol agents are likely to cause allergic reactions in sensitive humans. Some level of collateral impacts on nontarget organisms is inevitable even when highly specific biocontrol agents are used. For instance, biocontrol of an invasive weed may lead to a loss of habitat for some fauna and microflora dependent on the weed species.
7. Biological control products often are not economically viable in the marketplace. Unlike economically successful chemical pesticides [e.g., glyphosate (Roundup) and other products], biocontrol products are typically used on a very small scale, with a typical return of <$1 million per year per agent. An exception is *Bacillus thuringiensis*–based products (e.g., Dipel) used for the control of various insects. Bt products, as they are commonly referred to, have a collective worldwide market value of about $80–100 million [3].
8. Acceptance of biological control in the marketplace is often poor owing to the prevailing reliance on chemical pesticides for quick-fix solutions for the deep-seated problems of pest and disease outbreaks. Farmers and the general public are used to the quick action, high level of efficacy, convenience, and affordable cost of chemical pesticides despite their environmental drawbacks. The chemical pesticide industry has a well-established sales and promotional network. It is difficult to compete against this market force to sell biocontrol agents that have many limitations, as summarized in this list.
9. Finally, biological control agents, particularly those used as biopesticides, may cause the development of resistance in the biocontrol target, either by allowing naturally resistant host biotypes to become dominant or through selection for resistance genes in the host target population.

3 ECOLOGICAL BASIS OF BIOLOGICAL CONTROL

Biological control is in fact a practical application of the ecology of the host (cultivated or desired plant species or a habitat invaded by a pest), pests and diseases that attack the desired host or habitat (biocontrol target), the multitude of beneficial and antagonistic organisms that live on or around the target, and the environment that impacts the target, pathogen/pest complexes, and the biocontrol agents. It is generally agreed that agricultural and urban plant communities are ecologically disturbed communities that are subjected to pest and disease outbreaks. These outbreaks often result from practicing unsustainable forms of agriculture. However, with increasing need to feed the growing human population in the world, it is unrealistic to expect a return to a totally "sustainable" form of agriculture. Nonetheless, attempts should be made to balance the unsustainable tendencies of modern agriculture with ecologically beneficial pest and disease control methods. In this context, biological control is recognized as an ecologically beneficial strategy. However, because biological control has its limitations, it can never be the sole and permanent solution to pest or disease problems, although it should be the foundation for sustainable IPM programs [4]. Indeed, biological control is likely to be most successful when used as a component of IPM rather than as the sole method of control.

4 SCOPE OF THIS CHAPTER

We have attempted to present a brief review of biological control of plant diseases and weeds, with emphasis on *microbiological control* approaches. In line with our definition of biological control (see above), we discuss the use of agents (live organisms) as well as microbial genes and gene products. We have chosen examples of microbiological control agents that, in our view, best illustrate different biocontrol principles and application strategies. It is not our intention to suggest that these are the sole examples or the most suitable products and strategies. Clearly, there are numerous successful and elegant examples of biological control in use (e.g., classical biocontrol of insect pests, other microbial products in the market, etc.) that fall outside of the small number of cases we have chosen to present. For a more comprehensive examination of biological control in all its facets, which is beyond the scope of this chapter, the readers are referred to recent comprehensive treatises on biological control [5–9].

5 BIOCONTROL STRATEGIES BASED ON BIOCONTROL TARGET–BIOCONTROL AGENT INTERACTIONS

Biological control can occur naturally without direct human effort. Compared to natural biological control, the use of specific agents that are isolated, processed in several ways to ensure efficacy, and reintroduced to provide biological control is called introduced biocontrol. The latter can be further categorized as classical (inoculative; one-time or a limited number of introductions) or inundative (biopesticide) strategies. In some cases, periodic releases of a biocontrol agent may be necessary to augment a previously established or a naturally occurring level of the biocontrol agent. Density-dependent relationships between the biocontrol target and the biocontrol agent can be used to describe and distinguish these strategies, although the distinction will be arbitrary in some cases. The modes of biocontrol actions involved in these biological control systems can include one or more of the following: antibiosis, competition, hyperparasitism, hypovirulence, induced resistance, pathogenicity, and toxicity.

5.1 Naturally Occurring Biological Control

The term "suppressive soil" was coined to explain the phenomenon of natural suppression of potato scab observed following the addition of green manure [10,11]. The disease, characterized by conditions ranging from superficial lesions to deep pits on tubers, is caused by *Streptomyces scabies*, a filamentous bacterium. The disease can severely reduce tuber quality and result in unmarketable tubers. Natural disease suppression has been shown to be brought about by an increase in saprophytic organisms in the soil, including nonpathogenic *S. scabies* strains that are antagonistic toward the pathogen. A disease-suppressive soil shows low incidence of disease severity in spite of the presence of a high density of pathogen inoculum, a susceptible host plant, and favorable environmental conditions for disease development. In contrast, a disease-conducive soil shows high disease severity even in the presence of low inoculum density of the pathogen [12]. Every soil possesses the ability for some microbiological disease suppression and a continuous range of suppressiveness from a high degree of disease suppression through intermediate degrees of suppressiveness/conduciveness to the extreme of no disease suppression. In general, strains that are selected from suppressive soils are ready-made biocontrol agents because they are adapted to the plant or plant part where they must function [13].

Suppressive soils have been described from many countries, and fusarium wilt–suppressive soils are among the most extensively studied. Research carried out mainly in soils of the Châteaurenard region (Bouches-du-Rhône) of France [14–16] and the Salinas Valley of California [17–19] has established that disease suppressiveness of these soils is expressed against all formae speciales of *Fusarium oxysporum* but not against diseases caused by other soilborne pathogens and nonvascular *Fusarium* species. In most cases, disease suppressiveness could be transferred easily in previously heat-treated, disease-conducive soil by mixing in a small portion of disease-suppressive soil [20]. The level of soil suppressiveness, however, is correlated with physicochemical characteristics of the soil.

Fusarium wilt–suppressive soils typically have a large population of nonpathogenic *Fusarium* spp. (mainly nonpathogenic *Fusarium oxysporum*), bacteria (mainly *Pseudomonas fluorescens* and *P. putida*), and actinomycetes that contribute to biological control of fusarium wilts [21–23]. Moreover, the incidence of fusarium wilts appears to be related to the relative proportion of the pathogen population within the total population of *Fusarium* rather than to the absolute density of the pathogen population in soils.

Disease suppression by nonpathogenic *F. oxysporum* has been attributed to several mechanisms: (1) saprophytic competition for nutrients [15,16,24,25], (2) parasitic competition for infection sites at the root surface [26], and (3) induced systemic resistance (discussed in Sec. 6.1) [27–29]. Competition for nutrients determines the level of activity of the pathogen in soils and consequently plays an important role in the mechanism of soil suppression. Competition for carbon is another mechanism, because addition of glucose provided energy for *Fusarium* and caused an increase in disease incidence in both conducive and suppressive soils. However, a higher concentration of glucose was needed, indicating that competition for carbon is more intense in suppressive soils than in conducive soils [30]. Competition occurred simultaneously for both carbon and iron in the suppressive soil from Châteaurenard, but carbon appeared to be the first limiting factor in this soil. Competition for iron, a key element required by both the plant and microorganisms, is a mechanism shown to substantially influence suppressiveness of soils [19,30,31]. For instance, disease control afforded by strains of *Pseudomonas fluorescens* has been related to the ability of these bacteria to successfully compete for iron and nutrients and through antibiosis by the production of antimicrobial metabolites [32,33] such as 2,4-diacetylphloroglucinol, pyoluteorin, and hydrogen cyanide [34]. Direct correlation exists between siderophore (iron chelator) production by various fluorescent pseudomonads and their inhibition of chlamydospore germination of *Fusarium oxysporum* f.sp. *cucumerinum* [19].

Duffy and Défago [35] found that zinc and copper significantly improved the biocontrol activity of *P. fluorescens* CHA0 against *F. oxysporum* f.sp. *radicis-lycopersici* in soilless tomato culture. The authors suggested that zinc amendment improved biocontrol activity by reducing fusaric acid production by the pathogen, which resulted in increased antibiotic production by the biocontrol agent.

Practical use of antagonistic microorganisms recognized to be involved in the mechanisms of soil suppressiveness has been attempted. Extensive research has been carried out with the nonpathogenic *F. oxysporum* strain Fo47, a strain isolated from a suppressive soil in the Châteaurenard region of France that has been shown to induce resistance to fusarium wilt in tomato [36]. This strain is able to control fusarium wilt of several plants under well-defined conditions, especially in carnation (*Dianthus caryophyllus*) grown in steamed soil [37], cyclamen (*Cyclamen europaeum*) [38], flax (*Linum usitatissimum*) [14,39], and tomato (*Lycopersicon esculentum*) [36].

Other examples of natural disease control brought about by soil suppressiveness include control of common scab of potato by nonpathogenic *S. scabies* and other *Streptomyces* spp. [40], fusarium wilt of watermelon in Florida by nonpathogenic *F. oxysporum* and other *Fusarium* spp. [41], root rot of *Eucalyptus marginata* and avocado (*Persea gratissima*) caused by *Phytophthora cinnamomi* by a complex of antagonists [10,42], Pythium and Rhizoctonia damping-off of several plants by various soil microorganisms [10,42], and take-all disease of wheat (*Triticum aestivum*) by antagonistic microorganisms including *P. fluorescens* [10].

5.2 Introduced Biological Control Agents

5.2.1 Agents Used by Means of a Limited Number of Introductions

Some biological control agents are applied in the field through small releases to establish infection foci from which the agents spread further. Alternatively, the agents are released periodically to augment a background level of naturally occurring biocontrol agents. Agents that have the capacity for self-propagation and self-dissemination within the released area are most suitable for this method.

Control of Sclerotinia minor *by* Sporidesmium sclerotivorum. Mycoparasites (= hyperparasites of fungi) have been recognized as potential biocontrol agents since 1932, and intensive research has been carried out on numerous pathogen–hyperparasite systems. One such system is the control of lettuce drop disease caused by *Sclerotinia minor* by the mycoparasite *Sporidesmium sclerotivorum* [43].

Lettuce drop is an economically important disease of all types and cultivars of lettuce (*Lactuca sativa*). Disease incidence on romaine lettuce has been shown to be decreased significantly in fields treated with the biological control agent *S. sclerotivorum*. The biocontrol agent is a dematiaceous hyphomycete that parasitizes the sclerotia of several pathogens including *Botrytis cinerea*, *Claviceps purpurea*, *Sclerotinia sclerotiorum*, *S. minor*, *S. trifoliorum*, and *Sclerotium cepivorum* [43,44]. It has been reported from the continental United States, Australia, Canada, Finland, Japan, and Norway [45]. It produces multiseptate macroconidia, a *Selenosporella* state bearing microconidia, a few chlamydospores, microsclerotia, and mycelium in culture [44]. Macroconidia of *S. sclerotivorum* germinate within 3–5 days on the surface of host sclerotia and penetrate the rind and cortex without forming specialized penetration structures. The fungus develops intercellularly, and multiple infections may occur in the sclerotium. Sporulation may occur on the sclerotial surface and extend into the surrounding soil, where it can infect healthy sclerotia within a radius of 3 cm [44]. Approximately five macroconidia per gram of soil are needed to successfully infect sclerotia and bring about their decay. Each infected sclerotium produces about 15,000 new macroconidia in soil regardless of the initial inoculum density of the host [46]. Laboratory experiments with field soil have revealed that inoculum of *S. sclerotivorum* completely destroys sclerotia of *S. minor* within about 10 weeks at 20–25°C, pH of 5.5–7.5, and soil water potentials of −8 bars and higher. Under optimal field conditions, parasitized sclerotia may decay at all depths to at least

14 cm [43]. The fungus derives its energy for growth and sporulation from glucose that is released from sclerotial glucans released by glucanases produced by the host fungus [44].

A field study demonstrated that single applications of 100 and 1000 conidia of *S. sclerotivorum* per gram of soil caused control of lettuce drop of 40–83% in four successive crops over a 2-year period compared to the control plots. The number of sclerotia of the plant pathogen was significantly reduced by the mycoparasitic activity. The mycoparasite became established in the field and even increased its number of infective units over the experimental period [47]. Various alternatives to the addition of large quantities of *S. sclerotivorum* to soil to obtain biological control have been examined [43,48]. In field studies carried out in 1987–1989, it was demonstrated that lettuce drop could be controlled with rates as low as 0.08 macroconidium per gram of soil [49]. Thus, when properly applied and managed, this biocontrol agent can provide effective and economical biological control of lettuce drop.

Port Jackson Willow. Another highly successful inoculative biocontrol program, one directed at a weedy tree species, is taking place in South Africa. A gall-forming rust fungus, *Uromycladium tepperianum*, was imported from Australia and released into South Africa to control the alien invasive tree species *Acacia saligna* (Port Jackson willow) [50]. This tree is regarded as the most troublesome weed in the Western Cape Province of South Africa. It is difficult and costly to control by chemical and mechanical methods and therefore became a target for biological control. The fungus causes extensive gall formation on branches and twigs, accompanied by a significant energy loss. Heavily infected trees are eventually killed (Fig. 1).

The rust fungus was introduced into South Africa between 1987 and 1989, and in about 8 years the disease became widespread in the province and the tree density declined by at least 80% in rust-established sites. The number of seeds in the soil seed bank has also stabilized at most sites. Large numbers of trees have begun to die, and this process is continuing. Thus, *U. tepperianum* is providing very effective biocontrol following its inoculative release, which relied on a simple, low-input, manual inoculation of a small number of tree branches at each release site [50].

5.2.2 Agents Used as Bioprotectants

It is well known that certain naturally antagonistic microorganisms can be used to protect sites on plant surfaces and plant products from invading microbial pathogens [10,12]. Presently, some such microorganisms are being used as bioprotectants based on their capacity for competitive exclusion of pathogens at the infection site, lysis of pathogenic hyphae, production of pathogen-active antibiotics, and/or induction of systemic resistance that protects the plant against invading

FIGURE 1 Biological control of Port Jackson willow (*Acacia saligna*) by an introduced rust fungus, *Uromycladium tepperianum*. (A) Rust galls on a branch of *A. saligna*. (B) A heavily infected and galled *A. saligna* tree. (C) A "before-and-after" picture illustrating the success of this biocontrol program. (Photos courtesy of Plant Protection Research Institute, South Africa.)

pathogens. Generally, these organisms are selected from common, rhizosphere-resident bacteria with plant growth–promoting activities (i.e., plant growth–promoting rhizobacteria) or from microbial epiphytes of aerial plant surfaces. Some yeasts found on the surfaces of sugar-rich fruits are also considered. Root diseases caused by a variety of soilborne pathogens and postharvest diseases of fruits and vegetables are among the diseases controlled by this method [10,51,52].

Bacillus subtilis. *Bacillus* species are common, soil-inhabiting, spore-forming, rod-shaped, usually gram-positive, motile bacteria. Generally, they have relatively simple nutritional requirements and are aerobic or facultatively anaerobic. They form endospores within cells that may remain dormant for long periods. The endospores enable these bacteria to withstand adverse conditions such as high temperature and desiccation. The mechanisms of biocontrol by *Bacillus* spp. may include one or more of the following: antibiosis, competition for sites and nutrients, and hyperparasitism. A *Bacillus*-based product that is registered for commercial use in the United States is Kodiak™ (produced and marketed by Gustafson, Inc., Dallas, TX) [53].

Kodiak is registered by the U.S. Environmental Protection Agency (EPA) as a biofungicide for use in seed treatment [54]. It is used in combination with chemical seed treatments to give longer protection of plant roots against attack by soilborne and seedborne pathogens, mainly *Rhizoctonia solani* and *Pythium ultimum*. It is commonly used to protect cotton and legume seedlings, although it could be used to protect against a variety of other soilborne pathogens. Unlike the protective effect of chemical fungicides that diminish over time due to breakdown of the chemical in the soil, Kodiak offers extended protection because it consists of a living organism that can grow and multiply along with the growing plant roots.

Kodiak contains endospores of the bacterium *Bacillus subtilis* strain GB03. The endospores are produced under optimal conditions using liquid fermentation, concentrated, dried, and milled to a fine powder. The powder formulation of the product can be used as either a liquid or a dry blend with other chemicals used for seed treatment. The shelf life of the product is at least 2 years when stored at a temperature of ≤30°C. Kodiak provides yield increases by reducing the pathogen's inoculum level and the associated adverse effects on the crop plant's root system. The duration of control depends on the cultivar, the level of disease pressure present, and environmental factors. Cotton is the first crop in the United States in which Kodiak has been used on a large scale. Most of the cotton seed planted in the United States in 1998–1999 is said to have been treated with Kodiak for suppression of seedling diseases caused by soilborne pathogens. Other crops have also been known to show positive yield responses when Kodiak-treated seeds are used [53].

Postharvest Disease Control Agents. Postharvest disease control is emerging as an important area where microbial agents could have a significant role as bioprotectants. Fresh fruits and vegetables are highly disease-susceptible and therefore require specific measures to prevent postharvest losses. Harvested produce undergoes a perilous trip from the production fields to the consumers' tables during which it is exposed to numerous opportunities for disease development. It is harvested in the field, often by methods that can cause injury, handled in packinghouses (more chance for damage), subjected to time delays when shipped over long distances to markets, and again handled and left on shelves for several days before finally being delivered to the users. Wounds, improper handling, and time delay are therefore important factors that contribute to losses due to postharvest diseases. Second, because of its rich water and nutrient contents, fresh produce is naturally susceptible to attack by several pathogenic fungi and bacteria. Finally, during the ripening process, fruits and vegetables lose their intrinsic resistance that protects them during their development while attached to the plant.

An array of chemical agents, including synthetic fungicides; nonspecific, broad-spectrum chemicals such as chlorine; waxes and other polymers; and coloring agents, among others, are used on many fruits and vegetables to protect them against diseases, improve handling and visual qualities, improve shelf life, etc. These materials and treatments are coming under increasing scrutiny by the public, often resulting in their rejection, and biological control is being looked upon as an alternative. Other factors that promote the use of biocontrol include the development of fungicide resistance by postharvest pathogens, the lack of adequate new fungicides to replace older fungicides that are taken off the market, and the public's opposition to the use of irradiation as a protective measure.

Since the early 1980s, many antagonists have been isolated and shown to be effective in controlling numerous postharvest pathogens. Generally, epiphytic microorganisms isolated from plant surfaces are screened for antibiotic and disease-suppressive activity in a variety of in vitro assays. Although microorganisms from any source, such as soil, water, and plant surfaces, may possess antagonistic properties against postharvest pathogens, a preferred source is the plant or the plant organ (fruit or vegetable) itself. Conceptually, organisms that are preadapted for life on fruits and vegetables are more likely to be capable of affording bioprotection than microbes from unrelated habitats.

Various groups of microorganisms such as gram-negative and gram-positive bacteria, yeasts, and yeastlike filamentous fungi have been shown to be effective in protecting against postharvest pathogens. Major emphasis is placed on selecting agents that are effective in situ (at the site where protection is required); able to survive, colonize, and afford protection throughout the holding period of the produce; and compatible with various postharvest treatments and additives. Generally, in vitro assays are conducted as a necessary first step, but most often the activity seen in in vitro screenings does not hold out in subsequent in situ assays or under packinghouse conditions. Typically, these laboratory screenings are followed by tests under "real-life" or "field" conditions of the packinghouse and markets.

At least five mechanisms of action have been shown or postulated to be involved in the biocontrol of postharvest diseases: (1) colonization of the wounds by an antagonist capable of excluding the pathogen by competition for nutrients and space (niche competition); (2) inhibition of pathogen spore germination, growth, and sporulation; (3) direct lytic action on the pathogen; (4) antibiosis; and (5) induced resistance in the fruit or vegetable.

Use of microorganisms on produce that can be consumed raw poses some special considerations for risk analysis. Of particular concern are (1) nontarget effects of the biocontrol agent, including pathogenicity to the fruit and vegetable meant to be protected, potential toxicity and allergenicity to humans, and adverse effects of chronic exposure, determined from animal models; (2) production of metabolites that may have adverse human effects; and (3) potential of the biocontrol agent to grow at human body temperature (this is of concern when using yeasts and certain bacteria such as *Pseudomonas* spp.). Not all of these concerns may need to be addressed; a strategy of case-by-case analysis is followed by the EPA.

Three postharvest disease protectants are registered in the United States, including Bio-Save™ 10, Bio-Save™ 11, and Aspire™ (Table 1). These products are used to provide coatings on fruits through bin-drench or in-line application. Bio-Save is a line of postharvest disease preventatives based on naturally occurring bacteria and yeasts originally isolated from fruit surfaces [55]. These products are effective against multiple pathogens, preventing infection of fruit by outcompeting pathogens at the wound sites on fruit surfaces. Bio-Save 10 and Bio-Save 11 consist of *Pseudomonas syringae* strains ESC10 and ESC11, respectively. Bio-Save 10 is used to control green mold (*Penicillium digitatum*), blue mold (*P. italicum*), and sour rot (*Geotrichum candidum*) on citrus fruits. Bio-Save 11 is used against blue mold, benzimidazole-resistant strains of *P. expansum*, gray mold (*Botrytis cinerea*), and mucor rot (*Mucor pyriformis*) on pome fruits. Bio-Save products are produced and sold by EcoScience Produce Systems Division, Orlando, FL.

Aspire is a postharvest biofungicide composed of *Candida oleophila* isolate I-182 (Table 1). This naturally occurring yeast antagonist, isolated from tomato fruit, is effective against a wide range of postharvest pathogens, including *Penicillium* and *Botrytis* species on citrus and pome fruits [56]. The mode of action of this yeast is said to be through competition and is not known to produce antibiotics.

5.2.3 Agents Used as Biopesticides

Biopesticide is defined here as a biological control agent that is applied in an inundative manner (i.e., inundative biological control strategy) to control a target pest. Unlike the EPA's definition of biopesticides [57], which includes many naturally derived materials such as plant oils and baking soda in addition to living and nonliving biological agents, our definition is limited to living biocontrol agents that are applied inundatively to ensure a high initial level of attack on the biocontrol target. According to our definition, biopesticides may consist of bacteria, fungi, viruses, or protozoa as active ingredients. Biopesticides must be registered by the EPA under the rules and regulations of the Federal Insecticide, Fungicide and Rodenticide Act. Table 1 lists biopesticides that are currently registered by the EPA for the control of plant diseases and weeds.

Trichoderma-*Based Biofungicides.* *Trichoderma* spp., notably *T. harzianum, T. polysporum*, and *T. viride*, have been studied as potential biocontrol agents for nearly 50 years. About 40 different pathogenic fungi and diseases have been shown to be controlled by *Trichoderma* spp., which are soilborne, generally

Table 1 U.S. Environmental Protection Agency Approved Biopesticide Active Ingredients for the Control of Plant Diseases and Weeds

Active ingredient (agent)	Product name (registrant, if known) and Use
Bacteria	
Agrobacterium radiobacter K84	Norbac 84-C (New BioProducts, Inc., Corvallis, OR); Galltrol-A (AgBioChem, Inc., Orinda, CA); bioprotectants against crown gall disease (caused by *A. tumefaciens*) on various fruit crops
Bacillus subtilis GB03	System 3 (Helena Chemical Co., Memphis, TN); a bioprotectant against seedling pathogens on barley, beans, cotton, peanut, pea, rice, and soybeans
B. subtilis MBI 600	Kodiak line of biofungicides (Gustafson, Inc., Dallas, TX); soilborne root pathogens of cotton and legumes
Burkholderia cepacia type Wisconsin IsoJ82	Deny (Blue Circle) biofungicide (Stine Microbial Products, Shawnee, KS); root diseases caused by *Fusarium*, *Monosporascus*, *Pythium*, *Rhizoctonia*, and *Sclerotinia* species on greenhouse and field-grown crops such as vegetables, fruits, nuts, herbs and spices, ornamental flowers and bulbs, trees, shrubs, and grains
B. cepacia type Wisconsin M36	Deny (Blue Circle) bionematocide (Stine Microbial Products, Shawnee, KS); root knot, lesion, sting, spiral, needle, and lance nematodes on greenhouse and field-grown crops such as vegetables, fruits, nuts, herbs and spices, and grains
Pseudomonas aureofaciens strain Tx-1	Spot-Less biofungicide (Eco Soils Systems, Inc., San Diego, CA); dollar spot (caused by *Sclerotinia homeocarpa*), anthracnose (*Colletotrichum graminicola*), pythium (*Pythium aphanidermatum*), and pink snow mold (*Microdochium nivale*) on turf grass
P. fluorescens A506	BlightBan A506 (Plant Health Technologies, Fresno, CA); frost damage caused by ice-nucleating bacteria, fire blight caused by *Erwinia amylovora*, and russet-inducing bacteria
P. syringae ESC 10	Bio-Save 10 line of bioprotectants (EcoScience Produce Systems Division, Orlando, FL); green mold, blue mold, and sour rot on citrus fruits
P. syringae ESC 11	Bio-Save 11 line of bioprotectants (EcoScience Produce Systems Division, Orlando, FL); benzimidazole-resistant *Penicillium expansum*, gray mold, and mucor rot on pome fruits

Streptomyces griseoviridis K61	Mycostop biofungicide (Kemira Agro Oy, Helsinki, Finland); seed rots, root and stem rots, and wilt diseases of ornamental crops caused by *Alternaria*, *Fusarium*, and *Phomopsis* species; *Botrytis* gray mold and *Pythium* and *Phytophthora* root rots in greenhouse-grown ornamentals
Fungi	
Ampelomyces quisqualis M10	AQ10 biofungicide (Ecogen, Inc., Langhorne, PA); a fungal hyperparasite for the control of powdery mildew on various crops caused by *Uncinula necator* or *Oidium tuckeri* (in the conidial state)
Candida oleophila isolate I-182	Aspire bioprotectant (Ecogen, Inc., Langhorne, PA); postharvest fruit decay caused by various pathogens
Colletotrichum gloeosporioides f.sp. *aeschynomene* ATCC 20358	Collego bioherbicide (Encore Technologies, Minnetonka, MN); control of the weed northern jointvetch (*Aeschynomene virginica*)
Gliocladium catenulatum strain J1446	Primastop biofungicide (Kemira Agro Oy, Helsinki, Finland); for greenhouse and indoor use for the control of damping-off, seed rot, root and stem rot, and wilt diseases on various food and ornamental plants caused by various fungi
G. virens G-21	SoilGard, formerly GlioGard (Thermo Trilogy, Columbia, MD); damping-off and root rot pathogens, especially *Rhizoctonia solani* and *Pythium* spp. on ornamental and food crop plants grown in greenhouses, nurseries, homes, and interiorscapes
Puccinia canaliculata ATCC 40199	Dr. BioSedge (no known producer); a bioherbicide for yellow nutsedge, *Cyperus esculentus*
Trichoderma harzianum ATCC 20476	Binab T (Bio-Innovation AB, Sweden); a biofungicide to control wilt, take-all, and root rot diseases of plants, internal decay of wood products, and decay of tree wounds
T. harzianum KRL-AG2 and *T. polysporum* ATCC 20475	RootShield and T-22 lines of biofungicides (BioWorks, Inc., Geneva, NY); root diseases in nursery and greenhouse crops and as a seed treatment for beans, cabbage, corn, cotton, cucumbers, peanuts, sorghum, soybeans, sugar beets, tomatoes, all ornamental crops, and vegetatively propagated crops such as potatoes and bulbs

Table 1 Continued

Active ingredient (agent)	Product name (registrant, if known) and Use
Virus or viral gene derived	
Potato leafroll virus replicase protein as produced in potato plant	New Leaf potato (registered by Monsanto Company, St. Louis, MO) has resistance to infection by PLRV and prevents feeding by Colorado potato beetle. New Leaf Plus potato is genetically engineered to express Cry III protein from *B. thuringiensis* subsp. *tenebrionis* and the *orf1/orf2* gene from PLRV as the active ingredients.
The following viral coat proteins have been granted tolerance exemptions:	
Papaya ringspot virus coat protein	Protection against severe strains of papaya ring spot virus in papaya
Potato leafroll virus coat protein as produced in potato plant	Protection against potato leafroll virus in potato
Potato virus Y coat protein	Protection against some viruses in the potato virus Y group
Watermelon mosaic virus coat protein in squash	Protection against watermelon mosaic virus in squash
Watermelon mosaic virus 2 and zucchini yellow mosaic virus coat protein in Asgrow ZW20 squash	Squash cultivar with protection against watermelon mosaic virus 2 and zucchini yellow mosaic virus (Asgrow Seed Company)
Zucchini yellow mosaic virus coat protein	Protection against zucchini yellow mosaic virus

Source: Based on EPA compilation dated June 3 and 4, 1999 [57]. This list may be incomplete due to lack of full or up-to-date registration or availability of records.

saprophytic fungi found in moist, organic, slightly basic soils throughout the world. They are acidophilic; their growth and biocontrol activities are more pronounced under acidic conditions. They are also commonly found on root surfaces, decaying plant matter in soil, and sclerotia of other fungi. They are generally less affected by soil chemical and heat treatments and can quickly colonize chemical- and heat-treated soils, being efficient colonizers of empty ecological niches created by the elimination of other competing microbes. They also sporulate abundantly in culture and on natural and artificial substrates and produce both conidia and chlamydospores.

The modes of action of biocontrol by *Trichoderma* spp. include competition for nutrients and sites, antibiosis, enzymatic action, and hyperparasitism. The competitive action results from their capability to grow very rapidly and effectively colonize soil and plant surfaces. In this way, they effectively outcompete and exclude plant pathogens from infection sites. In addition, *Trichoderma* spp. are known to produce certain volatile and nonvolatile antibiotic metabolites in culture (in vitro) and at sites of interaction with plant pathogens (in situ). The metabolites reported to be produced by *Trichoderma* spp. include gliotoxin, gliovirin, viridin, trichodermin, peptide-containing antibiotics, and possibly several other unknown antibiotics. Moreover, several enzymes, including cellobiase, chitinase, exo- and endoglucanases, lipase, and protease, which are involved in the mechanism of biocontrol activity, are produced by *Trichoderma* spp. Finally, many workers have provided conclusive evidence of the involvement of mycoparasitism in several biocontrol systems involving *Trichoderma* isolates. The mycoparasitic activity involves several steps: (1) chemotropic growth of *Trichoderma* toward the host pathogen's mycelium; (2) recognition of the pathogen's mycelium by *Trichoderma* mycelium; (3) coiling of the pathogen's mycelium around the fungal mycelium's (4) excretion of extracellular enzymes by the *Trichoderma* mycelium; and (5) lysis of the host mycelium. Some degree of plant growth promotion has also been found with some *Trichoderma* treatments [58].

Despite the general capability for rapid colonization, individual biocontrol isolates of *Trichoderma* must be carefully selected for their ability to survive, multiply, and establish on developing plant root surfaces and in the rhizosphere. The term "rhizosphere competence" is applied collectively to denote the ability of a microbe to colonize, establish, and effectively compete with other microbes in the rhizosphere, a zone of increased microbial activity compared to soil areas farther from this zone.

Several *Trichoderma* preparations have been tested, and some registered for use, against soilborne, foliar, and fruit-infecting pathogens. *Trichoderma* preparations alone and in combination with chemical fungicides have been found to be effective and economically viable alternatives to disease management based solely on chemical control [59]. Use of *Trichoderma* spp. in combination with chemical fungicides can also help slow the development of pathogen strains that are resistant to chemical fungicides and improve the predictability and effectiveness of the biocontrol agent.

Several *Trichoderma*-based biofungicides are registered and sold in the United States and abroad [57,60]. Three active ingredients—*T. harzianum* ATCC 20476, *T. harzianum* KRL-AG2, and *T. harzianum* ATCC 20475—are currently registered by the EPA (Table 1). Bio-Trek™, Rootshield™, and T-22™ Planter Box are three products based on *T. harzianum* KRL-AG2 (strain T-22) that are sold by Bio Works, Inc. of Geneva, NY. They are used in a variety of ways: Bio-Trek 22G as granules that are broadcast for control of diseases of turf grasses and new turf seedlings; RootShield granules for application to greenhouse planting mix and soil for control of soilborne pathogens and root diseases caused by *Fusarium*, *Pythium*, and *Rhizoctonia* spp.; RootShield drench for control of root diseases in nursery and greenhouse crops; and T-22 Planter Box as a seed treatment for beans, cabbage, corn, cotton, cucumbers, peanuts, sorghum, soybeans, sugar beets, tomatoes, all ornamental crops, and vegetatively propagated crops such as potatoes and bulbs. Strain T-22 actively colonizes growing plant roots and competes with pathogens for nutrients and biological niche. T-22 is an aggressive colonizer of roots and a strong microbial competitor. It directly attacks and kills pathogenic fungi through mycoparasitism.

DeVine. DeVine™ is the first bioherbicide registered in the United States for control of milkweed vine, *Morrenia odorata*, a major problem weed in the citrus groves of Florida [61]. It is produced and sold by Encore Technologies, Minnetonka, MN. The vine climbs onto the citrus trees and covers the canopy, interfering with light availability for citrus and hindering cultural practices and harvesting. The bioherbicide product consists of a liquid concentrate of chlamydospores of a pathotype of *Phytophthora palmivora* originally isolated from dying milkweed vines found in central Florida. The pathogen infects the roots, causes a root rot, and completely wilts the milkweed vine plants. It is capable of killing vines of all ages. On the basis of extensive host range and efficacy studies, the *P. palmivora* pathotype was determined to be a safe biocontrol agent for use in citrus and was registered in 1981. DeVine is produced and sold as a made-to-order product and is shipped as fresh, ready-to-use liquid spore concentrate. DeVine is highly effective; one postemergent, directed application of the product provides more than 90% weed control that lasts for at least 18 months [62,63].

Collego. Collego®, a bioherbicide based on *Colletotrichum gloeosporioides* f.sp. *aeschynomene*, an anthracnose-causing fungal pathogen, has been in use since its EPA registration in early 1982 to control northern jointvetch (*Aeschynomene virginica*) in rice and soybean crops in Arkansas and the neighboring rice-producing states in the United States. The weed is an indigenous leguminous plant. In addition to competition with rice and soybean crops, it produces hard-textured seeds that tend to contaminate harvested rice and soybeans, reducing their market value. The bioherbicide pathogen causes foliar and stem lesions (an anthracnose disease). Stem lesions girdle the stem, causing complete plant death.

College was developed by scientists of the University of Arkansas and the U.S. Department of Agriculture [64] and is now produced and sold by Encore Technologies. The commercial product is a wettable powder formulation of dried spores produced by liquid fermentation. Collego is applied postemergence with fixed-wing aircraft or land-based sprayers. It is capable of killing northern joint-vetch plants of all ages. Collego has provided consistently high levels of weed control (>85%), and it is well accepted by rice and soybean growers. During nearly two decades of commercial use of this bioherbicide agent, no environmental or human health hazards have been encountered. The effectiveness of Collego has been attributed to its ability to cause rapid disease onset followed by rapid secondary disease spread within infected fields [64].

Bioherbicides for Weedy Grasses, Purple Nutsedge, and Pigweeds (Amaranths). The most problematic weeds in citrus groves in Florida are annual and perennial weedy grasses, some of which are also considered serious weeds in many crops in several countries [65]. These include bahiagrass (*Paspalum notatum*), bermudagrass (*Cynodon dactylon*), large crabgrass (*Digitaria sanguinalis*), crowfootgrass (*Dactyloctenium aegyptium*), goosegrass (*Eleusine indica*), guineagrass (*Panicum maximum*; tall and short biotypes), johnsongrass (*Sorghum halepense*), napiergrass (*Pennisetum purpureum*), natalgrass (*Rhynchelytrum repens*), southern sandbur (*Cenchrus echinatus*), Texas panicum (*Panicum texanum*), torpedograss (*Panicum repens*), vaseygrass (*Paspalum urvillei*), and yellow foxtail (*Setaria glauca*). These grasses are difficult to control, either because of their tolerance to available chemical herbicides or due to their growth habits that enable them to overcome other control measures. Narrow-leaf guineagrass, in particular, poses a major weed problem in citrus in Florida because of its capacity for prolific spread and tolerance to chemical herbicides.

Development of host-specific fungal plant pathogens as bioherbicides may provide a nonchemical option for managing these weedy grasses. However, to be successfully adopted by citrus growers, a bioherbicide with broad-spectrum biocontrol activity against the major grass weeds is preferable to several individual bioherbicides, each capable of controlling a single weed species. Such a broad-spectrum bioherbicide should also provide a high level of control. These problems may be overcome by using a mixture of host-specific pathogens that are mutually compatible, have similar requirements for disease development, and, in a mixture, are capable of controlling several grass species. Accordingly, we have attempted to develop a multiple-pathogen bioherbicide system using three host-specific pathogens that are combined and applied simultaneously to control several weeds [66].

The bioherbicide system is based on three fungal pathogens—*Drechslera gigantea*, *Exserohilum longirostratum*, and *Exserohilum rostratum*—that were isolated respectively from large crabgrass, crowfootgrass, and johnsongrass in Florida (Fig. 2). In trials conducted in a greenhouse, these pathogens, when used individually or as a mixture, caused severe foliar blighting and killed large crabgrass, crowfootgrass, guineagrass, johnsongrass, southern sandbur, Texas panicum, and yellow foxtail. The fungi were tested, each at 2×10^5 spores/mL or as a 1:1:1 (v/v) mixture. Four-week-old plants of the grass species were almost completely killed (85% control) by each pathogen or the pathogen mixture. The fungi were nonpathogenic to many nontarget crop species, including citrus [66].

The multiple-pathogen approach has been field tested. An emulsion-based inoculum preparation (40% oil concentration) of each pathogen and a pathogen mixture gave almost complete control of the seven weedy grasses mentioned. The control lasted for 14 weeks without any significant regrowth of the grasses. The bioherbicidal control of a natural population of guineagrass with the pathogen mixture was also field tested. Again, an emulsion-based inoculum preparation of individual pathogens and a mixture of the three pathogens controlled guineagrass almost completely, and the control lasted for at least 10 weeks without regrowth. Presently, these fungi are undergoing further development for possible registration as a bioherbicide system to control weedy grasses in tree crops such as citrus and for landscape maintenance.

Purple nutsedge (*Cyperus rotundus*) is considered the world's worst weed [65]. Despite various control attempts, it continues to increase in importance under current agricultural practices. Although various management strategies are

FIGURE 2 Effect of inoculation with a pathogen mixture on selected weedy grasses. Left to right (in each picture): Crowfootgrass, Texas panicum, yellow foxtail, guineagrass, southern sandbur, johnsongrass, and large crabgrass. (A) Uninoculated control and (B) a pathogen mixture (1:1:1 v/v).

available to control purple nutsedge, none is entirely satisfactory when used alone. The main reasons for the difficulty in controlling this weed are the weed's ability for rapid growth, its proliferation from rhizomes and tubers, and its production of dormant tubers. A fungus, *Dactylaria higginsii*, a dematiaceous hyphomycete isolated from diseased purple nutsedge plants collected in Florida, has shown promise as a bioherbicide agent for this weed [67–70]. It causes a severe foliar blight characterized by typical eye-shaped, pale brown spots surrounded by a dark border (Fig. 3). In greenhouse and field trials, purple nutsedge plants were killed when *D. higginsii* was applied at an inoculum concentration of 10^6 conidiospores/mL (= 10^{12} spores in 1000 L/ha). The fungus was highly pathogenic to younger plants (four- to six-leaf stage) compared to older plants (>six-leaf stage). A temperature range of 20–30°C and a 12 hr exposure to dew period (100% relative humidity) were ideal for disease development.

Dactylaria higginsii is capable of reducing purple nutsedge growth by nearly 90% when applied at the rate of 10^6 spores/mL under tomato and pepper cropping systems. This translated into effective suppression of competition from purple nutsedge and prevention of losses in crop yield. Further studies on inoculum production and formulation, large-scale efficacy trials, and integration with pest management and crop protection systems are under way to develop and register *D. higginsii* for commercial use.

Phomopsis amaranthicola, a newly described species that is the causal agent of a leaf and stem blight of *Amaranthus* species [71,72] (Fig. 4), has been shown to have potential as a broad-spectrum bioherbicide for several pigweeds and amaranths [71]. Pycnidiospore suspensions of this fungus were most effective in causing high levels of plant mortality compared to mycelial suspensions, under both greenhouse and field conditions. Fungal suspensions (consisting of spores and/or mycelia) amended with a hydrophilic mucilloid, Metamucil, were effective in causing plant mortality even in the absence of dew, a condition necessary for fungal infection of aerial plant parts. Spore suspensions ranging from 1.5×10^6 to 1.5×10^7 spores/mL were most effective in killing pigweeds at two- to four-leaf stages. Temperatures of 25–35°C were conducive to disease development and plant mortality [71]. The fungus penetrates its hosts directly within 20 hr after inoculation. Appressorium formation and intracellular colonization could not be observed, but cell necrosis was seen 6 days after inoculation [73].

Several species of *Amaranthus* are susceptible to the fungus, but susceptibility does not lead to mortality in all cases. Species in which at least one biotype was highly susceptible (80–100% mortality) included *A. acutilobus*, *A. lividus*, *A. powellii*, A. retroflexus, and *A. viridus*. Plants within the family Amaranthaceae but outside the genus *Amaranthus*, several important species of crop plants, and a substantial number of plant species that are reported to be attacked by another *Phomopsis* sp. were also tested. Significantly, no plant outside the genus *Amaranthus* was susceptible, and there was no evidence of infection on any of

FIGURE 3 Biological control of purple nutsedge (*Cyperus rotundus*) by *Dactylaria higginsii*. Effect of spores of *D. higginsii* (10^6 mL^{-1}) suspended in different carriers on disease severity and mortality of purple nutsedge. Left to right: 0.05% N-Gel + spores; 0.02% Silwet L-77 + spores; control, 0.5% Metamucil only; water + spores; and 0.5% Metamucil + spores.

the nontarget plants by *P. amaranthicola* as determined by microscopic examination and isolation techniques [71,72].

Phomopsis amaranthicola has been successfully field tested in Florida against *A. hybridus*, *A. lividus*, *A. spinosus*, *A. retroflexus*, and *A. viridus*. In addition, a triazine-resistant accession of *A. hybridus* was screened. As in greenhouse trials, spore suspensions were most effective in causing high levels of plant mortality, although *A. lividus* and *A. viridus* were effectively controlled with spore or mycelial treatments. The results indicated that this fungus could be developed as a bioherbicide for integrated management of pigweeds and amaranths [71].

5.3 System Management Approach

The term "system management approach" was proposed by Müller-Schärer and Frantzen [74] to describe the concept of weed management based on manipulation of an existing biological control system. It replaces the older terms "augmentative approach" and "conservation approach," which are difficult to define

Figure 4 Symptoms of foliar and stem lesions on redroot pigweed (*Amaranthus retroflexus*) caused by *Phomopsis amaranthicola*.

in practice. The system management approach excludes methods such as the introduction of exotic organisms (inoculative control) or the mass release of inoculum (inundative control), which, from the perspective of this proposed approach, are considered to cause disruptive events. The approach envisages the control of a single weed species and focuses on the use of native natural enemies, especially those that cannot be produced in large quantities (e.g., biotrophs such as rust fungi). The aim of the system management approach to weed control is not to eradicate plant species but to manipulate the weed pathosystem by shifting the balance between the host and an indigenous pathogen population in favor of the pathogen. Weed control is achieved by stimulating the buildup of a disease epidemic on the target weed population, thus reducing the competitiveness of the weed. The strategy calls for a fundamental knowledge of the underlying mechanisms of crop production systems and is compatible with modern agroecological concepts. Frantzen and Hatcher [75] reviewed several interactions of the plant–natural enemy–environment–human system as they relate to the system management approach.

5.3.1 Management of Common Groundsel with a Rust Fungus

The fundamental research required to validate the feasibility of the system management approach has been done with common groundsel (*Senecio vul-*

garis) and a rust pathogen, *Puccinia lagenophorae* [74,76,77]. Since the 1980s, the autoecious *P. lagenophorae* has been seriously considered as a biological control agent for common groundsel, an annual weed in Europe and parts of North America [78,79]. Paul and coworkers have contributed substantially to the current knowledge about the physiological consequences of the rust infection on groundsel [80] and made a distinction between the pathogen's ability to provide initial kill versus effective suppression of groundsel's growth [81].

Common groundsel is a problematic weed in horticultural crops owing to its short generation time, high seed production, and rapid germination throughout the year. Groundsel plants compete strongly with crops for resources. Furthermore, the occurrence of populations of groundsel resistant to *s*-triazine herbicides and partially resistant to phenylurea herbicides, coupled with the use of herbicides having limited effectiveness against groundsel, have contributed to the weed's dominance in some agroecosystems. The rust pathogen *P. lagenophorae*, widely distributed in Australia, was first detected in Europe on *S. vulgaris* in the early 1960s, and it is now common throughout Europe [82]. It infects leaves, stems, and capitula by aeciospores and causes severe malformations and distortions. It overwinters as mycelia within the host plant. The aeciospores lose their germinative capacity over winter and cannot serve as a fresh inoculum source in the following year. Moreover, groundsel plants infected by *P. lagenophorae* in early autumn generally die. Plants infected late in autumn are more likely to survive. A few isolated pustules within a weed population are normally enough to start an epidemic in the spring. However, the epidemic starts slowly from the overwintering inoculum source [83].

Because stimulation of epidemics and reduction in competitiveness of the target weed are the key objectives of the system management approach, Frantzen and Müller-Schärer [84] emphasized establishment of infection foci as a way to reduce the competitiveness of a target weed. It is assumed that the epidemic starts from these foci and that the pathogen's inoculum sources needed to control the weed can be calculated from the number of inoculum sources (infection foci) and their spatial distribution within a weed population. Disease epidemics should progress sufficiently rapidly to reduce the weed's competitiveness before the crop enters the critical period when it is sensitive to competition.

The presence of resistant biotypes in the weed population could be a complicating factor. Resistant weed biotypes may slow down or delay the onset of epidemics [85]. Experiments by Wyss and Müller-Schärer [86], conducted under controlled conditions and probed by means of component analysis, confirmed the existence of race-nonspecific quantitative resistance in this pathosystem. All host plant line–pathogen line interactions were compatible, but the plant lines tested showed variation in susceptibility to the rust fungus. The highest level of resistance for which differences between plant lines were detected occurred at the penetration-peg stage. Resistance was also detected during the formation of primary hyphae and sori, but impacts of the rust on the host and spore production still occurred in some host line–rust line interactions. Disease severity increased on individual genotypes infected by an aggressive rust line. On a long-term basis, consequences of differences in disease level on individual plants and selection by more aggressive pathogens could favor the buildup of less susceptible weed populations. Buildup of host resistance, previously unknown and originating as a genetic response to the disease pressure from a weed biocontrol agent, has not yet been recorded. However, an increase has been seen in the abundance of preexistent resistant weed biotypes following the control of a dominant susceptible biotype by a pathogen. In the case of groundsel and *P. lagenophorae*, other factors may influence the host plant fitness and alleviate or override the effects imposed by the rust fungus [85]. If this should occur, other strains of the rust fungus aggressive with respect to the resistant biotypes may be introduced to supplement the previous strain.

A preliminary field study designed as a small-scale experiment under simulated crop production practices was carried out in *Apium graveolens* var. *rapaceum* (celeriac) to monitor the epidemic buildup and to quantify the impact of the rust fungus [87]. In the absence of rust infection on groundsel, the fresh weight of the celeriac bulbs was reduced by 28% by weed competition. However, the introduction of the rust fungus strongly reduced competition from groundsel and reduced crop loss. Groundsel biomass was also reduced, but rust infection only weakened and did not kill the plants. The weakened plants still contributed to soil cover and thus may help to suppress subsequent germination of other weed species. Further research is under way to determine the level of disease necessary to sufficiently impact the host plant, the population dynamics of common groundsel and the rust, effects of the rust on weed competition, and the effect of pesticides on the infection process and on groundsel.

6 USE OF GENES AND GENE PRODUCTS

Successful biological control using microbial agents requires several complex and often specific interactions between the biocontrol target and the biocontrol agent. These interactions are the primary reason for the inconsistency and unpredictability of biocontrol systems. Understanding these interactions at the genetic and molecular level should render the biocontrol system more predictable and manageable. Hence, it is logical to search for genes and gene products involved in the mode of action of biocontrol agents. Once the traits involved in the modes of action are identified, they can be used as markers to search for effective strains. The genes encoding these traits could be cloned, expressed, and used to engineer biocontrol agents for improved performance or to render crop plants resistant to pests and diseases. It may also be possible to disrupt the signal transduction involved in host response(s) to the pathogen. Both susceptibility and resistance responses of plants to pathogens may be disrupted in this manner.

6.1 Systemic Acquired Resistance

Exploitation of the phenomenon of systemic acquired resistance (SAR) is an example of a disease management technology based on molecular and genetic mechanisms of plants. SAR refers to the phenomenon by which plants acquire an ability to defend themselves against pathogens through induction of host resistance. After the formation of a necrotic lesion, either as part of a hypersensitive response (HR) or as a susceptible disease symptom, a distinct signal transduction pathway, the SAR pathway, is activated. SAR is a host-activated resistance response that occurs in response to an invading pathogen. The SAR response may be triggered by necrosis initiated by a pathogen, microbial extracts, certain chemical treatments, and some epiphytic microorganisms such as plant growth–promoting rhizobacteria [88]. SAR may be expressed as local or systemic response in the plant. Many comprehensive reviews on SAR have been published within the past decade [89–92].

The following are some biological characteristics of SAR [93]:

1. It is induced by biotic or abiotic agents or pathogens causing necrosis (e.g., local lesions).
2. There is a delay of several days between induction and full expression of SAR.
3. Tissues not exposed to the inducing agent are also protected.
4. SAR is expressed as reductions in the disease levels (e.g., in lesion number and size) or in the reproduction of the pathogen (e.g., spore production, pathogen multiplication, etc.)
5. Protection is long-lasting, from a few weeks to several months.
6. Protection is nonspecific and is effective against pathogens unrelated to the inducing agent (broad-spectrum control).
7. Signal for SAR is translocated and is graft-transmissible.
8. Protection is not passed on to progeny.
9. Accumulation of pathogenesis-related proteins (PR proteins) is associated with SAR response.
10. PR proteins such as chitinase and β-1,3-glucanase may inhibit the advance of pathogens at a local site.

Some synthetic chemical compounds can activate a systemic resistance response against plant pathogens by inducing endogenous SAR pathways. In this category, some benzothiadiazole compounds have been found to be effective activators of SAR [94]. One such compound, Actigard™ (benzo[1,2,3]thiadiazole-7-carbothioic acid-*S*-methyl ester; common name, acibenzolar-*S*-methyl), has been developed by Novartis for possible commercial use. Actigard confers broad-spectrum control of plant diseases by mimicking the natural SAR response found in most plant species. It has been evaluated on approximately 50 crop species and has proven effective in conferring broad-spectrum control of both bacterial and fungal plant pathogens. It is applied to foliage at rates in the range of 17–70 g active ingredient per hectare per application—rates that are extremely low compared to the rates of currently registered disease control products in the market. Actigard is rapidly taken up and translocated throughout the entire plant. The resulting plant defense response interferes with the pathogen's life cycle and may slow the rates of penetration and colonization by fungal pathogens [95]. Actigard has been shown to confer resistance to three fungi, two bacteria, and one virus in tobacco [96]. This broad-spectrum disease control may be due to the induction of several genes that encode PR proteins; for instance, in tobacco a set of nine gene families are induced [97].

6.2 Transgenic Crops Expressing Viral Coat Protein Genes

It has been known for more than a quarter of a century that resistance to certain plant viruses could be generated in plants with the aid of virus-derived, resistance-inducing proteins, nucleic acids, and genes [98]. Viral coat proteins, replicases, movement proteins, defective interfering RNAs and DNAs, and nontranslated RNAs are capable of inducing resistance in transgenic plants [99]. Viral coat proteins in particular have been used to engineer broad-spectrum tolerance to plant viruses in some plants.

Coat protein–mediated resistance was first reported in tobacco infected by tobacco mosaic virus [100]. Since then, coat protein–mediated resistance has been used to confer resistance to a number of viruses in several plant species. Coat protein–mediated resistance can provide either broad- or narrow-spectrum protection against viruses, and the reason for this differential ability is not totally clear. However, it has been possible to achieve broad-spectrum resistance to different virus strains of tomato spotted wilt virus by combining genes encoding the nucleoprotein from several strains in a single construct [101]. The first commercial use of coat protein–mediated resistance resulted in a virus-resistant squash produced by Asgrow Co. of the United States. Presently, viral coat proteins from the following six plant viruses have been approved for tolerance exemptions by the EPA: papaya ringspot virus coat protein, potato leafroll virus coat protein as produced in potato, potato virus Y coat protein, watermelon mosaic virus coat protein in squash, watermelon mosaic virus 2 and zucchini yellow mosaic virus coat proteins in Asgrow ZW20, and zucchini yellow mosaic virus coat protein [57]. A few virus-resistant cultivars transformed with some of these coat protein genes have been introduced for commercial use, including Prelude II squash (Asgrow) and zucchini cultivars Independence II, Declaration II, and

Sensation (Asgrow) and Dividend and Revenue (Novartis/Rogers Seed Company). A transgenic plant cultivar containing resistance to infection by potato leafroll virus and feeding by Colorado potato beetle has been approved as a plant pesticide (Table 1).

The mode of action of coat protein–mediated resistance is not fully clear, but there is significant evidence to indicate that viral disassembly in the initially infected plant cells, a necessary step in virus replication, is inhibited [98]. However, several lines of evidence suggest that other mechanisms may also be involved. For instance, it has been suggested that decreased viral titers in some virus-resistant transgenic plants may be due to the formation of a stable RNA duplex that inhibits viral replication [102].

7 EPILOGUE

As in the past, pest control practices are rapidly changing, dictated by market-driven and societal forces. Among the changes that we anticipate in the next several decades is a greater reliance on biologically based pest control methods [2,103]. Both traditional biological control based on classical and biopesticidal methods and newer controls grounded in molecular genetic mechanisms are likely to be used to a greater extent than now. Major new technological innovations are expected in the area of biopesticides, especially with respect to genetic improvement of biocontrol agents. The current level of emphasis on research and development of biologically based pest control agents and methods supports this prognosis.

REFERENCES

1. R Charudattan, HK Kaya, WJ Lewis, CE Rogers. Editorial: Biological control: theory and application in pest management. Biol Control 1:1, 1991.
2. U.S. Congress, Office of Technology Assessment. Biologically Based Technologies for Pest Control. OTA-ENV-636. Washington, DC; US Govt Printing Office, 1995.
3. BA Federici. A perspective on pathogens as biological control agents for insect pests. In: TS Bellows, TW Fisher, eds. Handbook of Biological Control. San Diego: Academic Press, 1999, pp 517–548.
4. BJ Jacobson, PA Backman. Biological and cultural plant disease controls: alternatives and supplements to chemicals in IPM systems. Plant Dis 77:311–317, 1993.
5. D Rosen, FD Bennett, JL Capinera. Pest Management in the Subtropics. Biological Control—A Florida Perspective. Andover, UK: Intercept, 1994.
6. TS Bellows, TW Fisher. Handbook of Biological Control. San Diego: Academic Press, 1999.
7. JR Ruberson. Handbook of Pest Management. New York: Marcel Dekker, 1999.

8. CH Pickett, RL Bugg. Enhancing Biological Control: Habitat Management to Promote Natural Enemies of Agricultural Pests. Berkeley, CA: Univ California Press, 1998.
9. DA Andow, DW Ragsdale, RF Nyvall. Ecological Interactions and Biological Control. Boulder, CO: Westview Press, 1997, p 334.
10. RJ Cook, KF Baker. The Nature and Practice of Biological Control of Plant Pathogens. St. Paul, MN: Am Phytopathol Soc, 1983, p 539.
11. JD Menzies. Occurrence and transfer of biological factor in soil that suppresses potato scab. Phytopathology 49:648–652, 1959.
12. KF Baker, RJ Cook. Biological Control of Plant Pathogens. San Francisco: WH Freeman, 1974, p 433.
13. RJ Cook. Making greater use of introduced microorganisms for biological control of plant pathogens. Annu Rev Phytopathol 31:53–80, 1993.
14. C Alabouvette, P Lemanceau, C Steinberg. Recent advances in the biological control of fusarium wilts. Pest Sci 37:365–373, 1993.
15. P Lemanceau. Role of competition for carbon and iron in mechanisms of soil suppressiveness to fusarium wilts. In: EC Tjamos, CH Beckman, eds. Vascular Wilt Diseases of Plants. Berlin: Springer-Verlag, 1989, pp 385–396.
16. J Louvet. Microbial populations and mechanisms determining soil suppressiveness to fusarium wilts. In: EC Tjamos, CH Beckman, eds. Vascular Wilt Diseases of Plants. Berlin: Springer-Verlag, 1989, pp 367–384.
17. FM Scher, R Baker. Mechanism of biological control in a fusarium-suppressive soil. Phytopathology 70:412–417, 1980.
18. FM Scher, R Baker. Effect of *Pseudomonas putida* and a synthetic iron chelator on induction of soil suppressiveness to fusarium wilt pathogens. Phytopathology 72:1567–1573, 1982.
19. B Sneh, M Dupler, Y Elad, R Baker. Chlamydospore germination of *Fusarium oxysporum* f.sp. *cucumerinum* as affected by fluorescent and lytic bacteria from fusarium-suppressive soil. Phytopathology 74:1115–1124, 1984.
20. J Louvet, F Rouxel, C Alabouvette. Recherches sur la résistance des sols aux maladies. I. Mise en évidence de la nature microbiologique de la résistance d'un sol au dévelopment de la fuasariose vasculaire du melon. Ann Pytopathol 8:425–436, 1976.
21. TC Paulitz, CS Park, R Baker. Biological control of fusarium wilt of cucumber with nonpathogenic isolates of *Fusarium oxysporum.* Can J Microbiol 33:349–353, 1987.
22. B Sneh, D Pozniak, D Salomon. Soil suppressiveness to fusarium wilt of melon, induced by repeated croppings of resistant varieties of melons. J Phytopathol 120: 347–354, 1987.
23. CS Park, TC Paulitz, R Baker. Biocontrol of fusarium wilt of cucumber resulting from interactions between *Pseudomonas putida* and nonpathogenic isolates of *Fusarium oxysporum.* Phytopathology 78:190–194, 1988.
24. C Alabouvette. Fusarium-wilt suppressive soils from the Châteaurenard region: review of a 10-year study. Agronomie 6:273–284, 1986.
25. Y Couteaudier. Competition for carbon in soil and rhizosphere, a mechanism involved in biological control of fusarium wilts. In: EC Tjamos, GC Papavizas, RJ

Cook, eds. Biological Control of Plant Diseases. New York: Plenum Press, 1992, pp 99–104.
26. RW Schneider. Effects of nonpathogenic strains of *Fusarium oxysporum* on celery root infection by *F. oxysporum* f.sp. *apii* and a novel use of the Lineweaver-Burke double reciprocal plot technique. Phytopathology 74:646–653, 1984.
27. A Matta. Induced resistance to fusarium wilt diseases. In: EC Tjamos, CH Beckman, eds. Vascular Wilt Diseases of Plants. Berlin: Springer-Verlag, 1989, pp 175–196.
28. Q Mandeel, R Baker. Mechanisms involved in biological control of fusarium wilt of cucumber with strains of nonpathogenic *Fusarium oxysporum*. Phytopathology 81:462–469, 1991.
29. C Gessler, J Kúc. Induction of resistance to fusarium wilt in cucumber by root and foliar pathogens. Phytopathology 72:1439–1441, 1982.
30. P Lemanceau, C Alabouvette, Y Couteaudier. Modifications du niveau de réceptivité d'un sol résistant et d'un sol sensible aux fusarioses vasculaires en réponse à des apports de fer ou de glucose. Agronomie 8:155–162, 1988.
31. Y Elad, R Baker. The role of competition for iron and carbon in suppression of chlamydospore germination of *Fusarium* spp. by *Pseudomonas* spp. Phytopathology 75:1053–1059, 1985.
32. LS Thomashow, DM Weller. Current concepts in the use of introduced bacteria for biological disease control: mechanisms and antifungal metabolites. In: G Stacey, NT Keen, eds. Plant-Microbe Interaction. New York: Chapman & Hall, 1996, pp 187–235.
33. CR Howell, RD Stipanovic. Suppression of *Pythium ultimum*–induced damping-off of cotton seedlings by *Pseudomonas fluorescens* and its antibiotic, pyoluteorin. Phytopathology 70:712–715, 1980.
34. C Voisard, CT Bull, C Keel, J Laville, M Maurhofer, U Schnider, G Défago, D Haas. Biocontrol of root diseases by *Pseudomonas fluorescens* CHA0: current concepts and experimental approaches. In: F O'Gara, DN Dowling, B Boesten, eds. Molecular Ecology of Rhizosphere Microoganisms: Biotechnology and the Release of GMO's. Weinheim, Germany: VCH, 1994, pp 66–89.
35. BK Duffy, G Défago. Zinc improves biocontrol of fusarium crown and root rot of tomato by *Pseudomonas fluorescens* and represses the production of pathogen metabolites inhibitory to bacterial antibiotic biosynthesis. Phytopathology 87: 1250–1257, 1997.
36. JG Fuchs, Y Moënne-Loccoz, G Défago. Nonpathogenic *Fusarium oxysporum* strain Fo47 induces resistance to fusarium wilt in tomato. Plant Dis 81:492–496, 1997.
37. J Postma, H Rattink. Biological control of fusarium wilt of carnation with a non-pathogenic isolate of *Fusarium oxysporum*. Can J Bot 70:1199–1205, 1992.
38. P Mattusch. Biologische Bekämpfung von *Fusarium oxysporum* an einigen gärtnerischen Kulturpflanzen. Nachrichtenbl Deut Pflanzenschutzd 42:148–150, 1990.
39. P Lemanceau, C Alabouvette. Biological control of fusarium diseases by fluorescent *Pseudomonas* and non-pathogenic *Fusarium*. Crop Prot 10:279–286, 1991.
40. AD Ryan, LL Kinkel. Inoculum density and population dynamics of suppressive

and pathogenic *Streptomyces* strains and their relationship to biological control of potato scab. Biol Control 10:180–186, 1997.
41. RP Larkin, DL Hopkins, FN Martin. Effect of successive watermelon plantings on *Fusarium oxysporum* and other microorganisms in soil suppressive and conducive to fusarium wilt of watermelon. Phytopathology 83:1097–1105, 1993.
42. D Hornby. Suppressive soils. Annu Rev Phytopathol 21:65–85, 1983.
43. PB Adams. The potential of mycoparasites for biological control of plant diseases. Annu Rev Phytopathol 28:59–72, 1990.
44. PB Adams, WA Ayers. Histological and physiological aspects of infection of sclerotia of two *Sclerotinia* species by two mycoparasites. Phytopathology 73:1072–1076, 1983.
45. PB Adams, WA Ayers. The world distribution of the mycoparasites *Sporidesmium sclerotivorum*, *Teratosperma oligocladum* and *Laterispora brevirama*. Soil Biol Biochem 17:583–584, 1985.
46. PB Adams, JJ Marois, WA Ayers. Population dynamics of the mycoparasite, *Sporidesmium sclerotivorum*, and its host, *Sclerotinia minor*, in soil. Soil Biol Biochem 16:627–633, 1984.
47. PB Adams, WA Ayers. *Sporidesmium sclerotivorum*: distribution and function in natural biological control of sclerotial fungi. Phytopathology 71:90–93, 1981.
48. DR Fravel, PB Adams, WE Potts. Use of disease progress curves to study the effects of the biocontrol agent *Sporidesmium sclerotivorum* on lettuce drop. Biocontrol Sci Technol 2:341–348, 1992.
49. PB Adams, DR Fravel. Dynamics of *Sporidesmium*, a naturally occurring fungal mycoparasite. In: RD Lumsden, JL Vaughn eds. Pest Management: Biologically Based Technologies. Beltsville Symp XVIII, Washington, DC: Am Chem Soc, 1993, pp 189–195.
50. MJ Morris. Impact of the gall-forming rust fungus *Uromycladium tepperianum* on the invasive tree *Acacia saligna* in South Africa. Biol Control 10:75–82, 1997.
51. WJ Janisiewicz. Control of postharvest diseases of fruits with biocontrol agents. In: H Komada, ed. The Biological Control of Plant Diseases. Teipei, RC Taiwan: Food and Fertilizer Technology Center for the Asian and Pacific Region, 1991, pp 56–68.
52. CL Wilson, ME Wisniewski. Biological control of postharvest diseases of fruits and vegetables: an emerging technology. Annu Rev Phytopathol 27:425–441, 1989.
53. Gustafson, Inc. Kodiak™ Biological Fungicide. Tech Bull. Gustafson, Inc., 1400 Preston Road, Suite 400, Plano, TX 75093, 1993.
54. PM Brannen, DS Kenney. Kodiak: A successful biological control product for suppression of soilborne plant pathogens of cotton. J Ind Microbiol Biotechnol 19: 169–171, 1997.
55. EcoScience Produce Systems Division. Bio-Save™ Postharvest Disease Preventatives. Tech Data Sheet, 2 pp, undated. EcoScience Corporation, 4300 LB McLeod Road, Suite C, PO Box 3228, Orlando, FL.
56. Ecogen Inc. Aspire™ Biofungicide. Specimen Label. Ecogen Inc, 2005 Cabot Blvd West, PO Box 3023, Langhorne, PA, 1996.
57. US Environmental Protection Agency. What are Biopesticides? Office of Pesti-

cide Programs, Washington, DC. http://www.epa.gov/pesticides/biopesticides/what_are_biopesticides.htm, 1999.
58. I Chet. *Trichoderma*: Application, mode of action, and potential as a biological control agent of soilborne plant pathogenic fungi. In: I Chet, ed. Innovative Approaches to Plant Disease Control. New York: Wiley, 1987, pp 137–160.
59. GE Harman, B Latorre, E Agosin, R San Martin, DG Riegel, PA Nielson, A Tronsmo, RC Pearson. Biological and integrated control of Botrytis bunch rot of grape using *Trichoderma* spp. Biol Control 7:259–266, 1996.
60. RD Lumsden, JA Lewis, JC Locke. Managing soilborne plant pathogens with fungal antagonists. In: RD Lumsden, JL Vaughn, eds. Pest Management: Biologically Based Technologies. Beltsville Symp XVIII. Washington, DC: Am Chem Soc, 1993, pp 196–203.
61. R Charudattan. The mycoherbicide approach with plant pathogens. In: DO TeBeest, ed. Microbial Control of Weeds. New York: Chapman and Hall, 1991, pp 24–57.
62. DS Kenney. DeVine: The way it was developed—An industrialist's view. Weed Sci (suppl) 34:15–16, 1986.
63. WH Ridings. Biological control of stranglervine in citrus: a researcher's view. Weed Sci (suppl) 34:32, 1986.
64. DO TeBeest, GE Templeton. Mycoherbicides: progress in the biological control of weeds. Plant Dis 69:6–10, 1985.
65. LG Holm, DL Plucknett, JV Pancho, JP Herberger. The World's Worst Weeds: Distribution and Biology. Honolulu, Hawaii: Univ Press of Hawaii, 1977.
66. S Chandramohan. Multiple-pathogen strategy for bioherbicidal control of several weeds. PhD Dissertation, University of Florida, Gainesville, FL, 1999.
67. JB Kadir, R Charudattan. *Dactylaria higginsii*, a fungal bioherbicide agent for purple nutsedge (*Cyperus rotundus*). Biol Control 17:113–124, 2000.
68. JB Kadir, R Charudattan, RD Berger. Effects of some epidemiological factors on levels of disease caused by *Dactylaria higginsii* on *Cyperus rotundus*. Weed Sci 48:61–68.
69. JB Kadir, R Charudattan, WM Stall, TA Bewick. Effect of *Dactylaria higginsii* on interference of *Cyperus rotundus* with *L. esculentum*. Weed Sci 47:682–686, 1999.
70. JB Kadir, R Charudattan, WM Stall, BJ Brecke. Field efficacy of *Dactylaria higginsii* as a bioherbicide for the control of purple nutsedge (*Cyperus rotundus*). Weed Technol 14:1–6, 2000.
71. EN Rosskopf. Evaluation of *Phomopsis amaranthicola* sp. nov. as a biological control agent for *Amaranthus* spp. PhD Dissertation, University of Florida, Gainesville, FL, 1997.
72. EN Rosskopf, R Charudattan, YM Shabana, GL Benny. *Phomopsis amaranthicola*, a new species from *Amaranthus* sp. Mycologia 92:114–122, 2000.
73. GS Wyss, R Charudattan. Infection process of *Phomopsis amaranthicola* on *Amaranthus* spp.: its implications for biological control. Phytopathology 89:S86, 1999.
74. H Müller-Schärer, J Frantzen. An emerging system management approach for biological weed control in crops: *Senecio vulgaris* as a research model. Weed Res 36: 483–491, 1996.
75. J Frantzen, PE Hatcher. A fresh view on the control of the annual plant *Senecio vulgaris*. Integrated Pest Manage Rev 2:77–85, 1997.

76. H Müller-Schärer. Biological control of weeds in crops: a proposal of a new COST action. Communications of the 4th Intl Conf on "Non-chemical Weed Control," Dijon, France, 1993, IFOAM, pp 181–185.
77. H Müller-Schärer, PC Scheepens. Biological control of weeds in crops: A co-ordinated European research programme (COST-816). Integrated Pest Manage Rev 2: 45–50, 1997.
78. H Müller-Schärer, GS Wyss. Das Gemeine Kreuzkraut (*Senecio vulgaris* L.): Problemunkraut und Möglichkeiten der biologischen Bekämpfung. Zeitschrift für Pflanzenkrankheiten und Pflanzenschutz, Sonderheft 14:201–209, 1994.
79. GS Wyss, H Müller-Schärer. *Puccinia lagenophorae* as a classical biocontrol agent for common groundsel (*Senecio vulgaris*) in the United States? (abstr). Phytopathology 88(suppl 9):S99, 1998.
80. ND Paul, PG Ayres, SG Hallett. Mycoherbicides and other biocontrol agents for *Senecio* spp. Pest Sci 37:323–329, 1993.
81. ND Paul, PG Ayres. The impact of a pathogen (*Puccinia lagenophorae*) on populations of groundsel (*Senecio vulgaris*) overwintering in the field. II: Reproduction. J Ecol 74:1085–1094, 1986.
82. G Viennot-Bourgin. La Rouille australienne du Séneçon. Rev Mycol 29:241–258, 1964.
83. J Frantzen, H Müller-Schärer. Wintering of the biotrophic fungus *Puccinia lagenophorae* within the annual plant *Senecio vulgaris*: implications for biological weed control. Plant Pathol 48:483–490, 1999.
84. J Frantzen, H Müller-Schärer. A theory relating focal epidemics to crop–weed interactions. Phytopathology 88:180–184, 1998.
85. JJ Burdon. Diseases and Plant Population Biology. Cambridge: Cambridge Univ Press, 1987.
86. GS Wyss, H Müller-Schärer. Infection process and resistance in the weed pathosystem *Senecio vulgaris L.-Puccinia lagenophorae* Cooke, and implications for biological control. Can J Bot 77:1–9, 1999.
87. H Müller-Schärer, S Rieger. Epidemic spread of the rust fungus *Puccinia lagenophorae* and its impacts on the competitive ability of *Senecio vulgaris* in celeriac during early development. Biocontrol Sci Technol 8:59–72, 1998.
88. JW Kloepper, S Tuzun, L Liu, G Wei. Plant growth-promoting rhizobacteria as inducers of systemic disease resistance. In: RD Lumsden, JL Vaughn, eds. Pest Management: Biologically Based Technologies. Beltsville Symp XVIII. Washington, DC: Am Chem Soc 1993, pp 156–165.
89. ZX Chen, J Malamy, J Henning, U Conrath, P Sanchezcasas, H Silva, J Ricigliano, DF Klessig. Induction, modification, and transduction of the salicylic acid signal in plant defense responses. Proc Natl Acad Sci USA 92:4134–4137, 1995.
90. M Hunt, J Ryals. Systemic acquired resistance signal transduction. Crit Rev Plant Sci 15:583–606, 1996.
91. U Neuenschwander, K Lawton, J Ryals. Systemic acquired resistance. In: G Stacey, NT Keen, eds. Plant–Microbe Interactions, Vol 1. New York: Chapman and Hall, 1996, pp 81–106.
92. L Sticher, B Mauch-Mani, JP Métraux. Systemic acquired resistance. Annu Rev Phytopathol 35:235–270, 1997.

93. JA Ryals, UH Neuenschwander, MG Willits, A Molina, H-Y Steiner, MD Hunt. Systemic acquired resistance. Plant Cell 8:1809–1819, 1996.
94. J Gorlach, S Volrath, G Knauf-Beiter, G Hengy, M Oostendorp, T Staub, E Ward, H Kessman, J Ryals. Benzothidiazole, a novel class of inducers of systemic acquired resistance activates gene expression and disease resistance in wheat. Plant Cell 8: 629–643, 1996.
95. FA Ruess, K Mueller, G Knauf-Beiter, T Staub. Plant activator CGA-245704: an innovative approach for disease control in cereals and tobacco. Brighton Crop Protection Conference—Pests and Diseases. Farnham, Surrey, UK: British Crop Protection Council, 1996, pp 53–60.
96. L Friedrich, K Lawton, W Ruess, P Masner, N Specker, M Gut Rella, B Meier, S Dicher, T Staub, S Uknes, JP Metraux, H Kessmann, J Ryals. A benzothiadiazole derivative induces systemic acquired resistance in tobacco. Plant J 10:61–70, 1996.
97. Novartis Crop Protection, Inc. Actigard™. Tech Bull. Greensboro, NC: Novartis Crop Protection, Inc., 1998.
98. DC Baulcombe. Mechanisms of pathogen-derived resistance to viruses in transgenic plants. Plant Cell 8:1833–1844, 1996.
99. RN Beachy. Mechanisms and applications of pathogen-derived resistance in transgenic plants. Curr Opin Biotechnol 8:215–220, 1997.
100. AP Powell, RS Nelson, B De, N Hoffmann, SG Rogers, RT Fraley, RN Beachy. Delay of disease development in transgenic plants that express the tobacco mosaic virus coat protein gene. Science 232:738–743, 1986.
101. M Prins, P De Haasn, R Luyten, M Van Veller, M Van Grinsven, R Goldbach. Broad resistance to tospoviruses in transgenic tobacco plants expressing three tospoviral nucleoprotein gene sequences. Mol Plant Microbe Interact 8:85–91, 1995.
102. SZ Pang, JL Slightom, D Gonsalves. Different mechanisms protect transgenic tobacco against tomato spotted wilt and impatiens necrotic spot tospoviruses. BioTechnol 11:819–824, 1987.
103. RWF Hardy, RN Beachy, WH Browning, JD Caulder, R Charudattan, P Faulkner, FL Gould, MK Hinkle, BA Jaffee, MK Knudson, WJ Lewis, JE Loper, DL Mahr, NK Van Alfen. Ecologically Based Pest Management: New Solutions for a New Century. Washington, DC: Natl Acad Press, 1996, p 144.

3

Economic Issues of Agricultural Pesticide Use and Policy in the United States

Craig D. Osteen and Merritt Padgitt*
Economic Research Service
U.S. Department of Agriculture
Washington, D.C., U.S.A.

1 INTRODUCTION

The development and growing use of synthetic organic pesticides have been an integral part of a technological revolution in U.S. agriculture that increased productivity by 2.5-fold between 1948 and 1994 [1]. Synthetic organic pesticide use grew dramatically from the late 1940s to the early 1980s before stabilizing and increased at a much slower rate through the 1990s.† Major factors affecting the trend since 1980 have been the development and use of new pesticides with reduced application rates and of genetically modified crops that reduce or modify the use of conventional pesticides.

Growth in pesticide use has created many controversies about potential effects of pesticide use on food safety, water quality, worker safety, wildlife

* The authors are agricultural economists with the Resource Economics Division, Economic Research Service, U.S. Department of Agriculture. The views presented are those of the authors and do not represent the official views of any agency or organization.

† The discussion of pesticide use trends is based on data collected through 1997, which were available when this chapter was written.

mortality, and pest control. These controversies reflect two major themes that have influenced the evolution of pesticide and pest management policy [2,3]:

1. Increasing pesticide use may be counterproductive for pest control, resulting in higher pest damages or control costs.
2. Undesirable health or environmental effects of the use of some pesticides may outweigh production benefits.

Increased public concern about the dietary risks of pesticides during the 1980s and 1990s led to a major change in pesticide law. New public concerns about the potential effects of genetically modified crops on pest control, human health, and the environment are emerging. The current focus of pesticide policy is on reducing dietary and other pesticide risks to meet safety standards rather than weighing risks and benefits and on mitigating adverse impacts by finding "safer" alternatives. Integrated pest management (IPM) has become a policy tool for reducing the risks of pesticide use as well as an approach for improving the effectiveness of pest control. This chapter discusses major pesticide use trends in the United States; the effects of such factors as pesticide productivity, farm programs, and pesticide regulations on use; and changing law and policy.

2 PESTICIDE USE TRENDS

Effective chemical control of agricultural pests became prevalent in the 1800s [4]. Paris green (copper acetoarsenite) was developed in the United States in the 1870s to combat the potato beetle, and Bordeaux mixture (quicklime and copper sulfate) was developed in France in the 1880s to control disease in grape culture. Prior to World War II, arsenicals, sulfur compounds, and oils were commonly used. However, the development of synthetic organic materials, such as 2,4-D and DDT, during World War II heralded the modern age of chemical pesticides. Pesticide expenses as a portion of farm production expenses (excluding operator dwellings) rose from 0.2% in 1920 to 4.8% in 1997 [5].

2.1 Aggregate Trends

Synthetic organic pesticide use grew rapidly from the late 1940s to the early 1980s as the percentage of crop acreage treated with pesticides increased. By the late 1970s, growth of pesticide use had slowed, because high proportions of crop acreages were being treated annually. Trends in pesticide use since 1980 have been heavily influenced by changes in crop acreage and the replacement of older compounds with new ones applied at lower per-acre rates. Synthetic organic pesticide use increased during the 1990s, but more slowly than before 1980. The U.S. Environmental Protection Agency (USEPA) published estimates that agricultural pesticide use grew from 366 million lb of active ingredient (a.i.) in 1964 to 843

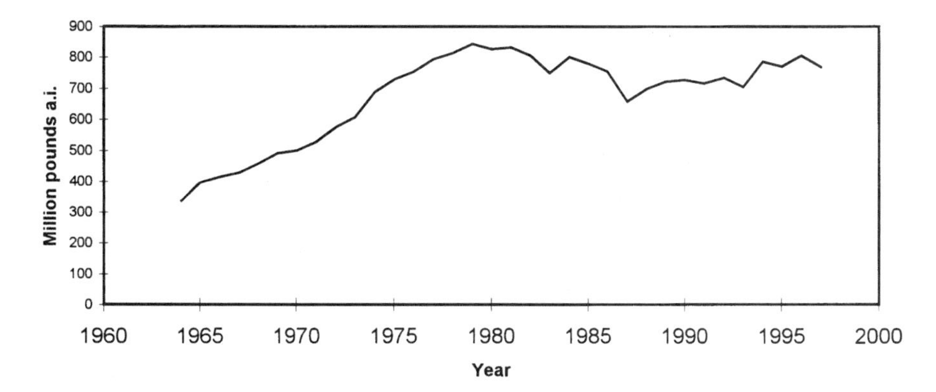

FIGURE 1 Quantity of agricultural pesticides used in the United States. (Data from Ref. 6.)

million lb in 1979, fell to 658 million lb in 1987, but rose to 770 million lb in 1997 (Fig. 1) [6]. (Estimates exclude sulfur, petroleum oil, wood preservatives, biocides, and other nonconventional chemicals.)

Some economists developed quality-adjusted indices that show larger long-term increases in pesticide use than the USEPA quantity estimates, because the materials used and their properties, such as toxicity and persistence, have changed over time. In particular, pesticides applied at rates of a fraction of a pound per acre have replaced pesticides applied at rates of several pounds per acre to control the same pests. Ball et al. [1] and Fernandez-Cornejo and Jans [7] developed quality-adjusted indices that showed that use increased by about threefold from 1968 to 1992, while unadjusted USEPA quantity estimates increased by 1.6 times.

Padgitt and others [8,9] developed aggregate use estimates for major crops from 1964 to 1997 from U.S. Department of Agriculture (USDA) pesticide surveys.* Use on these crops grew from 215 million lb a.i. in 1964 to 572 million

* Estimates in Table 1 and Figure 2 were constructed for corn, soybeans, wheat, cotton, potatoes, other vegetables, citrus fruit, apples, and other fruits and berries from USDA surveys conducted between 1964 and 1997. In years when the surveys did not include all states producing the crop, the estimates assume use rates similar to those of surveyed states. These estimates account for 52–56% of cropland acres for the 1964, 1966, and 1971 estimates and 67–70% of cropland acres for the 1982–1997 estimates. These estimates exclude use on such major crops as peanuts, rice, sorghum, barley, oats, rye, other grains, tobacco, alfalfa, hay, pasture, and nuts, because they were not surveyed or were surveyed only in a few years after 1982, making estimation of use after that date difficult. The excluded crop uses contribute to the differences between these estimates and the USEPA estimates [6]. These estimates also exclude sulfur, oils, and other nonconventional pesticides as well as postharvest pesticide use.

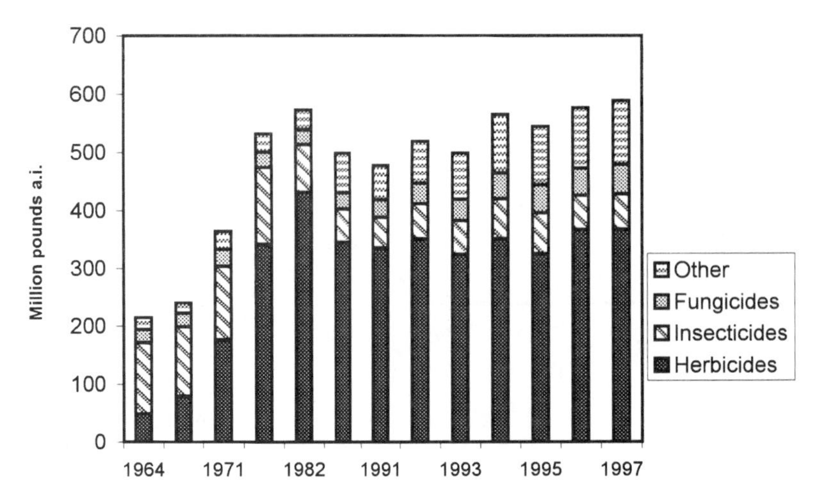

FIGURE 2 Pesticide use on major crops. (Data from Refs. 8 and 9.)

lb in 1982, fell to 478 million lb in 1991, and rose to a high of 588 million lb in 1997 (Fig. 2 and Table 1). Major components in that trend were:

1. An increase in pesticide use on corn and soybeans from 50 million lb a.i. in 1964 to 421 million lb a.i. in 1982, and then a decline to 312 million lb a.i. in 1997.
2. An increase in pesticide use on potatoes and other vegetables from 27 million lb a.i. in 1964 to 139 million lb in 1997.
3. An increase in pesticide use on cotton from 95 million lb a.i. in 1964 to 112 million lb a.i. in 1971 and then a decline to 68 million lb a.i. in 1997—a trend heavily influenced by changes in insecticide ingredients applied.
4. An increase in herbicide use on major crops from 48 million lb a.i. in 1964 to 430 million lb a.i. in 1982 and then a decline to 366 million lb a.i. in 1997.
5. An increase in insecticide use from 123 million lb a.i. in 1964 to 132 million lb a.i. in 1976, a dramatic fall to 83 million lb a.i. in 1982, and a continuing decline to 50–60 million lb in the 1990s.
6. An increase in fungicide use from 22 million lb a.i. in 1964 to 51 million lb a.i. in 1997.
7. An increase in use of "other pesticides" from 21 million lb a.i. in 1964 to 110 million lb a.i. in 1997.
8. A change in the mix of pesticides used over time, which reduced average application rates per acre, especially for herbicides and insecti-

cides. Also, during the 1990s, the number of pesticide treatments and ingredients applied per acre increased and an increasing proportion of treatments were made after planting rather than before or at planting.

2.2 Insecticides

In the 1950s, insecticides were widely used on a variety of high value crops including cotton, tobacco, fruits, potatoes, and other vegetables (Table 2) [10–15]. Somewhat later, insecticide use on other major field crops, particularly corn, increased rapidly. Insecticides were applied to less than 10% of corn acreage during the mid-1950s but to 35–40% by 1976. Since the mid-1980s, the proportion of corn acres treated fell from 45% to 25–30% in the 1990s. The proportion of cotton, potatoes, and many fruit and vegetable acres treated with insecticides remained high in the 1990s (Tables 2–4) [16,17].

The quantity of insecticide applied to major crops increased from 1964 to 1976 but in 1997 declined to less than 50% of that in 1976 (Table 1). Cotton and corn accounted for most of that decline. Cotton insecticide quantity fell from 73 million lb a.i. in 1971 to 64 million lb in 1976 and to 19 million lb in 1982, and varied between 10 and 30 million lb from 1982 to the late 1990s. Corn insecticide quantity declined from 30 million lb a.i. in 1982 to less than 21 million lb a.i. in the 1990s.

The decline in insecticide use reflects the changes in the compounds used, with reduced per-acre application rates. In the 1960s and 1970s, organophosphates and carbamates replaced organochlorines (Table 5) [12,18–21].* (See footnotes to Table 5 for examples of pesticides in the major classes.) Synthetic pyrethroids were rapidly adopted after their introduction in the late 1970s and accounted for over 20% of insecticide acre-treatments by 1982.† However, insecticide groups used in the 1960s—the organochlorines, organophosphates, and carbamates—still accounted for over 90% of insecticide quantity, and many active ingredients used in the 1960s continued to be widely used in the 1990s. The use of other new, low-rate insecticides, including abamectin (an antibiotic), diflubenzuron (a benzoylphenyl urea), and imidacloprid (a chloronicotinyl), increased during the 1990s. Synthetic pyrethroids and newer insecticide groups accounted for less than 5% of insecticide quantity in 1997 but because of their low rates of application, accounted for about one-third of insecticide acre-treatments.

The adoption of genetically modified crops may influence future insecticide use trends, but emerging concerns about their pest control, environmental, and

* The estimates for insecticide and herbicide families are restricted to use on corn, cotton, soybeans, wheat, and potatoes, which were surveyed in more years than the other major crops.

† Acre-treatments are the number of acres treated with a pesticide multiplied by the average number of treatments per acre.

TABLE 1 Estimated Quantity (Millions of Pounds) of Pesticide Active Ingredients Applied to Selected U.S. Crops, 1964–1997[a]

Commodity	1964	1966	1971	1976	1982	1990	1991	1992	1993	1994	1995	1996	1997
Herbicides													
Corn	25.5	46.0	101.1	207.1	243.4	217.5	210.2	224.4	202.0	215.6	186.3	211.6	211.8
Cotton	4.6	6.5	19.6	18.3	20.7	21.1	26.0	25.8	23.6	28.6	32.9	27.7	29.2
Wheat	9.2	8.2	11.6	21.9	19.5	16.6	13.6	17.4	18.3	20.7	20.0	30.5	24.3
Soybeans	4.2	10.4	36.5	81.1	133.2	74.4	69.9	67.4	64.1	69.3	68.1	77.8	83.7
Vegetables	3.5	5.7	5.6	7.2	5.9	7.3	7.2	8.0	8.2	9.1	10.1	10.6	9.9
Fruit	1.2	2.6	1.3	6.0	7.4	7.8	8.2	7.6	7.3	7.4	7.5	7.5	7.5
Total	48.2	79.4	175.7	341.6	430.1	344.7	335.1	350.6	323.5	350.7	324.9	365.7	366.4
Insecticides													
Corn	15.7	23.6	25.5	32.0	30.1	23.2	23.0	20.9	18.5	17.3	15.0	16.1	17.5
Cotton	78.0	64.9	73.4	64.1	19.2	13.6	8.2	15.3	15.4	23.9	30.0	18.7	19.3
Wheat	0.9	0.9	1.7	7.2	2.9	1.0	0.2	1.2	0.2	2.0	0.9	2.3	1.2
Soybeans	5.0	3.2	5.6	7.9	11.6	0.0	0.4	0.4	0.3	0.2	0.5	0.4	0.8
Vegetables	9.8	11.2	11.1	9.0	8.3	8.3	8.1	9.0	9.2	10.1	8.8	7.9	8.6
Fruit	13.9	15.5	10.4	11.6	10.6	11.3	12.9	13.3	14.4	14.5	14.7	13.9	13.2
Total	123.3	119.3	127.7	131.8	82.7	57.4	52.8	60.1	58.0	68.0	69.9	59.3	60.6
Fungicides													
Corn	0.0	0.0	0.0	0.0	0.1	0.0	0.0	0.0	0.0	0.0	0.0	0.0	0.0
Cotton	0.2	0.4	0.2	0.0	0.2	1.0	0.7	0.8	0.7	1.1	1.0	0.5	0.9
Wheat	0.0	0.0	0.0	0.9	1.1	0.2	0.1	1.2	0.7	1.0	0.5	0.2	0.1
Soybeans	0.0	0.0	0.0	0.2	0.1	0.0	0.0	0.1	0.0	0.0	0.0	0.0	0.0
Vegetables	7.7	7.6	9.8	9.3	10.7	15.7	16.3	20.9	23.1	29.6	32.4	32.2	35.2
Fruit	14.3	15.3	19.3	16.3	13.1	10.9	12.3	12.0	12.1	12.9	13.5	13.9	14.4
Total	22.2	23.3	29.3	26.7	25.3	27.8	29.4	35.0	36.6	44.6	47.4	46.8	50.6

Other pesticides													
Corn	0.1	0.5	0.4	0.5	0.1	0.0	0.0	0.0	0.0	0.0	0.0	0.0	0.0
Cotton	12.4	14.2	18.7	12.7	9.3	15.2	15.5	15.8	12.7	15.6	19.7	18.7	18.5
Wheat	0.0	0.0	0.2	0.0	0.0	0.0	0.0	0.0	0.0	0.0	0.0	0.0	0.0
Soybeans	0.0	0.0	0.1	2.0	2.4	0.0	0.0	0.0	0.0	0.0	0.0	0.0	0.0
Vegetables	5.9	0.6	9.8	13.7	21.4	52.4	44.2	56.5	67.2	84.6	79.7	81.6	85.5
Fruit	2.9	3.4	2.4	1.9	0.9	0.4	0.4	0.4	0.1	0.9	1.6	3.8	6.2
Total	21.3	18.7	31.6	30.8	34.1	68.0	60.1	72.7	80.0	101.1	101.0	104.1	110.2
All pesticides													
Corn	41.2	70.1	127.0	239.5	273.7	240.7	233.2	245.2	220.5	233.0	201.3	227.7	229.3
Cotton	95.3	86.0	111.9	95.2	49.5	50.9	50.3	57.6	52.3	69.1	83.7	65.6	68.0
Wheat	10.1	9.2	13.6	30.0	23.5	17.8	13.8	19.7	19.1	23.8	21.5	32.9	25.7
Soybeans	9.2	13.7	42.2	91.1	147.4	74.4	70.4	67.8	64.4	69.5	68.7	78.1	84.5
Vegetables	26.9	25.0	36.2	39.0	46.3	83.6	75.9	94.4	107.8	132.4	131.1	132.3	139.2
Fruit	32.4	36.6	33.4	35.8	32.0	30.2	33.9	33.4	34.0	35.6	37.1	39.1	41.2
Total	215.1	240.6	364.3	530.6	572.4	497.6	477.5	518.1	498.1	563.4	543.4	575.7	587.9

[a] Estimates include preharvest use of synthetic organic pesticides on corn, soybeans, wheat, cotton, potatoes, other vegetables, citrus fruit, apples, and other fruits and berries. They cover 52–56% of cropland for 1964, 1966, and 1971 and 67–70% for the 1982–1997 estimates. In years when the surveys did not include all states producing the crop, the estimates assume similar use rates for those states. Estimates exclude sulfur, oils, and other nonconventional pesticides. See footnote in Section 2.1 for more details.

Source: Refs. 8 and 9.

TABLE 2 Share of Crop Acres (Percent) Treated with Insecticides

Year	Corn	Cotton	Soybeans	Wheat	Sorghum	Apple	Citrus	Other deciduous	Other fruits/nuts	Potatoes	Other vegetables	Tobacco	Peanuts	Rice
1952	1	48	NA	NA	NA	—[a]	—[a]	—[a]	1	75	61	47	NA	NA
1958	6	66	NA	NA	NA	NA	NA	NA	NA	80	74	58	NA	NA
1966	33	54	4	2	2	92	97	72	59	89	56	81	70	10
1971	35	61	8	7	39	91	88	87	71	77	56	77	87	35
1976	38	60	7	14	27	NA	NA	NA	NA	NA	NA	76	55	11
1979	NA	48	NA	NA	NA	NA	NA	NA	NA	94	74	NA	NA	NA
1980	43	NA	11	NA	4	NA	NA	NA	NA	NA	NA	NA	NA	NA
1982	37	36	12	3	26	NA	NA	NA	NA	NA	NA	85	48	16
1984	42	63	8	NA	NA	NA	NA	NA	NA	NA	NA	NA	NA	NA
1985	45	65	7	5	NA	NA	NA	NA	NA	NA	NA	NA	NA	NA
1986	41	NA	4	7	NA	NA	NA	NA	NA	NA	NA	NA	NA	NA
1987	41	61	3	7	17	NA	NA	NA	NA	NA	NA	NA	NA	NA
1988	35	61	8	4	NA	NA	NA	NA	NA	89	NA	NA	NA	18
1989	32	68	3	11	NA	NA	NA	NA	NA	91	NA	NA	NA	22
1990	31	NA	NA	4	NA	NA	NA	NA	NA	88	NA	NA	NA	10
1991	31	66	2	8	16	NA	NA	NA	NA	92	NA	NA	56	16
1992	29	65	1	6	NA	NA	NA	NA	NA	88	—[c]	NA	NA	11
1993	28	65	2	3	NA	99	—[b]	—[b]	—[b]	86	NA	NA	NA	NA
1994	27	71	1	13	NA	NA	NA	NA	NA	83	—[c]	NA	NA	NA
1995	26	75	2	7	NA	98	—[b]	—[b]	—[b]	85	NA	NA	NA	NA
1996	29	79	1	13	NA	NA	NA	NA	NA	92	—[c]	96	NA	NA
1997	30	74	2	7	NA	96	—[b]	—[b]	—[b]	91	NA	NA	NA	NA

NA = Not available.

[a] Individual crop estimates not available; but Eichers et al. [12] presented estimates of the percent of total fruit and nut acres treated with insecticides: 82% in 1952, 81% in 1958, 87% in 1966, and 90% in 1971.

[b] See Table 3 for more detailed fruit information.

[c] See Table 4 for more detailed vegetable information.

Source: Refs. 10–15.

TABLE 3 Fruit-Bearing Acreage Treated with Pesticides, Major Producing States, 1993–1997

			Percent of planted area receiving applications											
			Herbicide			Insecticide			Fungicide			Other		
Fruit	Planted acres (1000s)	No. of states surveyed[a]	1993	1995	1997	1993	1995	1997	1993	1995	1997	1993	1995	1997
Grapes, all types	894	6	64	74	75	64	67	60	75	90	87	21	27	22
Oranges	833	2	94	97	91	90	94	88	57	69	65	14	13	14
Apples	351	10	43	63	60	99	98	96	88	93	90	56	59	56
Grapefruit	159	2	93	92	91	93	89	91	85	86	71	5	3	4
Peaches	136	9	49	66	54	99	97	82	98	97	84	3	4	6
Prunes	101	1	40	46	48	93	73	71	84	84	58	—[b]	4	4
Avocados	64	2	50	29	44	12	15	33	10	9	12	20	—[b]	20
Pears	68	4	44	65	57	98	96	90	92	90	85	59	44	52
Lemons	48	1	45	83	78	94	73	73	87	64	66	39	42	56
Cherries, sweet	48	4	71	61	61	88	92	84	14	93	80	34	48	45
Plums	44	1	49	48	74	98	75	85	99	71	69	—[b]	—[b]	8
Olives	37	1	70	54	53	89	14	16	79	30	30	—[b]	—[b]	—[b]
Cherries, tart	32	4	67	67	78	27	94	98	33	98	99	59	68	75
Nectarines	38	1	84	82	73	98	97	82	95	96	79	—[b]	—[b]	—[b]
Tangerines	39	1	84	83	80	87	90	79	59	73	56	4	21	3
Blueberries	34	4	75	73	67	91	86	83	81	87	88	2	8	14
Apricots	20	1	48	34	30	94	83	62	98	92	52	—[b]	—[b]	1
Figs	17	1	89	54	48	17	—[b]	1	—[b]	—[b]	—[b]	—[b]	—[b]	—[b]
Raspberries	13	2	83	92	90	80	83	90	92	90	95	—[b]	—[b]	5
Tangelos	13	1	95	99	96	97	96	97	89	82	91	6	8	27
Temples	7	1	99	99	96	98	98	98	92	97	94	2	—[b]	—[b]
Kiwi	6	1	63	65	41	11	13	20	—[b]	—[b]	15	—[b]	—[b]	—[b]
Dates	5	1	39	29	—[b]	75	12	4	40	54	18	—[b]	—[b]	—[b]

[a] Surveys were conducted in major producing states; the set of minor producing states surveyed was modified slightly between years.
[b] Insufficient reports to estimate.
Source: Ref. 16.

TABLE 4 Vegetable Acreage Treated with Pesticides, Major Producing States, 1992–1996

Vegetable	Planted acres (1000s)	Number of states surveyed[a]	Herbicide 1992	Herbicide 1994	Herbicide 1996	Insecticide 1992	Insecticide 1994	Insecticide 1996	Fungicide 1992	Fungicide 1994	Fungicide 1996	Other 1992	Other 1994	Other 1996
			Percent of acres											
Sweet corn, proc.	417	5	92	94	90	75	66	74	19	9	11	2	3	2
Tomatoes, proc.	318	1	90	76	78	81	71	71	92	86	90	27	41	48
Green peas, proc.	222	5	91	93	89	49	50	35	1	—[b]	2	—[b]	—[b]	—[b]
Lettuce, head	195	2	68	60	52	97	100	98	76	77	76	1	—[b]	1
Watermelon	164	6	37	41	43	53	45	41	71	64	65	4	4	6
Sweet corn, fresh	146	8	75	79	79	84	81	89	41	36	42	—[b]	—[b]	—[b]
Snap beans, proc.	134	4	95	91	90	68	58	72	55	41	49	—[b]	—[b]	—[b]
Onion	127	8	86	88	88	79	76	83	83	89	89	13	21	20
Cantaloupe	113	3	44	41	36[c]	78	82	85[c]	73	41	47[c]	5	10	15[c]
Honeydews	—[c]	—[c]	29	21	NA	84	88	NA	51	40	NA	10	12	NA
Carrots	108	6	67	72	89	37	34	40	79	71	78	13	12	21
Broccoli	106	1	58	67	64	95	96	96	31	36	37	1	2	1
Tomatoes, fresh	89	6	75	52	54	95	94	93	86	91	90	NA	58	56
Lettuce, other	74	2	59	46	52	92	89	86	72	60	73	—[b]	—[b]	1
Cucumbers, proc.	72	6	74	77	76	34	48	36	32	30	34	2	4	11
Asparagus	72	3	86	91	88	64	70	56	28	23	33	—[b]	—[b]	—[b]
Snap beans, fresh	67	7	52	60	49	77	79	75	62	63	73	3	—[b]	—[b]
Peppers, bell	65	5	65	57	67	85	92	88	66	73	75	34	36	43
Cabbage, fresh	64	7	49	55	62	96	97	94	53	60	57	1	2	3
Cucumbers, fresh	49	8	54	45	60	75	74	68	66	81	77	13	8	17
Strawberries	45	7	39	41	37	86	88	85	87	89	86	56	69	72
Cauliflower	44	2	44	62	31	94	99	97	21	51	18	2	2	1
Lima beans, proc.	31	5	NA	55	49	NA	84	60	NA	24	18	NA	—[b]	—[b]
Celery	26	2	82	64	68	100	100	97	98	99	86	12	3	—[b]
Spinach, fresh	12	3	57	52	56	73	75	72	63	46	49	5	—[b]	3

Percent of planted area receiving applications (Herbicide, Insecticide, Fungicide, Other).

[a] Surveys were conducted in major producing states; the set of minor producing states surveyed was modified slightly between years.
[b] Insufficient reports to estimate.
[c] Cantaloupes and honeydew melons included with other melons in 1996.
Source: Ref. 17.

TABLE 5 Shares (Percent) of Insecticide Use by Class[a]

Insecticide class	1964	1966	1971	1976	1982	1991	1997
Quantity							
Carbamates[b]	7	4	10	16	15	11	14
Organochlorines[c]	73	73	51	31	9	2	2
Organophosphates[d]	20	23	39	49	71	80	79
Pyrethroids[e]	0	0	0	0	4	3	3
Others	0	0	0	4	<1	5	1
Acre-treatments[f]							
Carbamates	NA	NA	NA	NA	14	11	13
Organochlorines	NA	NA	NA	NA	5	2	2
Organophosphates	NA	NA	NA	NA	60	57	56
Pyrethroids	NA	NA	NA	NA	21	27	25
Others	NA	NA	NA	NA	<1	3	4

NA = Not available.
[a] Estimated for corn, cotton, potatoes, soybeans, and wheat; excludes oils, sulfur, and other inorganics.
[b] Examples include aldicarb, carbaryl, carbofuran, formetanate, methomyl, and oxamyl.
[c] Examples include dicofol, endosulfan, methoxychlor, and many materials no longer registered: aldrin, chlordane, deldrin, DDT, and toxaphene.
[d] Examples include azinphos-methyl, chlorpyrifos, fonofos, malathion, methyl parathion, mevinphos, parathion, phorate, and terbufos.
[e] Examples include permethrin, cypermethrin, tralomethrin, deltamethrin, cyhalothrin, cyfluthrin, and esfenvalerate.
[f] Total acreage treated with a pesticide multiplied by average number of applications per acre.
Source: Refs. 12, 18–21.

health effects could limit further adoption. Crops that include a gene that produces the *Bacillus thuringiensis* (Bt) toxin to control Lepidopteran pests were introduced in the mid-1990s. This technology helps to control the European corn borer, a target for insecticides on a small portion of corn acreage, and bollworm, tobacco budworm, and pink bollworm, major targets for cotton insecticide use. USDA surveys showed that Bt-treated seed was planted on 19% of corn acreage and 17% of cotton acreage in the surveyed states in 1998 [22]. Bt-treated seed was planted on 35% of cotton acreage in the Mississippi Delta states, where a major portion of insecticide treatments is for bollworms and budworms.

2.3 Herbicides

Herbicide quantity increased rapidly from the late 1950s before stabilizing in the 1980s. Approximately 10% of corn and wheat and 5% of cotton acres were treated with herbicides in 1952 (Table 6). Herbicide use on corn, cotton, and soybeans (for which there are no data before 1966) stabilized at 90–97% of acres planted since 1980. Winter wheat herbicide use has varied in the range of 30–60% of planted acreage since 1986, while spring wheat use has varied between 80% and 95%. Limited data show similar increases for potatoes, peanuts, rice, and sorghum as well as for other fruits and vegetables (Tables 3, 4, and 6).

Herbicide quantity on the major crops increased dramatically between 1964 and 1982 (by 8.9-fold), but in the 1990s was 15–20% lower than estimated for 1982 (Table 1). The quantity applied to corn and soybeans, which account for the major portion of herbicide use, grew from 30 million lb a.i. in 1964 (62% of use on the major crops) to 377 million lb a.i. in 1982 (88%), before falling to 296 million lb (81%) in 1997. The quantity of herbicides used on cotton, wheat, vegetables, and fruit generally increased between 1964 and 1997, but these crops accounted for a declining share of herbicide use.

Much of the decline in quantity since 1982 was due to reduced crop acreage, particularly during the 1980s because the proportion of acreage treated with herbicides remained high, and to lower application rates for commonly used herbicides such as atrazine. But the change in the herbicide compounds used, which also reduced average application rates per acre, contributed (Table 7) [12,18–21]. Shares of total herbicide quantity declined for phenoxys, phenyl ureas, and benzoics between 1964 and 1997 and for carbamates since 1982. (See footnotes for Table 7 for examples of herbicides in each class.) During this time, shares grew significantly for amides and anilines. The share for triazines increased until 1976, then declined, but still exceeded 20% in the 1990s. New families of herbicides introduced since the 1970s account for increasing shares of use and include phosphinic acids, bipyridyls, benzothiadiazoles, benzoxazoles, oximes, pyridazinones, pyridines, sulfonyl ureas, and imidazolinones. Herbicide groups reported in the 1960s accounted for over 80% of herbicide applied in 1997, but families not reported before 1976 accounted for about 40% of acre-treatments. In particular, the shares for phosphinic acids and sulfonyl ureas have grown dramatically since 1982.

The adoption of genetically modified, herbicide-tolerant crops may influence future herbicide use trends by encouraging the application of specific herbicides, which might otherwise kill the crop, to control weeds. Emerging concerns about environmental and health effects and the development of herbicide-resistant weed species could limit further adoption. Currently, herbicide-tolerant corn, cotton, soybeans, and canola have been developed. The most commonly planted are glyphosate-tolerant, but glufosinate ammonium–tolerant corn and bromoxynil-tolerant cotton are also available. USDA surveys showed that herbicide-tolerant seed was planted on 18% of corn, 44% of soybean, and 26% of cotton acreage in surveyed states in 1998 [22]. These are large increases from 3% of corn, 7% of soybeans, and less than 1% of cotton acreage in 1996. The increased acreage

TABLE 6 Share (Percent) of Crop Acres Treated with Herbicides

Year	Corn	Cotton	Soybeans	Winter wheat	Spring wheat	Sorghum	Apples	Citrus	Other deciduous	Other fruit/nuts	Potatoes	Vegetables	Tobacco	Peanuts	Rice
1952	11	5	NA	12	—[a]	NA	NA	NA	NA	NA	NA	NA	NA	NA	NA
1958	27	7	NA	20	—[a]	NA	NA	NA	NA	NA	NA	NA	NA	NA	NA
1966	57	52	27	28	—[a]	30	16	29	13	18	59	28	2	63	52
1971	79	82	68	41	—[a]	46	35	22	19	34	51	40	7	92	95
1976	90	84	88	38	—[a]	51	NA	NA	NA	NA	NA	NA	55	93	83
1979	NA	91	NA	NA	NA	NA	NA	NA	NA	NA	73	NA	NA	NA	NA
1980	93	NA	92	NA	NA	61	NA	NA	NA	NA	NA	NA	NA	NA	NA
1982	95	97	93	42	—[a]	59	NA	NA	NA	NA	NA	NA	71	93	98
1984	95	93	94	NA	NA	NA	NA	NA	NA	NA	NA	NA	NA	NA	NA
1985	96	94	95	44	—[a]	NA	NA	NA	NA	NA	NA	NA	NA	NA	NA
1986	96	NA	96	53	86	NA	NA	NA	NA	NA	NA	NA	NA	NA	NA
1987	96	94	95	61	89	82	NA	NA	NA	NA	NA	NA	NA	NA	NA
1988	96	95	96	53	83	NA	NA	NA	NA	NA	NA	NA	NA	NA	98
1989	97	93	96	61	91	NA	NA	NA	NA	NA	77	NA	NA	NA	97
1990	95	95	95	34	89	NA	NA	NA	NA	NA	79	NA	NA	NA	98
1991	96	92	97	30	94	78	NA	NA	NA	NA	91	NA	NA	97	95
1992	97	91	98	35	91	NA	NA	NA	NA	NA	93	—[c]	NA	NA	97
1993	98	92	98	45	95	NA	43	—[b]	—[b]	—[b]	91	NA	NA	NA	NA
1994	98	94	98	50	96	NA	NA	NA	NA	NA	92	—[c]	NA	NA	NA
1995	97	97	98	59	95	NA	63	—[b]	—[b]	—[b]	94	NA	NA	NA	NA
1996	93	93	97	55	83	NA	NA	NA	NA	NA	91	—[c]	75	NA	NA
1997	97	96	98	47	82	NA	60	—[b]	—[b]	—[b]	88	NA	NA	NA	NA

NA = Not available.
[a] Spring wheat information combined with winter wheat information.
[b] See Table 3 for more detailed fruit information.
[c] See Table 4 for more detailed vegetable information.
Source: Refs. 10–15.

TABLE 7 Shares (Percent) of Herbicide Use by Class[a]

Herbicide class	1964	1966	1971	1976	1982	1991	1997
Quantity							
Arsenicals[b]	2	2	4	1	1	2	1
Phenoxys[c]	43	32	12	8	4	4	6
Phenyl ureas[d]	4	3	4	4	2	2	1
Amides[e]	0	4	24	30	31	35	35
Triazines[f]	23	30	32	32	26	29	26
Dintro group[g]	4	1	3	1	1	0	0
Carbamates[h]	10	9	5	11	17	9	3
Anilines[i]	2	7	8	9	11	12	13
Benzoics[j]	6	10	6	2	2	2	3
Phosphinic acids[k]	0	0	0	<1	1	2	6
Sulfonyl ureas[l]	0	0	0	0	<1	<1	<1
Other new families[m]	0	0	0	2	3	3	6
Others	6	2	2	<1	<1	<1	<1
Acre-treatments[n]							
Arsenicals	NA	NA	NA	NA	1	1	1
Phenoxys	NA	NA	NA	NA	13	10	11
Phenyl ureas	NA	NA	NA	NA	4	2	1
Amides	NA	NA	NA	NA	20	16	12
Triazines	NA	NA	NA	NA	26	24	17
Dinitro group	NA	NA	NA	NA	2	0	0
Carbamates	NA	NA	NA	NA	6	2	1
Anilines	NA	NA	NA	NA	15	13	10
Benzoics	NA	NA	NA	NA	5	6	7
Phosphinic acids	NA	NA	NA	NA	1	2	8
Sulfonyl ureas	NA	NA	NA	NA	<1	9	14
Other new families	NA	NA	NA	NA	7	15	18
Others	NA	NA	NA	NA	<1	<1	<1

NA = Not available.

[a] Estimated for corn, cotton, potatoes, soybeans, and wheat.

[b] DMSA, MSMA.

[c] 2,4-D, 2,4-DB, MCPA, MCPB.

[d] Diuron, linuron, fluometuron, terbacil.

[e] Alachlor, acetochlor, metolachlor, propachlor.

[f] Atrazine, cyanazine, propazine, simazine, metribuzin, ametryne.

[g] Dinoseb, DNBP.

[h] Butylate, EPTC, pebulate.

[i] Oryzalin, pendimethalin, ethalfluralin, trifluralin.

[j] Chloramben, dicamba, naptalam.

[k] Glyphosate, glufosinate-ammonium.

[l] Chlorsulfuron, halosulfuron, metsulfuron, nicosulfuron, primisulfuron.

[m] Includes bipyridyls (paraquat), benzothiadiazoles (bentazon), benoxazoles (fenaxaprop), imidizolinones (imazaquin, imazethapyr), diphenyl ethers (acifluorfen, diclofop, lactofen, oxyfluorfen), oximes (clethodim, clomazone, sethoxydim), pyridines (clorpyralid, fluazifop), pyridazinones (norfluorazon), and others that first appeared in pesticide use surveys since 1976.

[n] Sum of acreage treated with a pesticide multiplied by average number of applications per acre.

Source: Refs. 12, 18–21.

of herbicide-tolerant crops may be a factor in the dramatic increase of glyphosate (the primary phosphinic acid) use in the 1990s.

2.4 Fungicides

The estimated quantity of fungicides used on the major crops increased by about 2.3 times between 1964 and 1997 (Table 1). Fruits and vegetables, including potatoes, accounted for over 94% of fungicide use over that time period. Most of the increase occurred on potatoes and vegetables—more than 4.5-fold between 1964 and 1997. Potato acreage treated with fungicides increased steadily from 24% in 1966 to 85–98% in the 1990s (Table 8). An estimated 20% of the acres of "other vegetables" were treated with fungicides in 1966 and 1971, and by the 1990s much higher proportions of the acreage of many vegetables, such as celery, tomatoes, lettuce, melons, strawberries, and green peas, were treated (Table 4). By the early 1970s, a high proportion of fruit acreage was treated with fungicides, including about 70% of apple acreage and over 50% of citrus acreage. During the 1990s, somewhat higher proportions of apple, citrus, and other fruit crop acres were treated (Table 3).

As is the case for herbicides and insecticides, the change in fungicide compounds used over time contributed to lower per-acre application rates (Table 9) [12,18–21]. (See footnotes to Table 9 for more widely used fungicides in each class.) Shares of quantity declined for inorganics (primarily copper compounds) and dithiocarbamates since the 1960s but increased for phthalimides.* However, pthalimides, inorganic materials, and dithiocarbamates together accounted for over 90% of fungicide quantity in the 1960s and still accounted for almost 90% in 1997. The shares of newer groups, such as benzimidazoles, azoles, dicarboximides, metal organics, and acyclalanines accounted for about 10% of quantity but 35% of acre-treatments in 1997.

2.5 Other Pesticides

The estimated quantity of "other pesticides" used on the major crops increased by over fivefold between 1964 and 1997 (Table 1). This category includes soil fumigants, desiccants, harvest aids, and growth regulators. For the crops included, cotton, fruits, and vegetables accounted for virtually all of the quantity in the late 1990s.† Growth in the use of fumigants on potatoes and other vegetables and of sulfuric acid (a harvest aid) on potatoes accounts for much of the increased

* Estimates of shares of fungicide families include use on fruits and vegetables as well as corn, soybeans, cotton, wheat, and potatoes.

† Tobacco is a major use of "other pesticides" not included in these totals, but the proportional growth in use has not been large. Estimated use on tobacco was 18 million lb in 1964, 19 million lb in 1976, and 25 million lb in 1996.

TABLE 8 Share (Percent) of Crop Acres Treated with Fungicides (Excluding Seed Treatments)

Year	Corn	Cotton	Soybeans	Winter wheat	Spring wheat	Apples	Citrus	Other deciduous	Other fruits/nuts	Potatoes	Other vegetables	Tobacco	Peanuts	Rice
1966	<1	2	<1	<1	—[a]	72	73	58	39	24	20	7	35	<1
1971	1	4	2	<1	—[a]	67	58	54	46	49	18	7	85	<1
1976	1	8	3	<1	—[a]	NA	NA	NA	NA	NA	NA	30	76	<1
1979	NA	NA	NA	NA	NA	NA	NA	NA	NA	64	37	NA	NA	NA
1982	<1	2	1	1	—[a]	NA	NA	NA	NA	NA	NA	60	79	3
1988	NA	NA	NA	NA	NA	NA	NA	NA	NA	62	NA	NA	NA	14
1989	NA	NA	NA	NA	NA	NA	NA	NA	NA	69	NA	NA	NA	22
1990	NA	NA	NA	3	NA	NA	NA	NA	NA	67	NA	NA	NA	12
1991	<1	6	<1	1	3	NA	NA	NA	NA	69	NA	NA	89	24
1992	<1	7	<1	2	4	NA	NA	NA	NA	72	—[c]	NA	NA	21
1993	<1	6	<1	2	3	88	—[b]	—[b]	—[b]	76	NA	NA	NA	NA
1994	<1	10	<1	1	2	NA	NA	NA	NA	92	—[c]	NA	NA	NA
1995	<1	8	<1	1	3	93	—[b]	—[b]	—[b]	85	NA	NA	NA	NA
1996	<1	6	<1	1	<1	NA	NA	NA	NA	89	—[c]	49	NA	NA
1997	<1	7	<1	1	<1	90	—[b]	—[b]	—[b]	98	NA	NA	NA	NA

NA = Not available.

[a] Spring wheat information combined with winter wheat information.

[b] See Table 3 for more detailed fruit information.

[c] See Table 4 for more detailed vegetable information.

Source: Refs. 10–15.

Table 9 Share (Percent) of Fungicide Use by Class[a]

Fungicide class	1964	1966	1971	1997
Quantity				
Phthalimides[b]	23	29	21	31
Dithiocarbamates[c]	40	42	32	34
Inorganics[d]	28	23	40	24
Dinocap, dodine, quinones	4	5	4	0
Acyclalanines[e]	0	0	0	1
Azoles[f]	0	0	0	1
Benzimidazoles[g]	0	0	0	1
Dicarboximides[h]	0	0	0	2
Metal organics[i]	0	0	0	2
Other	5	2	4	4
Acre-treatments[j]				
Phthalimides	NA	NA	NA	27
Dithiocarbamates	NA	NA	NA	22
Inorganics	NA	NA	NA	15
Dinocap, dodine, quinones	NA	NA	NA	0
Acyclalanines	NA	NA	NA	5
Azoles	NA	NA	NA	4
Benzimidazoles	NA	NA	NA	13
Dicarboximides	NA	NA	NA	3
Metal organics	NA	NA	NA	11
Other	NA	NA	NA	10

[a] Includes use on fruit and vegetables as well as on corn, soybeans, cotton, wheat, and potatoes; excludes sulfur use.
[b] Includes captan, chlorothalonil.
[c] Includes maneb, mancozeb, metiram, thiram.
[d] Primarily copper compounds; excludes sulfur.
[e] Metalaxyl.
[f] Includes fenbuconazole, propiconazole, myclobutanil, triadimefon, and others.
[g] Includes benomyl, thiophanate-methyl, and thiabendazole.
[h] Includes iprodione, vinclozolin.
[i] Includes fosetyl-aluminum and triphenlytin hydroxide.
[j] Total acreage treated with a pesticide multiplied by average number of applications per acre.
Source: Refs. 12, 18–21.

quantity. These materials are used at very high per-acre rates and accounted for 85% of the quantity of other pesticides but less than 5% of the acres treated in 1997. In 1997, about 30 million lb of sulfuric acid, which was not reported in the early USDA surveys, was used on only 14% of potato acreage. The quantity of fumigants (methyl bromide, 1,3-D, chloropicrin, and metam-sodium) on the

TABLE 10 Share (Percent) of Crop Acres Treated with Other Pesticides

Year	Corn	Cotton	Soybeans	Wheat[a]	Sorghum	Apples	Citrus	Other deciduous	Other fruits/nuts	Potatoes	Other vegetables	Tobacco	Peanuts	Rice
1966	<1	26	<1	<1	<1	28	38	5	1	9	<1	69	<1	<1
1971	<1	36	<1	<1	<1	26	66	5	3	17	24	85	<1	<1
1976	1	34	1	<1	<1	NA	NA	NA	NA	NA	NA	86	6	NA
1979	NA	NA	NA	NA	NA	NA	NA	NA	NA	51	NA	NA	NA	NA
1982	<1	30	1	<1	<1	NA	NA	NA	NA	NA	NA	93	13	<1
1989	NA	50	NA	NA	NA	NA	NA	NA	NA	NA	NA	NA	NA	NA
1991	<1	58	<1	<1	<1	NA	NA	NA	NA	45	NA	NA	5	1
1992	<1	48	<1	<1	NA	NA	NA	NA	NA	43	—[c]	NA	NA	1
1993	<1	63	<1	<1	NA	56	—[b]	—[b]	—[b]	53	NA	NA	NA	NA
1994	<1	66	<1	<1	NA	NA	NA	NA	NA	60	—[c]	NA	NA	NA
1995	<1	56	<1	<1	NA	59	—[b]	—[b]	—[b]	57	NA	NA	NA	NA
1996	<1	60	<1	<1	NA	NA	NA	NA	NA	56	—[c]	98	NA	NA
1997	<1	73	<1	<1	NA	56	—[b]	—[b]	—[b]	65	NA	NA	NA	NA

NA = Not available.
[a] Spring wheat information combined with winter wheat information.
[b] See Table 3 for more detailed fruit information.
[c] See Table 4 for more detailed vegetable information.
Source: Refs. 10–15.

included crops increased from about 10 million lb during the 1964–1971 period to over 60 million lb in the 1990s. The use of growth regulators, desiccants, and harvest aids on cotton and other crops account for most of the acreage treated with "other pesticides."

Potatoes and vegetables have accounted for most of the increase in the quantity of "other pesticides" used (by almost 15 times). The proportion of potato acreage treated with such materials increased from 9% in 1966 to 55–60% in the late 1990s (Table 10). Limited information indicates that the acreage of other vegetable crops treated with these materials has also increased. The 1971 survey estimated that 24% of other vegetables were treated with such materials. Currently, a large proportion of tomato, strawberry, and pepper acres are treated with "other pesticides," including methyl bromide and other fumigants (Table 4). Cotton remains a major site for growth regulators and harvesting aids, but the quantity used increased only 50% from 1964 to 1997. The percent of cotton acreage treated increased from 26% in 1966 to over 60% in the late 1990s (Table 10). The increase in percent of acreage treated has been offset by changes from older materials, such as arsenic acid, sodium chlorate, and tribufos, to new ones applied at lower per-acre rates, such as ethephon, mepiquat chloride, thidiazuron, paraquat, and dimethepin. Growth regulators are also used on various fruit crops, including apples, pears, lemons, and tart cherries (Table 3).

3 ECONOMIC FACTORS AFFECTING PESTICIDE USE

Various economic factors affect farmers' choices of pest control practices and how intensively they use them. According to economic efficiency criteria, producers should choose the combination of pest control methods that maximizes the difference between the value of pest damage reductions and control costs. They should increase the use of pest control inputs until the marginal value of damage reduction (the value of the last unit used) equals the marginal cost. As a result, the prices of crops, pesticides, and other practices should influence the use of pesticides and other pest control practices. Fruits and vegetables for fresh markets often bring higher prices than those for processing markets, and market-driven quality standards can encourage pesticide use to prevent rots, surface blemishes, or other quality defects to increase returns. There can also be price incentives for postharvest pesticide use to protect the quality of stored grains, fresh fruits, dried fruits, and nuts.

Financial risk (variability of returns) and uncertainty (incomplete information about outcomes) are also important considerations in farmers' pest control decisions. Risk results from variations in yields and returns that are affected by changes in market conditions and natural variations in weather, pest infestations, and other factors affecting output. Uncertainty, which increases perceived risk, results from imperfect information about how these factors vary. Farmers do not know the precise value of pest damage without control or the reductions in damage from using control practices. They must develop expectations of crop value and potential yield savings from control. Rational decisions will subsequently appear suboptimal if pest infestations or crop values were different than expected. Because reducing the risk of large financial losses is important to many producers, some may find it rational to apply pesticides or other inputs in excess of profit-maximizing levels. Crop insurance for pest damage has been suggested as an alternative way to reduce risk without increasing pesticide use, but some research indicates that crop insurance encourages pesticide use [23]. Uncertainty about pest damage can be reduced by information about pest infestation levels from scouting or monitoring; models predicting yield losses from pests, weather, and other factors; and information about the effectiveness of pest control practices.

Pest mobility may create externalities, which are costs and damages not considered by the grower because another grower bears some of the impact of the decision. The more mobile a pest species is, the greater the externalities can be. Mobile pests can reinfest a treated area from an untreated area. From the viewpoint of a group of farmers, the most effective strategy might be for all to treat. However, a single farmer might underestimate potential pest damage, because some of it occurs elsewhere, and decide not to treat or to treat less than is desirable. Mobile pests can also spread resistance to pesticides and reduce their effectiveness. The response of a grower in that case might be to increase pesticide use to increase control. But, from the viewpoint of the group, the most effective strategy might be for the grower to help manage resistance by reducing application rates, eliminating treatments, or using nonchemical practices. Large area control programs can coordinate grower actions and more effectively control more mobile and damaging pests and manage resistance [24]. They may also create economies of scale for monitoring or controlling pests. Government pest eradication programs for such pests as the boll weevil may require grower participation and/or provide subsidies for participation to improve effectiveness of the program and to prevent nonparticipating growers from benefiting [25].

3.1 Pesticide Cost Efficiency

One argument for the increase in synthetic pesticide use from the end of World War II through 1980 is that pesticides often cost less and contributed to higher, less variable yields than previously used methods. Fernandez-Cornejo et al. [26] reviewed pesticide productivity studies that account for the yield-increasing effects of pesticides as well as the effects of pesticide and crop prices. They said that many of the studies, but not all, showed pesticides to be cost-efficient inputs from the farmer's perspective because marginal return to pesticide use exceeded cost [27–37]. They also said that some studies indicate that the marginal return of pesticide use is declining over time, which is to be expected as pesticide use increases.

Relative price trends may have influenced the cost effectiveness of pesticides and the amount used. Ball et al. [1] estimated that pesticide prices generally rose relative to crop and fuel prices between 1948 and 1997, which would tend to discourage pesticide use, but that pesticide prices fell relative to wages, which would tend to encourage more pesticide use to reduce labor use (Fig. 3). However, pesticide prices fell relative to wages, fuel prices, and crop prices from the late 1960s to 1980, a period of rapid growth in pesticide use. Price trends during that period would have reduced the costs of pesticides relative to other control methods and encouraged the substitution of pesticides for labor, fuel, and machinery used in pest control [38]. The increase in crop prices relative to pesticide prices would have increased the returns to pesticides and other yield-increasing inputs and encouraged greater use. These trends also may have induced technological change to take advantage of relatively cheap pesticides [39]. However, pesticide price trends since 1980 have returned to the longer trend, with pesticide prices rising relative to crop prices and fuel prices but continuing to fall relative to wages. Rising relative pesticide prices may have contributed to the stabilizing

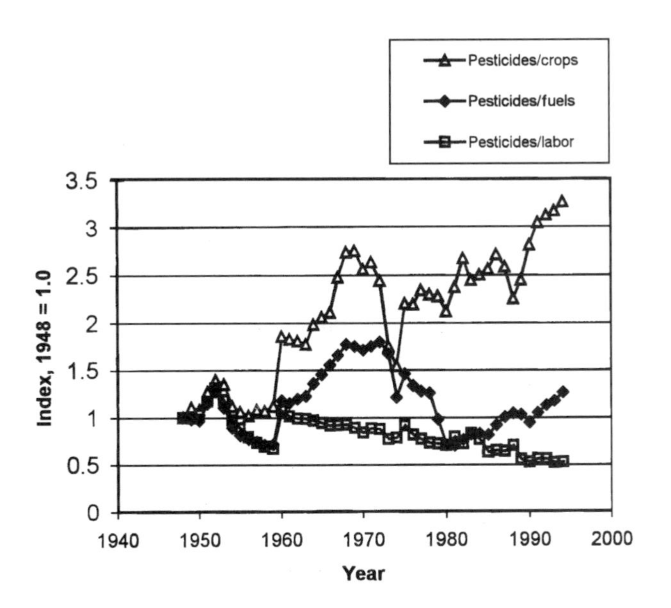

FIGURE 3 Relative price of pesticides compared to fuels, labor, and crops. (Data from Ref. 1.)

of pesticide use since 1980 and also may have resulted from a high level of demand for pesticide use in crop production.

3.2 Effect of Farm Programs

Many economists argue that commodity programs encouraged more pesticide use than would have been optimal under free markets [36,40]. The combination of target prices, loan rates, acreage restrictions, and inflexible base acreage encouraged greater per-acre use of pesticides and other yield-increasing inputs and more continuous cropping instead of rotation. Ribaudo and Shoemaker [41] found that participants in federal commodity programs used higher nitrogen fertilizer and herbicide application rates than did nonparticipants. By increasing returns and reducing financial risk for program crops, the programs may have encouraged more program crop acreage and greater pesticide use. However, acreage restrictions reduced total pesticide use in comparison to previous years by reducing acreage planted to program crops.

Pesticide use grew rapidly during the 1960s when farm programs restricted crop acreage. From the mid-1970s to the early 1980s, when pesticide use grew to market saturation, acreage restrictions were relaxed, export demand for U.S. commodities was high, and crop prices and acres increased. During the 1980s, low crop prices, acreage diversion, and land retirement contributed to reductions in pesticide use. During the 1990s, pesticide use increased, and increased acreage of planted crops, relaxation of acreage restrictions, greater planting flexibility within the programs, and higher crop prices may have contributed.

Farm program incentives for pesticides or other yield-enhancing inputs were steadily decreased through changes in farm legislation in 1977, 1985, 1990, and 1996. These changes steadily reduced restrictions on farmers' planting decisions and the relation between current production and program payments. Under the Federal Agriculture Improvement and Reform Act of 1996, producers were permitted to plant 100% of their total base acreage plus additional acreage to any crop (with some exceptions for fruits and vegetables) without loss of federal subsidy. However, producers' greater planting flexibility could lead to increased pesticide use when idled land returns to production.

4 COUNTERPRODUCTIVE PESTICIDE APPLICATIONS

Despite the apparent contribution to production efficiency, increased pesticide use is not a panacea for all pest problems. Scheduled or prophylactic treatments when pest infestations are low may have little effect on yield, and the value of damage reduction might not exceed cost. Some applications destroy beneficial organisms and natural enemies to pests. As a result, secondary outbreaks could require additional treatments, while species that were adequately controlled by natural enemies become pests. Continued exposure of pest populations to a chemical often leaves the most resistant individuals, which reduces the effectiveness of the chemical, creates the potential for pest outbreaks, and encourages further counterproductive pesticide use. Continuous plantings of some crops can encourage pest population growth and greater use of pesticides than the rotation of several crops would. A monoculture of genetically uniform, high-yielding varieties and high use of pesticides without regard for beneficial species or pest resistance can create the potential for damaging pest outbreaks. As a result, reducing pesticide use could lower pest damage and control costs in some circumstances. Stern et al. [42] discussed the economic threshold and integrated control concepts as ways to address the problems of counterproductive pesticide applications. These concepts have had a significant influence on the science and economics of pest management.

4.1 Economic Thresholds

The concept of the economic threshold is based on the notion that pests should be controlled only when the value of damage reduction exceeds the cost of control [42–44]. Treatments are economically justified when infestations exceed the threshold or pest population level where damage reduction equals control cost. Pest monitoring information and damage projections, which incur costs, are needed to implement thresholds. If they eliminate uneconomic applications and reduce pesticide use, thresholds can reduce pest control costs, the destruction of beneficial species and natural enemies to pests, the development of pest resistance, and adverse health, safety, and environmental effects.

According to economic theory, thresholds and pesticide application rates will respond to economic factors. Higher crop prices or lower control costs increase optimal rates or lower thresholds. With some exceptions, economists generally argue that risk and uncertainty encourage more pesticide use through higher rates or lower thresholds [45–48]. In contrast, some studies indicate that growers may use nonpesticide practices to reduce risk from pest damage [49–51]. Improved monitoring information about pest damage can reduce uncertainty and thus reduce dosages or increase thresholds [52]. However, the benefits of monitoring must be compared to the costs. One study showed that premature insecticide applications on soybeans in Georgia had little effect on net returns compared with strict threshold compliance, allowing farmers to maintain a high level of crop protection without incurring the costs of a scouting program [53]. Economists also examined the impacts of dynamics on economic thresholds with optimal control models [54,55].

4.2 Integrated Pest Management

Integrated pest management (IPM) is an approach that can reduce counterproductive pesticide applications. Stern et al. [42] originally defined integrated control as "applied pest control which combines and integrates biological and chemical control." IPM focuses on optimizing the use of chemical, biological, and cultural controls, including varietal resistance to pests, trap crops, augmentation of natural enemies, and crop rotation, to manage pest problems rather than relying solely on chemical use [56]. IPM programs often include pest monitoring and economic thresholds. Methods of biological control that can be included in an IPM program include the use of pest predators, parasites, and other beneficial organisms and can also include pheromones or microbial organisms that are regulated as pesticides. Organic production and sustainable agriculture are approaches to crop production that can incorporate various pest management techniques to reduce or eliminate pesticide use. Certification of organic production often excludes the use of synthetic organic pesticides.

Integrated pest management was originally developed as an approach to control pests more cost effectively over time, and it has influenced the science and practice of pest control. More recently, IPM has become a policy tool to reduce the use and risks of pesticides. In the late 1980s, there was an emerging interest by some groups in the United States in restricting or reducing the total amount of pesticides used, and one goal was to reduce the adverse environmental and health effects. Many proponents argued that some pesticides were overused and that more efficient application technology, nonchemical practices, pest monitoring, and economic thresholds or crop rotations can reduce pesticide use with relatively small economic losses, while adverse environmental and health effects would be reduced significantly [57]. Some European countries, including Denmark and Sweden, instituted programs to reduce pesticide use by 50%. Pettersson [58] said that the quantity of active ingredient used in Sweden was reduced by 50% between 1985 and 1990 with little effect on acreage treated, which was attributed to the reduction of application rates, the use of more efficient application technology, and a change to new, lower application rate pesticides. More recently, some groups have argued that the practice of IPM has become overly oriented to using pesticides to control pests rather than reducing pesticide use [59]. In response, the concepts of bio-intensive IPM and ecologically based IPM have been developed [60,61]. These concepts focus on reducing the use of synthetic organic pesticides, increasing the emphasis on reduced risk pesticides and nonchemical practices, and understanding crop and pest ecology.

The United States has instituted a policy of implementing IPM to help reduce health and environmental risks from pesticides but has not adopted a goal of reducing pesticide use by a specified percentage. In September 1993, the Clinton Administration called for reducing the use of high-risk pesticides, particularly through increased use of IPM techniques, and set a goal that by 2000 75% of all farms would use IPM techniques that reduce pesticide use. In August 1994, the USDA and the USEPA signed a Memorandum of Understanding for an IPM Initiative to develop IPM techniques and pursue this goal. The Food Quality Protection Act of 1996, which is discussed in more detail in Section 5.2, also requires the USDA, in cooperation with the USEPA, to conduct research and education programs to support the adoption of IPM.

5 PESTICIDE REGULATORY POLICY

Pesticide use has increased within the context of regulatory law and policy, which have been shaped by changing public attitudes and political pressure. One important issue has been the balance of production benefits against the health and environmental hazards of pesticide use [62]. There have been major public reactions to the alleged health and environmental hazards of increased pesticide use since the 1960s. Important issues include farm worker safety, cancer risks, birth defects, wildlife mortality, water quality, endangered species, and food safety. Unless they directly affect returns, adverse health and environmental effects might not affect the grower's decision to apply a pesticide. One major idea that changed regulatory policy is that the hazards of using some pesticides might outweigh their benefits. In recent years, some interest groups have argued that benefits should have no role in regulatory decision-making.

The regulatory process defines what pesticides and use practices are legal. Under the Federal Insecticide, Fungicide, and Rodenticide Act (FIFRA) and the Federal Food, Drug, and Cosmetic Act (FFDCA), the USEPA decides whether or not to register new uses of previously registered or unregistered pesticides, modify existing pesticide registrations, and cancel some or all registered uses of pesticides on the market.* The Clean Air Act, Clean Water Act, Endangered Species Act, and the Occupational Safety and Health Act also affect the use of pesticides.

Current pesticide regulatory policy recognizes a role for pesticides in crop production but emphasizes protection from hazards of use. The regulatory approach is to mitigate the risks of using pesticides by modifying use rates and practices, cancel uses of pesticides that do not meet safety standards, and register

* Before a pesticide can be used in the United States, it must be registered under FIFRA, currently administered by the USEPA. Registrations specify sites (such as specific crops or livestock) where pesticides can be applied, application rate, methods of use, or locations of use for pesticide products. For a pesticide to be registered for use on a food crop, FFDCA requires residue tolerances or exemptions from tolerance for the raw commodity and all processed foods and feeds, rotational crops, and livestock where residues can be found. The USEPA establishes residue tolerances; the FDA monitors residues and enforces the tolerances.

"reduced risk" pesticides. The focus is on meeting safety standards, especially for dietary risks, rather than on weighing risks and benefits.

Over time, the regulatory process has influenced aggregate quantities of pesticides used by affecting what pesticides can be used and their use practices. The requirements of the regulatory process have influenced the innovation of pest control products that are considered for registration and the structure of markets for those products. Ollinger and Fernandez-Cornejo [63] estimated that the research and development of a new pesticide takes 11 years and can cost manufacturers between $50 and $70 million. Their results indicated that regulation encourages the development of less toxic pesticide materials, such as biological pesticides, discourages new chemical registrations, encourages firms to abandon pesticide registrations for minor crops, and favors large firms over smaller ones. An important result has been the development and registration of pesticides with low application rates. However, the extent of overall pesticide use has primarily responded to such economic factors as input and output markets and commodity programs.

5.1 A Review of Changing Policy

From the early 1900s, before pesticide use was widespread, until the 1960s, when it was increasing rapidly, U.S. pesticide legislation encouraged adoption of the new technology by regulating product effectiveness, labeling contents, and warning users about acutely toxic ingredients [62,64]. (See Table 11 for a summary of important pesticide legislation.) Concerns about the presence and safety of chemical residues in food emerged in the 1950s, which resulted in FFDCA amendments in 1954 and 1958 to require pesticide residue tolerances for raw food and feed commodities and processed products. The 1958 amendment included the Delaney Clause, which prohibited food additives found to induce cancer in humans or animals.

Public concern about potential hazards of chemical use to the environment emerged in the 1960s, when pesticide use was growing rapidly. FIFRA amendments in the 1960s and 1970s focused the regulatory process on protection from health and environmental hazards. These laws created a role for balancing the risks and benefits in decisions to register new uses of pesticides or to modify, cancel, or suspend existing ones. The result was a series of formal reviews on the risks and benefits of pesticides.

Reregistration became a major focus in the 1980s and 1990s. The review of previously registered pesticides was identified as reregistration in the 1978 amendments, and in 1988 amendments were passed to speed the process and provide additional financial resources through fees. During this process, the USEPA raised many pesticide risk issues with registrants, who, in many cases, voluntarily changed labels or canceled uses to meet safety standards without going through a costly formal review. An important impact was to focus the regulatory process on the data and procedures of risk assessment and to reduce the role of formal risk and benefit comparisons.

5.2 Food Quality Protection Act of 1996

Several important issues led to the Food Quality Protection Act of 1996 (FQPA), which amended FIFRA and FFDCA. Among those issues were (1) increasing public concern about the safety of pesticide residues in food, especially in food consumed by children; (2) enforcement of the Delaney Clause; and (3) concerns voiced by producers of fruits, vegetables, and other small-acreage crops that the regulatory process resulted in an inadequate number of pesticides being available for use on those crops.

5.2.1 Food Safety and the Delaney Clause

The National Academy of Sciences (NAS) highlighted concerns about the unique sensitivity of children and suggested changes to the USEPA's risk assessment process [65]. NAS also described the regulatory confusion created by the "Delaney paradox" where a no carcinogenic risk rule applied to residue tolerances for pesticides that concentrate in processed food and a benefit–risk rule applied to those that do not concentrate [66]. Under its interpretation, the USEPA would revoke or deny tolerances for a raw commodity if the tolerance for a processed product was revoked or denied under the Delaney Clause, leading to the cancelation of the pesticide's registration for those crops. NAS argued that rigorous application of the Delaney Clause would reduce the USEPA's flexibility to reduce dietary cancer risks. It would prevent the USEPA from registering new pesticides with a slight cancer risk even if they would displace the use of more hazardous materials. It would also require the USEPA to address negligible dietary risks instead of other, more significant health risks.

The USEPA attempted to apply a negligible risk rule to the Delaney Clause, but the Natural Resources Defense Council filed a lawsuit to prevent that. In 1992, the Ninth Circuit U.S. Court decided that a negligible risk rule could not be applied under existing law. As a result of this decision and other lawsuits, the USEPA wrote rules to revoke tolerances under the Delaney Clause. The decision created a strong incentive for agricultural interests to seek a legislative resolution to Delaney Clause issues and prevent the loss of pesticide tolerances and registrations. Two legislative approaches to resolving the paradox were proposed: a risk–benefit rule for all food tolerances and a negligible risk rule, which considered only risk, for all food tolerances.

The FQPA resolved the Delaney paradox, created new dietary risk standards, and required a reassessment of residue tolerances against those standards. As a result, pesticides are no longer subject to the Delaney Clause but to a new

Table 11 Important Pesticide Legislation

The Insecticide Act of 1910 Prohibited the manufacture, sale, or transport of adulterated or misbranded pesticides; protected farmers and ranchers from the marketing of ineffective products.

Federal Food, Drug, and Cosmetic Act of 1938 (FFDCA) Provided that safe tolerances be set for residues of unavoidable poisonous substances, such as pesticides, in food.

Federal Insecticide, Fungicide and Rodenticide Act of 1947 (FIFRA) Required pesticides to be registered before sale and that the product label specify content and whether the substance was poisonous.

Miller Amendment to FFDCA of 1954 Amended the Federal Food, Drug, and Cosmetic Act (FFDCA) to require that tolerances for pesticide residues be established (or exempted) for food and feed (Section 408). Allowed consideration of risks and benefits in setting tolerances.

Food Additives Amendment to FFDCA of 1958 Amended FFDCA to give authority to regulate food additives against a general safety standard that does not consider benefits (Section 409); included the Delaney Clause, which prohibited food additives found to induce cancer in humans or animals. Pesticide residues in processed foods were classified as food additives, whereas residues on raw commodities were not. When residues of a pesticide applied to a raw agricultural commodity appeared in a processed product, the residues in processed foods were not to be regulated as food additives if levels were no higher than sanctioned on the raw commodity.

FIFRA Amendments of 1964 Increased authority to remove pesticide products from the market for safety reasons by authorizing denial or cancelation of registration and the immediate suspension of a registration, if necessary, to prevent an imminent hazard to the public.

Federal Environmental Pest Control Act (FEPCA) of 1972 Amended FIFRA to significantly increase authority to regulate pesticides. Allowed registration of a pesticide only if it did not cause "unreasonable adverse effects" to human health or the environment; required an examination of the safety of all previously registered pesticide products within 4 years using new health and environmental protection criteria. Materials with risks that exceeded those criteria were subject to cancelation of registration. Specifically included consideration of risks and benefits in these decisions.

Table 11 Continued

FIFRA Amendment of 1975 Required consideration of the effects of registration cancelation or suspension on the production and prices of relevant agricultural commodities.

Federal Pesticide Act of 1978 Identified review of previously registered pesticides as "reregistration"; eliminated the deadline for reregistration but required an expeditious process.

FIFRA Amendments of 1988 Accelerated the reregistration process by requiring that all pesticides containing active ingredients registered before November 1, 1984, be reregistered by 1995; provided the USEPA with additional financial resources through reregistration and annual maintenance fees levied on pesticide registrations.

The Food Quality Protection Act of 1996 (FQPA) Amended FIFRA and FDCA. Set a consistent safety standard for risks from pesticide residues in foods to "ensure that there is a reasonable certainty that no harm will result to infants and children from aggregate exposure." Pesticide residues are no longer subject to the Delaney Clause of FDCA; both fresh and processed foods may contain residues of pesticides classified as carcinogens at tolerance levels determined to be safe. The USEPA is required to reassess existing tolerances of pesticides within 10 years, with priority to pesticides that may pose the greatest risk to public health. Benefits no longer have a role in setting new tolerances but may have a limited role in decisions concerning existing tolerances. Included special provisions to encourage registration of minor use and public health pesticides.

uniform safety standard for pesticide-related risks in raw and processed foods: "A reasonable certainty of no harm from aggregate exposure to the pesticide chemical residue." For carcinogens treated as nonthreshold effects, this standard means negligible risk, instead of no risk, for both raw and processed foods. For threshold effects, the standard is satisfied if exposure is lower by an ample margin of safety than the no-effect level.

In setting tolerances, the USEPA must consider dietary exposures to a pesticide from all food uses and from drinking water as well as nonoccupational exposure, such as homeowner use of a pesticide for lawn care. The USEPA must also consider increased susceptibility to infants and children or other sensitive subpopulations and the cumulative effects from other substances with a "common mechanism of toxicity."* FQPA directs the USEPA to use an additional tenfold

* Two or more pesticides have a common mechanism of toxicity if they cause a common toxic effect to human health by the same, or essentially the same, sequence of major biochemical events.

margin of safety in setting residue tolerances in some cases to protect infants and children. The USEPA must review all residue tolerances of currently registered pesticides against this new standard by 2006, with priority given to pesticides that may pose the greatest risk to public health. The timetable specifies 33% by 1999 (which was achieved), 66% by 2002, and the remainder by 2006. If risk of a pesticide exceeds the standard, the USEPA will reduce residue limits or revoke tolerances for uses of the pesticide until the standard is met. If a common mechanism of toxicity is identified for a group of pesticides, the acceptable risk for one pesticide can be reduced by risks from other pesticides.

In 1997, the USEPA gave a high priority to organophosphates, carbamates, and probable carcinogens in the tolerance reassessment process. The USEPA focused first on the organophosphates and coordinated the tolerance review with the ongoing reregistration process. Ecosystem and worker safety risks are being examined along with dietary, drinking water, and nonoccupational exposure risks. As a result of the review, registrants took actions to reduce risks from azinphos-methyl and methyl parathion in 1999 and chlorpyrifos and diazinon in 2000. The review resulted in a number of proposals to reduce worker safety and ecosystem risks of organophosphate pesticides. In addition, the USEPA is assessing the cumulative risks of organophosphates.

5.2.2 Minor Use Pesticides

Growers of minor crops such as fruits, nuts, and vegetables argued that the pesticide regulatory process created inadequate incentives to register and reregister pesticides for use on those crops. These small-acreage crops create relatively small markets for pesticides, except for fungicides and some "other pesticides," even though crop values and per-acre use of pesticides are often very high. Registrants have a financial incentive to register or reregister pesticides for major crops such as corn, soybeans, cotton, and wheat that create large markets for pesticides and to cancel minor uses as a cost-effective way to reduce risks to acceptable levels. A registrant might also decide not to incur the costs of conducting toxicology tests to retain registrations, so that minor use registrations are canceled for procedural reasons. Similarly, registrants often pursue new registrations for major crops but not for minor crops because of the cost of registration and small potential for sales.

The FQPA created incentives to register pesticides for minor uses by providing additional time to submit registration data, waiving data requirements in some cases, and extending the period of exclusive use of data by the registrant. Minor uses were defined as those crops grown on less than 300,000 acres or whose use provides insufficient financial incentive for registration (but other conditions must apply). Although these provisions help to reduce the costs of registering pesticides for minor uses, it is not clear that they will offset the loss of registrations under the tolerance reassessment process.

5.3 Implications of the FQPA for Risk Management

The FQPA changed the potential role of benefits in pesticide regulatory decisions and emphasized risk assessment procedures and risk reduction. Under the new standard, benefits of use cannot be considered when setting new residue tolerances for raw or processed foods. However, they can be considered in special circumstances when evaluating existing tolerances to justify, for a limited time, an aggregate dietary cancer risk for a pesticide that is slightly greater than negligible.* In general, it is expected that benefits of use will have a small role in justifying higher dietary risks [67]. But the USEPA could use benefits-of-use information to identify which tolerances could be modified or revoked to reduce the costs of meeting the standard for a pesticide or group of pesticides.

Within the context of benefit–cost analysis, the use of a pesticide would be regulated so that the marginal benefit of the regulation (costs avoided by reducing risks) equals marginal cost (lost economic benefits) [68,69]. Breyer [70] indicated that a major problem in regulation occurs if the marginal cost of risk reduction, the cost of preventing the last unit of adverse effect, such as saving a statistical life, is significantly greater for some hazard regulations than for others.

Risks could be reduced more cost-effectively if higher marginal cost regulations were made less stringent and lower marginal cost regulations more stringent so that marginal costs were closer or equal in magnitude. For example, if the cost of saving the last statistical life were greater for a pesticide regulation than for an automobile safety regulation, the impact of saving the same number of lives could be reduced if the pesticide regulation were less stringent and the automobile safety regulation were more stringent. The regulations could also be modified to save more lives for the same economic impact.

Although the FQPA standard eliminates the most costly tolerance revocations that the Delaney Clause would have caused, it does not eliminate the problem that Breyer described. The reason is that the USEPA can regulate the pesticide's uses to minimize the impacts of meeting the overall safety standard but the economic impacts of decisions generally cannot affect the acceptable amount of risk. The FQPA standard has many characteristics of what Harper and Zilberman [69] call a safety-fixed rule. Such rules allow an "efficient allocation of regulatory restrictions affecting a single chemical," but they do not "address

* Benefits can be considered if the pesticide protects consumers against adverse health effects greater than the risks of the pesticide or if the pesticide is needed to prevent a significant disruption in domestic production of an adequate, wholesome, and economical food supply.

directly the tradeoffs between aggregate economic benefits and environmental risks . . . , nor do they assure an efficient allocation of social resources among regulations affecting different chemicals." As a result, the marginal cost per life saved or illness prevented by tolerance decisions could vary from pesticide to pesticide, depending upon the relative cost-effectiveness of each one's alternatives. Similarly, the marginal costs could vary significantly between residue tolerance decisions and regulations of nonpesticide hazards. If the marginal costs of tolerance regulations were higher than the marginal costs of other regulations, the only way to equate marginal costs would be to make the other regulations stricter.

Nondietary pesticide risks, such as those to workers or wildlife, are still subject to a risk–benefit rule. Under such a rule, the impact of a regulation can affect the acceptable amount of risk: A high cost per unit of risk prevented can justify a less restrictive regulation, or a low cost, a more restrictive regulation. As a result of the two rules, the marginal cost of saving lives could be higher for dietary risks than for worker safety risks. However, Cropper et al. [71], in a study of USEPA cancelation decisions, found that the cost of a statistical life saved was several orders of magnitude higher for worker safety risks than dietary risks, which implies that stricter regulations for dietary risks were warranted. However, by focusing on decisions relating to currently registered pesticides, they did not address the economic impacts of the original tolerance-setting process, which might have reduced dietary risk to acceptable levels.

The FQPA rules restrict the USEPA's ability to consider the impacts of a decision on the availability of alternative pest control practices and on other pesticide risks. Although the FQPA requirement to address the highest risk pesticides earlier in the tolerance review process means that decisions should not increase dietary risk, it is unclear what the FQPA means for other risks. Conceivably, tolerance reassessment could force growers to use alternatives that increase risks to workers or wildlife, which in turn could force more regulatory actions. However, the USEPA has been assessing worker safety and ecosystem risks in conjunction with the tolerance reassessment process, which could prevent tolerance decisions from increasing other pesticide risks. Ultimately, the tolerance reassessment process could reduce the number of pest control alternatives, especially for minor crops, and increase the costs of preventing both dietary and nondietary risks unless alternative practices are developed.

Although the FQPA might not minimize the costs of risk reduction, it might be consistent with the policy preferences held by much of the public. Horowitz [72], in a survey of 1000 Maryland households, found a preference for regulations that would reduce pesticide risks, compared to a regulation that would reduce an alternative risk from automobile exhaust, regardless of the number of lives saved. This preference may reflect the fear of unknown pesticide residues on food and the involuntary nature of such risks, even if scientists may view them as insignificant. In addition, the FQPA's special risk assessment provisions and the additional margin of safety for children imply a willingness to incur higher marginal costs for protecting them from the dietary risks of pesticide residues than for protecting adults.

6 SUMMARY

Synthetic organic pesticide use grew dramatically from the late 1940s to the early 1980s and then stabilized, increasing at a much slower rate through the 1990s. Important components are a dramatic rise in herbicide and insecticide use until the early 1980s followed by a decline, although use on potatoes and other vegetables generally increased. Increased pesticide use is part of a larger technological change in agriculture that increased productivity by 150% between 1947 and 1994. Pesticide use grew to market saturation on many large-acreage crops by 1980, so that major factors affecting pesticide use since then have been changes in crop acreage and the replacement of older compounds with newer ones applied at lower per-acre rates. The extent of pesticide use has responded primarily to economic factors such as markets and farm programs, but the pesticide regulatory process has influenced the aggregate quantity by influencing the types of new pesticides developed, registering new materials, and removing others from the market.

Several cited studies indicate that, from the farmer's viewpoint, financial returns have justified increased pesticide use. During the post–World War II period, pesticide prices generally rose relative to crop and other input prices but fell relative to wages, which encouraged the substitution of pesticide use for labor. However, pesticide prices fell relative to crop and other input prices during 1965–1980, a period of rapid increases in pesticide use. There is also an argument, supported by economic theory, that farm programs encouraged more pesticide use per acre than is economically efficient, but in recent years acreage restrictions have helped to stabilize pesticide use. Also, changes in farm program legislation since 1977 have reduced incentives for pesticide use.

However, increased pesticide use has not solved all pest control problems. One concern is that pesticides are overused, resulting in overly rapid development of pest resistance and mortality of beneficial species including natural enemies of pests. The result may be that farmers spend too much on pesticides and have greater pest losses than would otherwise occur. A response has been the use of IPM and economic thresholds to eliminate unnecessary, counterproductive pesticide applications and encourage nonchemical practices where economically feasible. Encouraging the implementation of IPM has become a policy tool to reduce the undesirable health and environmental effects of pesticide use and to improve the cost-effectiveness of pest control.

Also important is the idea that, from society's viewpoint, the health and environmental effects of some pesticides, including food safety, water quality, worker safety, and wildlife mortality, outweigh their production benefits. Changing societal attitudes toward pesticide risks and benefits have had a profound effect on pesticide policy. Pesticide regulatory policy was at first a response to the availability of the new technology that encouraged adoption by attempting to ensure product quality. However, public concerns emerging in the 1960s changed policy to emphasize protection from various hazards, so that most regulatory decisions involved a risk–benefit comparison. Nevertheless, public concern about pesticide hazards, including dietary risks, and the USEPA's ability to resolve pesticide controversies continued. The Food Quality Protection Act of 1996 resolved the Delaney paradox and created new food safety standards, which made benefits of use a minor consideration in decisions related to dietary risk.

REFERENCES

1. E Ball, JC Bureau, R Nehring, A Somwaru. Agricultural productivity revisited. Am J Agric Econ 79:1045–1063, 1997.
2. C Osteen. Pesticide use trends and issues in the United States. In: D Pimentel, H Lehman, eds. The Pesticide Question: Environment, Economics, and Ethics. New York: Chapman and Hall, 1993, pp 307–336.
3. CD Osteen, PI Szmedra. Agricultural Pesticide Use Trends and Policy Issues. AER-622. USDA, Econ Res Serv, 1989.
4. W Klassen, PH Schwartz. ARS research program in chemical insect control. In: JL Hilton, ed. Agricultural Chemicals of the Future. Totowa, NJ: Rowman and Allenheld, 1985.
5. USDA. Farm business economics briefing room. Farm sector performance data, U.S. production agriculture expenses 1910–97. USDA, Econ Res Serv website (www.ers.usda.gov/briefing/farmincome/expense/expagprd.wk1), 2000.
6. AL Aspelin. Pesticide Industry Sales and Usage, 1996 and 1997 Market Estimates. US Environ Protect Agency, 733-R-99-001, 1999.
7. J Fernandez-Cornejo, S Jans. Quality-adjusted price and quantity indices for pesticides. Am J Agric Econ 77:645–659, 1995.
8. M Padgitt. Pesticides. In: M Anderson, R Magleby, eds. Agricultural Resources and Environmental Indicators, 1996–97. USDA, Econ Res Serv, Ag Handbook No. 712, 1997.
9. M Padgitt, D Newton, C Sandretto. Production Practices for Major Crops in U.S. Agriculture, 1990–97. SB-969 USDA, Econ Res Service, 2000.
10. A Fox, T Eichers, P Andrilenas, R Jenkins, H Blake. Extent of Farm Pesticide Use on Crops in 1966. AER-147 USDA, Econ Res Serv, 1968.
11. PA Andrilenas. Farmers' Use of Pesticides in 1971—Extent of Crop Use. AER 268. USDA, Econ Res Serv, 1975.
12. TR Eichers, PA Andrilenas, TW Anderson. Farmers' Use of Pesticides in 1976. AER-418. USDA, Econ Stat Coop Serv, 1978.

13. USDA, Econ Res Serv. Inputs Outlook and Situation Report. IOS-6, November 1984 and IOS-2, October 1983.
14. USDA, Econ Res Serv. Agricultural Resources: Inputs Situation and Outlook Report. AR-1, February 1986; AR-5, January 1987; AR-9, January 1988; AR-13, February 1989; AR-15, August 1989; AR-17, February 1990; and AR-20, October 1990.
15. USDA, Natl Agric Stat Serv. Agricultural Chemical Usage: 1990 Field Crops Summary. Ag Ch 1 (91), March 1991 (and subsequent issues).
16. USDA, Natl Agric Stat Serv. Agricultural Chemical Usage: Fruits Summary, 1993, 1995, 1997, Ag Ch 1(94), Ag Ch 1(96), Ag Ch 1(98).
17. USDA, Natl Agric Stat Serv. Agricultural Chemical Usage: Vegetables Summary, 1992, 1994, 1996, Ag Ch 1(93), Ag Ch 1(95), Ag Ch 1(97).
18. TR Eichers, PA Andrilenas, R Jenkins, A Fox. Quantities of Pesticides Used by Farmers in 1964. AER-131. USDA, Econ Res Serv, 1968.
19. TR Eichers, PA Andrilenas, R Jenkins, H Blake, A Fox. Quantities of Pesticides Used by Farmers in 1966. AER-179. USDA, Econ Res Serv, 1970.
20. PA Andrilenas. Farmers' Use of Pesticides in 1971—Quantities. AER-252. USDA, Econ Res Serv, 1974.
21. USDA, Econ Res Serv. Unpublished pesticide use data for 1982, 1991, 1997.
22. ERS Issues Center. Genetically engineered crops for pest management. June 25, 1999. USDA, Econ Res Serv website (www.ers.usda.gov/whatsnew/gmo/index.htm/).
23. JK Horowitz, E Lichtenberg. Insurance, moral hazard, and chemical use in agriculture. Am J Agric Econ 75:926–935, 1993.
24. CO Calkins. Areawide IPM as a tool for the future. In: S Lynch, C Greene, C Kramer-Leblanc, eds. Proceedings of the Third National IPM Symposum/Workshop, MP-1542. USDA, Econ Res Serv, 1997, pp 154–158.
25. DF Dumas, RE Goodhue. The cotton acreage effects of boll weevil eradication: a county level analysis. J Agric Appl Econ 31:475–497, 1999.
26. J Fernandez-Cornejo, S Jans, M Smith. Issues in the economics of pesticide use in agriculture: a review of the empirical evidence. Rev Agric Econ 20:462–488, 1998.
27. HF Campbell. Estimating the marginal productivity of agricultural pesticides: The case of tree-fruit farmers in the Okanagan Valley. Can J Agric Econ 24:23–30, 1976.
28. GA Carlson. Long-run productivity of pesticides. Am J Agric Econ 59:543–548, 1977.
29. C Carrasco-Tauber, LJ Moffitt. Damage control economics: Functional specificity and pesticide productivity. Am J Agric Econ 79:47–61, 1997.
30. M Duffy, M. Hanthorn. Returns to Corn and Soybean Tillage Practices. AER-508. USDA, Econ Res Serv, 1984.
31. J Fernandez-Cornejo, S Jans, M Smith. The economic impact of pesticide use in U.S. agriculture. Northeast Agricultural and Resource Economics Association meeting, Atlantic City, NJ, 1996.
32. DE Hawkins, WF Slife, ER Swanson. Economic analysis of herbicide use in various crop sequences. IL Agric Econ 17:8–13, 1977.
33. JC Headley. Estimating the productivity of agricultural pesticides. J Farm Econ 50: 13–23, 1968.

34. JY Lee, M Langham. A simultaneous equation model of the economic-ecologic system in citrus groves. Southern Journal of Agricultural Economies 5:175–180, 1973.
35. BH Lin, S Jans, K Ingram, L Hansen. Pesticide Productivity in Pacific Northwest Potato Production. In: Agricultural Resources: Inputs. AR-29. USDA, Econ Res Serv, 1993.
36. JA Miranowski. The demand for agricultural crop chemicals under alternative farm programs and pollution control solutions. PhD Dissertation. Harvard Univ, Cambridge, MA, 1975.
37. ML Teague, BW Brorsen. Pesticide productivity: what are the trends? J Agric Appl Econ 27:276–282, 1995.
38. S Daberkow, KH Reichelderfer. Low-input agriculture: trends, goals, and prospects for input use. Am J Agric Econ 70:1159–1166, 1988.
39. SM Capalbo, TT Vo. A review of the evidence on agricultural productivity and aggregate technology. In: S Capalbo, J Antle, eds. Agricultural Productivity Measurement and Explanation. Washington, DC: Resources for the Future, 1988.
40. JC Headley. Productivity of agricultural pesticides. In: Economic Research on Pesticides for Policy Decisionmaking. USDA, Econ Res Serv, 1971.
41. MO Ribaudo, RA Shoemaker. The effect of feedgrain program participation on chemical use. Agric Res Econ Rev 24:211–220, 1995.
42. VM Stern, RF Smith, R Van den Bosch, K Hagen. The integrated control concept. Hilgardia 29:81–101, 1959.
43. JC Headley. Defining the economic threshold. In: National Research Council, Pest Control Strategies for the Future. Washington, DC: Natl Acad Sci, 1972, pp 100–108.
44. PM Hillebrandt. The economic theory of the use of pesticides. J Agric Econ 13: 464–472, 1960.
45. RB Norgaard. The economics of improving pesticide use. Annual Review of Entomology 24:45–60, 1976.
46. G Feder. Pesticides, information, and pest management. Am J Agric Econ 61:97–103, 1979.
47. LJ Moffitt. Risk-efficient action thresholds for pest control decisions. J Agric Econ 37:69–75, 1986.
48. CD Osteen, LJ Moffitt, AW Johnson. Risk efficient action thresholds for nematode management. J Prod Agric 1:332–338, 1988.
49. CR Greene, RA Kramer, GW Norton, EG Rajotte, RM MacPherson. An economic analysis of soybean integrated pest management. Am J Agric Econ 67:567–572, 1985.
50. WF Lazarus, ER. Swanson. Insecticide use and crop rotation under risk: Rootworm control in corn. Am J Agric Econ 65:738–747, 1983.
51. PS Liapis, LJ Moffitt. Economic analysis of cotton integrated pest management strategies. Southern Journal Agric Econ 15:97–102, 1983.
52. PC Pingali, GA Carlson. Human capital, adjustments in subjective probabilities, and the demand for pest controls. Am J Agric Econ 67:853–861, 1985.
53. PI Szmedra, RW McClendon, ME Wetzstein. Risk efficiency of pest management strategies: A simulation case study. Trans ASAE 29:1642–1648, 1988.

54. D Hueth, U Regev. Optimal pest management with increasing pest resistance. Am J Agric Econ 56:543–552, 1974.
55. CR Taylor, JC Headley. Insecticide resistance and the evaluation of control strategies for an insect population. Can Entomol 107:237–242, 1975.
56. RF Smith, JL Apple, DG Bottrell. The origins of integrated pest management concepts for agricultural crops. In: JL Apple, RF Smith, eds. Integrated Pest Management. New York: Plenum Press, 1976, pp 1–16.
57. WJ Lewis, JC van Lenteren, SC Phatak, JH Tomlinson III. A total approach to sustainable pest management. Proc Natl Acad Sci USA 94:12243–12248, 1997.
58. O Pettersson. Swedish pesticide policy in a changing environment. In: D Pimentel, H Lehman, eds. The Pesticide Question: Environment, Economics, and Ethics. New York: Chapman and Hall, 1993, pp 182–205.
59. J Curtis. Fields of Change: A New Crop of American Farmers Finds Alternatives to Pesticides. New York: Natural Resources Defense Council, 1998.
60. CM Benbrook, E Groth, M Hansen, J Halloran, S Marquardt. Pest Management at the Crossroads. Yonkers, NY: Consumers Union, 1996.
61. National Research Council. Ecologically Based Pest Management: New Solutions for a New Century. Washington, DC: Natl Acad Press, 1996.
62. JD Conner Jr, SE Lawrence, CA O'Conner III, C Volz, KW Weinstein, JC Chambers, AA Kerester, SW Landfair, RH Rahinsky, EM Weaver. Pesticide Regulation Handbook. New York: Executive Enterprises Pub Co, 1987.
63. M Ollinger, J Fernandez-Cornejo. Regulation, Innovation, and Market Structure in the U.S. Pesticide Industry. AER-719. USDA, Econ Res Serv, 1995.
64. National Research Council. Regulating Pesticides. Washington, DC: Natl Acad Sci, 1980.
65. National Research Council. Pesticides in the Diets of Infants and Children. Washington, DC: Natl Acad Press, 1993.
66. National Research Council. Regulating Pesticides in Food: The Delaney Paradox. Washington, DC: Natl Acad Press, 1987.
67. M Phillips, LP Gianessi. An analysis of the economic benefit provisions of the Food Quality Protection Act. Review Agric Econ 20:377–389, 1998.
68. E Lichtenberg, RC Spear, D Zilberman. The economics of reentry regulation of pesticides. Am J Agric Econ 75:946–958, 1993.
69. C Harper, D Zilberman. Pesticides and worker safety. Am J Agric Econ 74:68–78, 1992.
70. S Breyer. Breaking the Vicious Circle: Toward Effective Risk Regulation. Cambridge, MA: Harvard Univ Press, 1993.
71. ML Cropper, WN Evans, SJ Berardi, MM Ducla-Soares, PR Portney. The determinants of pesticide regulation: a statistical analysis of EPA decision making. J Polit Econ 100:175–197, 1992.
72. JK Horowitz. Preferences for pesticide regulation. Am J Agric Econ 76:396–406, 1994.

4

Risk Assessment

Nu-may Ruby Reed
California Environmental Protection Agency
Sacramento, California, U.S.A.

1 PESTICIDE SAFETY

Safety regulation for pesticides has come a long way. In the United States, the Federal Insecticide Act of 1910 was the earliest law on consumer protection. It was designed mainly to protect farmers from substandard or fraudulent products. In 1938, the Pure Food Law of 1906 was amended to require that foods shipped in interstate commerce be pure and wholesome. Colors were added to white insecticides (sodium fluoride and lead arsenate) to distinguish them from flour or other cooking ingredients. Additionally, residue tolerances in foods were established for arsenic and lead.

The 1940s ushered in the era of discovery and an acceleration in pesticide use. Behind the apparent beneficial effects of chemical arsenals against pests loomed the potential hazards to humans and the environment [1–3]. Reports of bird and fish kills, the pollution of surface and ground waters, and cases of human poisoning (e.g., from organophosphates) were alarming. The toxicities of chemicals previously thought to be benign were revealed after some period of use (e.g.,

---

Views expressed do not necessarily represent those of the Department of Pesticide Regulation. Mention of trade names or commercial products is not an endorsement or opinion of their use.

DDT). The concerns about hazards to humans and the environment called for a systematic process to safeguard pesticide use. With the birth of the U.S. Environmental Protection Agency (USEPA), the 1970s were a decade of pesticide safety legislation and regulation [4]: the Federal Environmental Pesticide Control Act, which specified methods and standards of control; a further amendment of the 1947 Federal Insecticide, Fungicide and Rodenticide Act (FIFRA); and the Clean Water Act. Meanwhile, the 1970 Occupational Safety and Health Act provided safety standards for occupational exposures. Vested by FIFRA and the Federal Food, Drug, and Cosmetic Act (FFDCA), the USEPA regulates the registration and use of pesticides. Together with the Food and Drug Administration (FDA), the USEPA also establishes the tolerances for pesticides in food. A tolerance is the maximum residue level of a pesticide allowed in or on human food and animal feed. A similar standard established by the international Codex Alimentarius Commission is the "Maximum Residue Limit."

One significant change that accompanied the formation of the USEPA was the opening of the decision-making process to the public. In 1983, the National Academy of Science (NAS) published *Risk Assessment in the Federal Government* [5]. This report formally introduced risk assessment as the tool for characterizing and predicting the risk of toxic substances, building on the foundation of best available scientific knowledge. Pesticide laws have since been amended to better ensure safety and to address additional concerns as new scientific information becomes available. The most recent law that significantly affects the approach to risk assessment is the 1996 Food Quality Protection Act (FQPA), which established a tougher standard for setting pesticide tolerances in foods.

Food safety issues stemmed from two major concerns. One regards the adequacy of the enforcement efforts. The other regards the setting of standards that are adequate for the protection of health, including the health of those who may be more sensitive, such as infants and children. The latter concern was the emphasis of the 1993 NAS report [6] and the key scientific rationale behind the FQPA. To ensure a reasonable certainty of no harm from the use of pesticides, the FQPA requires that risk assessments for setting tolerances explicitly consider (1) the sensitivity and exposure of infants and children, including in utero exposures, (2) the aggregate exposure from multiple pathways and routes, and (3) the cumulative risk from multiple chemicals with a common mechanism of action. In addition, pesticides are to be tested for the potential of endocrine disruption.

2 FRAMEWORK OF HEALTH RISK ASSESSMENT

Human health risk assessment (HRA) is a scientific evaluation of the magnitude or probability of harm to human health posed by a single risk agent or substance or a mixture of such agents or substances. HRA provides health-based information on risks for risk management decisions, e.g., setting pesticide food tolerance

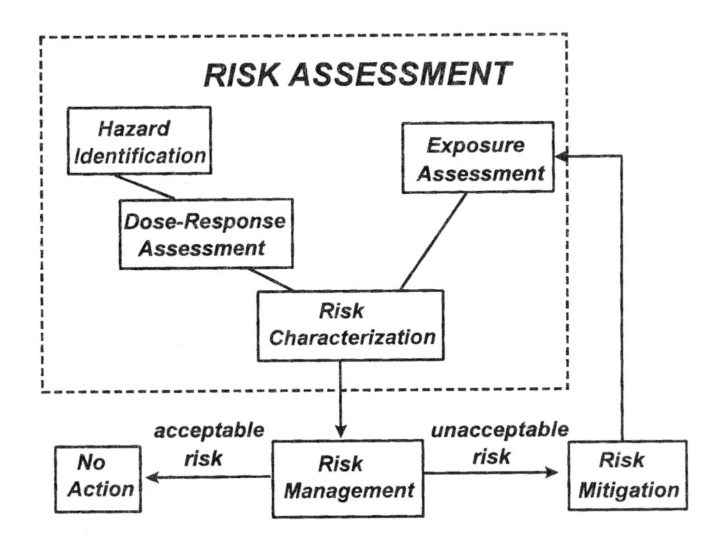

FIGURE 1 Health risk assessment framework.

and permissible concentrations in water and air, establishing public health policies, and determining the needs for risk mitigation.

According to the paradigm set forth by the NAS in 1983, HRA consists of four basic components: hazard identification, dose–response assessment, exposure assessment, and risk characterization. The flow of the HRA process, with its relation to risk management and mitigation, is illustrated in Figure 1. Hazard identification describes the inherent toxicity of a risk agent. Dose–response assessment describes the relationship between the dose and the magnitude, severity, or probability of a toxicological response. Exposure assessment estimates the level of current or anticipated human exposure. Risk characterization integrates the information from the previous three components and estimates the potential risk as the probability or likelihood of adverse effects on a population. Risk management decisions are then made regarding whether the estimated risk is acceptable. When the risk is judged to be above the level of concern, measures to reduce the exposure are explored. The risk associated with any feasible mitigation option is reassessed through iterating the risk assessment process until options that would result in acceptable risk are found.

2.1 Health Risk Assessment Guidelines

In the 1980s, the USEPA published the first series of risk assessment guidelines for various types of health hazards (e.g., cancer, developmental toxicity, reproductive toxicity). These guidelines provided scientific rationale and consistency

Table 1 Most Recent Versions of the USEPA Risk Assessment Guidelines[a]

Guideline	Federal Register publication	Year
Guidelines for neurotoxicity risk assessment	63(93):26926–26954	1998
Toxic Substances Control Act Test guidelines	62(158):43819–43864	1997
Reproductive toxicity risk assessment	61(212):56274–56322	1996
Proposed guidelines for carcinogen risk assessment[b]	61(79):17960–18011	1996
Principles of neurotoxicity risk assessment	59(158):42360–42402	1994
Cross-species scaling factor for carcinogen risk assessment based on equivalence of mg $kg^{-3/4}$ day^{-1}	57(109):24152–24173	1992
Guidelines for exposure assessment	57(104):22888–22938	1992
Guidelines for developmental toxicity risk assessment	56(234):63798–63826	1991
Guidelines for mutagenicity risk assessment	51(185):34006–34012	1986

[a] A similar list can also be found online in National Center for Environmental Assessment (NCEA).
[b] The 1999 review draft (NCEA-F-0644) was adopted as the interim guidelines in 2001.

in risk assessment methods and practices in all branches of government regulation. As more scientific information became available, many of these guidelines were revised in the 1990s, and new ones were added. A list of the most current risk assessment guidelines is given in Table 1. Many of these guidelines are available online through the USEPA's Office of Research and Development, National Center for Environmental Assessment [7]. In addition, scientific policies and guidance documents pertaining to nine FQPA focus issues specific to pesticide risk assessment are published by the USEPA's Office of Pesticide Programs [8] and are available online.

2.2 Data for Risk Assessment

In 1988, FIFRA was amended to require the USEPA to reregister those pesticides that had been in use before current scientific and regulatory standards were formally established. To ensure that the use of a pesticide would not adversely affect human health and the environment, the USEPA further expanded the testing requirements. Four categories of tests are currently required for food use pesticides: chemistry, environmental fate, toxicology, and ecological effects. Lists of specific tests for each category are published in the Code of Federal Regulation Title 40 (40 CFR), Part 158.

Data call-in (DCI) is additionally issued when sufficient data are not available to reliably characterize the risk of a pesticide. For example, in reassessing the existing tolerances under the FQPA, DCIs for developmental neurotoxicity studies have been issued for some organophosphates. DCIs have also been issued to address specific exposure scenarios, such as residential or drinking water exposures. In addition, registrants may also conduct studies to refine a risk assessment without any requirement to do so. For example, to better characterize the dietary exposure, registrants may conduct market basket surveys or studies on residues after food processing.

3 PESTICIDE HEALTH RISK ASSESSMENT

Regarding the approach and the practices, the HRA process for pesticides is no different from the process for other environmental risk agents. However, because of the requirements for toxicology tests, pesticide risk assessment is unique in having a standard and extensive set of data for the risk evaluation. This does not mean that the knowledge base is complete for predicting the current or future potential health risks of pesticides. In risk assessment, assumptions are routinely made to bridge the knowledge gaps. These generic "default" assumptions are highlighted in the following step-by-step presentation of risk assessment.

3.1 Hazard Identification

Health hazards are described by the type of toxicity and the condition of exposure under which these effects occurred. For obvious ethical reasons, the evaluation relies mainly on experiments conducted in laboratory animals. The toxicity database is inclusive, encompassing toxicities to all organs and systems after various durations of exposure or experimental treatment. A list of areas of toxicity data is summarized in Figure 2. Note that "oncogenicity" as used in this chapter refers to the potential to cause benign or malignant tumors. A somewhat interchangeable term frequently used in risk assessment is "carcinogenicity." Strictly speaking, the latter term refers specifically to the formation of carcinoma, a form of malignant tumor, or cancer.

3.1.1 Acute Toxicity Categories

The standard battery of acute toxicity studies required for pesticide registration includes oral, inhalation, and dermal toxicities, skin and eye irritations, and dermal sensitization (allergic response). Median lethal dose (LD_{50}) and concentration (LC_{50}) are determined from the route-specific toxicity studies. LD_{50} and LC_{50} are the dose levels that kill 50% of the animals in a test. These indices of lethality

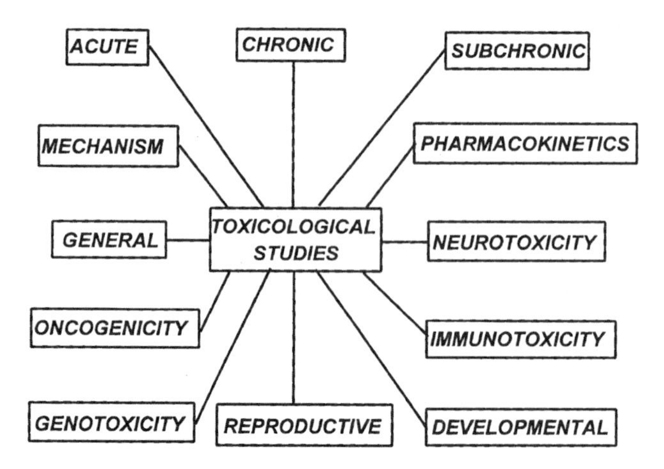

FIGURE 2 Toxicological database for risk assessment.

form the basis for classifying pesticides and their products into toxicity categories. This classification is then used to assign the human hazard signal word posted on the label of a marketed product. The criteria specified by the USEPA [9] for this simple application of acute toxicity data are summarized in Table 2. For example, if a pesticide product has an oral LD_{50} of 10 mg/kg, it is classified as a Category I (i.e., highest toxicity) substance, even if the categories for inhalation and dermal routes of toxicity are numerically higher (i.e., lower toxicity). The label on the container would bear the word "DANGER" as well as a distinctive "POISON" in red, accompanied by a skull and crossbones. Toxicity category classification for end use products is also used for determining the minimum personal protective equipment (PPE) for pesticide handlers.

3.1.2 Adverse Effect Identification

Designating hazard signal words to ensure safe use and handling of pesticide products is only one aspect of hazard identification. To address the short- and long-term effects of a chemical, risk assessment takes into account all aspects of toxicity, not just lethality. In this endeavor, all pertinent reports of toxicity are collected for identifying potential adverse health effects to humans. These include both the standard batteries of tests required for registration and all pertinent publications in the scientific literature.

It is recognized that increasing the dose level and/or prolonging and repeating the exposure to a risk agent would result in increasing severity of the toxicological response and/or number of affected target organs. This general pattern is illustrated in Figure 3. At a very low dose or for a short time of exposure,

TABLE 2 Toxicity Categories of Pesticides

Toxicity study	Toxicity category I	II	III	IV
	Toxicity data			
Oral LD_{50}	≤50 mg/kg	50–500 mg/kg	500–5,000 mg/kg	>5 g/kg
Inhalation[a] LC_{50}	≤0.05 mg/L	0.05–0.5 mg/L	0.5–2 mg/L	>2 mg/L
Dermal LD_{50}	≤0.2 g/kg	0.2–2 g/kg	2–5 g/kg	>5 g/kg
Eye irritation				
Corrosive	Irreversible	—	—	—
Corneal irritation	≥21 days	8–21 days	≤7 days	≤24 hr
Skin irritation[b]	Corrosive	Severe	Moderate	Mild
	Human hazard signal word[c]			
	Danger Poison[d]	Warning	Caution	Caution

[a] For a 4 hr exposure.
[b] Skin irritation: observation at 72 hr.
[c] Based on the highest category of the five studies.
[d] Based on the highest category for oral, inhalation, or dermal toxicities. The signal word is in red and is accompanied by a skull and crossbones.
Source: Ref. 9.

the initial manifestation of a risk agent may be detected as transient clinical signs (e.g., dizziness, nausea). With increasing dose and/or time, toxicity to the liver and kidneys may become evident through more thorough investigations. Ultimately, as the dose continues to increase, death can be expected. In this illustration, each target organ or toxicological effect is identified as a part of the inherent toxicity of the risk agent.

The identification of target organs and judgment on the adversity of toxicological effects are essential for setting risk assessment priorities. For example,

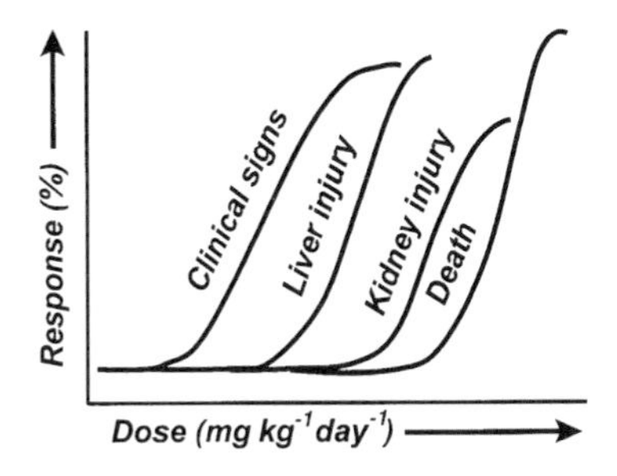

FIGURE 3 Illustrated toxicological responses.

considering the detrimental effects of cancer, a safety evaluation program may choose to place higher priority on evaluating the risk of cancer-causing chemicals. On the other hand, considering the neurological effects of organophosphates (OPs) that are immediately manifested, another safety evaluation program may elect to evaluate their risks first.

Categorizing a risk agent according to its type of toxicity is also essential for addressing mandates of laws and regulations for the protection of human health. Some national and state programs are mandated to focus on specific areas of toxicity. For example, Proposition 65 passed by California voters in 1986 requires the state to list chemicals known to cause cancer as well as those with reproductive or developmental effects. Public warning of the potential risk is required for the listed chemicals when the risk may be significant. The determination to list a chemical under a specific category of hazard is made at the conclusion of the hazard identification step.

3.2 Dose–Response Assessment

Philippus Aureolus Theophrastus Bombastus von Hohenheim-Paracelsus (1493–1541), widely recognized as the father of toxicology, is often quoted as stating, "All substances are poisons; there is none that is not a poison. The right dose differentiates a poison from a remedy." Following this basic principle, the toxicological hazards of a risk agent are described in the context of the exposure dose, route, and duration.

Two general dose–response models are used in HRA, the threshold and the nonthreshold models. Patterns for these two models are illustrated in Figure 4. The threshold model assumes that there is a threshold dose for a toxicological effect below which no effects are expected. Alternatively, the nonthreshold model assumes that a threshold dose does not exist. This means that any minute increase in the dose is expected to produce an increase in response. There is general agree-

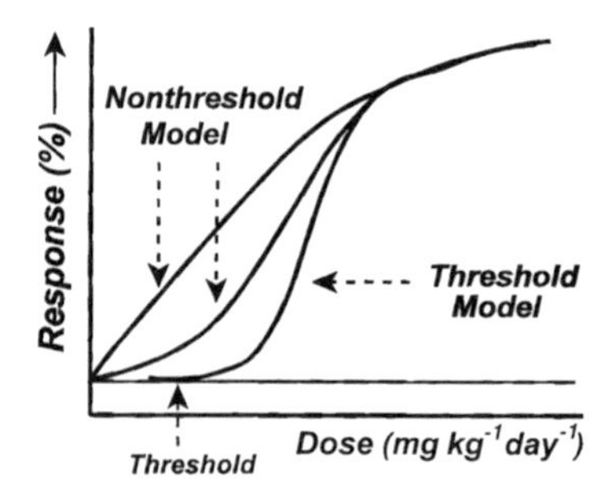

FIGURE 4 Dose–response models.

ment that a threshold exists for all toxicological effects other than cancer. However, policies differ regarding oncogenicity or cancer effects. European countries generally regard cancer effects as likely to have a threshold, whereas the United States considers that a threshold may not exist for a cancer, especially when it could be caused by a direct action on the genetic materials. Understandably, this nonthreshold assumption would present the risk in a more "alarming" way. In some cases, it is viewed as "conservative," tending to overstate the risk.

3.2.1 Threshold Model: No Observed Effect Level and Benchmark Dose

The main focus for describing the dose–response relationship in a threshold model is to establish the threshold. This threshold can then be used as a point of reference for gauging the "degree of safety" associated with human exposure. When human exposure is substantially below this threshold, the risk is judged as "not likely to occur." A conventional term for this threshold is the No Observed Effect Level (NOEL). NOEL is the highest dose in a toxicity study at which no effects are observed. A data point that provides the context for NOEL is the Lowest Observed Effect Level (LOEL). LOEL is the lowest dose in a toxicity study at which an effect is observed. The effect that occurred at LOEL is identified as the most sensitive or critical endpoint of toxicity. Theoretically, if the potential human exposure would not result in an unacceptable risk for this critical endpoint, and provided that the toxicity database is complete, there would be no concern for any other effects, however detrimental they might be.

Although the conventional NOEL approach is rather simple and straightforward, it has many apparent deficiencies [10]. It does not take into account all data points for the entire dose–response curve. Also, the value for NOEL is dictated by the dose selection predetermined by the study design, providing no consistent reference point for comparing NOELs between two studies. Moreover, this approach tends to define a higher NOEL from a study that shows a greater data variation or uses a smaller sample size, especially when the NOEL is delineated on the basis of statistical significance. The NOEL thus determined may be inadequate for protecting human health.

An alternative is the Benchmark Dose (BMD) approach [11]. It entails mathematically fitting a curve to the data points. Accordingly, the dose (the BMD) corresponding to a predetermined Benchmark Response (BMR) is estimated. The BMR is usually selected as a 1%, 5%, or 10% increase in response over the controls, corresponding to the level that can be statistically differentiated from the controls within the sample size commonly employed in a toxicity study. The BMD approach overcomes the deficiencies of the NOEL approach. Theoretically, two studies using different dose levels but similar in quality, conduct, and protocol should yield similar BMDs for the same toxicological endpoint, whereas they may have different NOELs. Unfortunately, the BMD approach is gaining its usefulness, although a standardized set of criteria and guidelines for its use are not yet established. One critical need is the availability of mathematical tools. Until 2000, user-friendly software programs had been costly and limited in accommodating the variety of data types (e.g., dichotomous incidence data, continuous data of physiological measurements, "nested" fetal data, within-the-litter effects). A bundle of BMD software is now available for download by the USEPA through the NCEA.

3.2.2 Adversity of Endpoints

It is generally implied that the critical endpoint for risk assessment is adverse. However, "adversity" is sometimes subjective or conditional. An example is the ongoing debate on the use of blood cholinesterase (ChE) inhibition as an endpoint for characterizing the health risk of organophosphates. ChEs are a family of enzymes that hydrolyze choline esters. Acetylcholinesterase (AChE) terminates impulses across nerve synapses by hydrolyzing the neural transmitter acetylcholine (Ach). Inhibition of AChE leads to accumulation of ACh, resulting in overstimulation of nerves followed by depression or paralysis of the central and peripheral nervous systems. AChE is highly selective for acetyl esters as substrates and is the predominant form of ChE in the central nervous system and neuromuscular junctions of peripheral tissues [12,13]. Butyrylcholinesterase (BuChE) is another form of cholinesterase that preferentially hydrolyzes butyryl and propionyl esters, although it will also hydrolyze a wider range of esters, including ACh [13]. Nonsynaptic AChE is essentially the only ChE present in the red blood cells (RBCs) of higher animals. BuChE is the predominant form of ChE in human plasma. With respect to being an endpoint for risk assessment, one opinion is that ChE inhibition in the blood (i.e., plasma and RBC) is an indicator and a reasonable surrogate for the AChE inhibition in the brain and peripheral tissues for which data are lacking [14]. On the other hand, an argument was made that since blood ChE has not been shown to consistently correlate with brain ChE, it should be used mainly as a biomarker of exposure, not an indicator of a health hazard [15]. However, the lack of consistency in accurately measuring the ChE activites further complicated the attempt to identify any correlations. Nevertheless, the term "No Observed Adverse Effect Level" (NOAEL), once interchangeable with NOEL, may now be useful for emphasizing the adversity of endpoints.

3.2.3 Nonthreshold Model: Potency Slope

The focus for a nonthreshold model is to estimate the slope of the dose–response curve. This is achieved through mathematical curve-fitting by maximizing the likelihood function [16,17]. Contrary to the BMD approach, the nonthreshold approach requires extensive downward extrapolation of the estimated slope into the low-response range. The extrapolation is necessary because of the expectation that increased oncogenic risk from environmental contaminants should not exceed a range of probability around one in a million, or five orders of magnitude below the BMR of 10%. For an experiment to detect a statistically significant increase in cancer incidence at this low range would require substantially more than 1 million animals in a test. A typical rodent bioassay that utilizes 50 animals per dose group is simply unable to verify the extrapolated slope. Thus, the approach for the slope extrapolation is a policy decision based on the best available scientific knowledge.

A weight-of-evidence approach is used to determine whether a chemical is likely to cause cancer in humans, whether the cancer-causing process is likely to be nonthreshold, and whether the dose–response relationship is probably "linear" in the low-response range. Several factors are included in this weight-of-evidence consideration. Among these are the evidence of oncogenicity in humans and in laboratory animals, the evidence of genotoxicity (causing changes in the genetic materials), and the mechanistic data regarding relevance of the cancer-causing process in humans. Tables 3–5 provide the three most frequently used carcinogen classification schemes. The classification in Table 3 follows the 1986 USEPA cancer risk assessment guidelines and is still in use. Table 4 is the scheme used by International Agency for Research on Cancer (IARC). Instead of the alphanumeric classification based mainly on the evidence gained from humans and animals, the 1996 USEPA proposed guidelines favor a descriptive classification that takes into account the genotoxic potential and the mechanistic data. The scheme in the 2001 interim guidelines (1999 review draft) is shown in Table 5.

For chemicals with sufficient weight of evidence (e.g., A, B1, and B2 carcinogens), showing genotoxic potential, and with a mechanism of oncogenicity relevant to humans, the general nonthreshold approach is to extrapolate the slope

Table 3 USEPA 1986 Carcinogen Classification[a]

	Animal evidence				
Human evidence	Sufficient	Limited	Inadequate	No data	No evidence
Sufficient	A	A	A	A	A
Limited	B1	B1	B1	B1	B1
Inadequate	B2	C	D	D	D
No data	B2	C	D	D	E
No evidence	B2	C	D	D	E

[a] Group A: Human carcinogen
[a] Group B: Probable human carcinogen
[a] Group C: Possible human carcinogen
[a] Group D: Not classifiable as to human carcinogenicity
[a] Group E: Evidence of noncarcinogenicity for humans
Source: U.S. Fed Reg 51(185):33993–34012.

Table 4 International Agency for Research on Cancer (IARC) Carcinogenicity Classification[a]

	Animal evidence				
Human evidence	Sufficient	Limited	Inadequate	No data	No evidence
Sufficient	1	1	1	1	1
Limited	2A	2A[b], 2B	2A[b], 2B	2A[b], 2B	2A[b], 2B
Inadequate	2A[b], 2B	2B[b]	3	3	4[b]
No data	2A[b], 2B	2B[b], 3	3	3	4[b]
No evidence	2A[b]	3	3	3	4

[a] Group 1: Carcinogenic to humans.
[a] Group 2: 2A, Probably carcinogenic to humans; 2B, possibly carcinogenic to humans.
[a] Group 3: Not classifiable.
[a] Group 4: Not carcinogenic to humans.
[b] Supporting evidence from other relevant data.
Source: IARC 1987 Monograph Preamble; updated January 1999 (http://www.iarc.frl).

Table 5 USEPA Interim Human Carcinogen Descriptors

Carcinogen to Humans
Causal evidence in humans, or evidence of association in humans (insufficient for showing causality) and having compelling evidence in animals with similar mode of action in humans.

Likely to Be Carcinogenic to Humans
The weight of evidence ranging between possitive association in humans plus strong evidence in animals, to no human data but the animal carcinogenicity mode of action is pertinent to humans.

Suggestive Evidence of Carcinogenicity, But Not Sufficient to Assess Human Carcinogenic Potential
Suggestive evidence in animals and humans, but insufficient for conclusion on human carcinogenic potential.

Data Are Inadequate for an Assessment of Human Carcinogenic Potential
Lack of pertinent data, or having conflicting evidence.

Not Likely to Be Carcinogenic to Humans
Such as: robust human data showing no basis of concern; negative evidence in animals; animal carcinogenicity not pertinent to humans or human route of exposure; carcinogenic effects not anticipated below a defined dose range.

Source: USEPA 1999 Review Draft Guidelines for Carcinogen Risk Assessment (NCEA-F-0644).

linearly from the observable range to the low-response range. Two estimates of the slope are usually made; the best estimate (maximum likelihood estimate; MLE) and its statistical 95th percentile upper bound (UB). The slope is the probability of response per unit dose, or milligrams per kilogram per day (expressed as $mg/kg/day)^{-1}$. These slope estimates derived from data in animal studies are adjusted to humans. For oncogenic effects, the adjustment is based on the dose equivalence between animals and humans when it is expressed as per kilogram body weight to the 3/4 power (e.g., $mg/kg^{3/4}/day$) [18]. Since the dose is usually expressed in milligrams per kilogram per day, the adjustment factor for the slope is the 1/4 power of the animal/human body weight ratio, or $(BWt_{animal}/BWt_{human})^{1/4}$. Ideally, both MLE and UB should be presented for bounding the estimated slope. However, "potency" is often given as a single number, referring only to the UB. The estimated cancer risk in a lifetime is then calculated by multiplying the potency with the lifetime average exposure. The "risk" is the estimated probability of occurrence above the background rate. A risk of 1×10^{-6} means a one in a million increase in probability.

3.3 Exposure Assessment

Human exposure is estimated on the basis of current and/or anticipated exposure patterns (e.g., frequency, duration) and levels (or concentrations) in the exposure media. Depending on the properties, use pattern, and persistence of a pesticide, human exposures may be a single day, short-term, intermediate-term, and/or long-term. Corresponding to the exposure duration of toxicity data, the following exposures are commonly assessed: absorbed daily dose (ADD) for acute effects, seasonal average daily dose (SADD) for subchronic effects, annual average daily dose (AADD) for chronic nononcogenic effects, and lifetime average daily dose (LADD) for oncogenic effects.

A general equation for calculating the exposure is

$$\text{Exposure} = \text{concentration} \times \frac{\text{intake rate}}{\text{BWt}}$$

A list of useful data for exposure assessment is given in Table 6. Standard intake rates such as water consumption, respiratory volume, and other physiological parameters for the U.S. population are available in the *Exposure Factor Handbook* [19]. The exposure calculation takes into account the time frame (e.g., acute, chronic), duration (e.g., hours per day, food consumption per eating session), and frequency (e.g., repeated pattern) of exposure. Judicious considerations of spatial, temporal, geographic, and individual patterns can yield more realistic exposure estimates. However, these complex analyses are time-consuming and resource-intensive. Thus, a tiered approach is often used to maximize the use of limited resources. The initial screening assessment usually uses maximum or extreme

TABLE 6 Parameters and Data for Pesticide Exposure Assessment

Exposure parameter	Examples
Properties	Physical and chemical properties, degradation, dissipation, octanal–water coefficient, Henry's law constant, solubility
Concentration	Residue in food and water, concentrations in air, amount on contact surface (e.g., soil, foliage, water, countertop, carpet), transfer factor
Intake rate	Food consumption, water ingestion, in respiratory volume, body surface, body weight
Pharmacokinetics	Absorption factor, biomarker, pharmacokinetic parameters
Exposure pattern	Exposure duration (e.g., hours per day), exposure frequency (e.g., days per week, years per lifetime)
Use pattern	Season, frequency, amount, applied by professional or homeowner
Human activities	Time spent indoors and outdoors, activity level (affecting respiratory volume, water intake), change of location (daily travel, move residence), proximity to agricultural farm(s)
Pesticide use	Lawn, home, and garden

high values for more than one exposure parameter. By multiplying these values, the resultant exposure becomes "worst case" or even unrealistic. If the estimated risk does not exceed the level of concern, no subsequent tiers of refinement may be necessary. When needed, a probabilistic (distributional) analysis is conducted in the refining tiers. This analysis captures the range and variation of exposure parameters instead of using a single value for the parameters as in a point estimate (deterministic) analysis. A general guide for probabilistic analysis using the Monte Carlo technique was published by the USEPA in 1997 [20].

The dose that enters the system circulation is estimated by multiplying the exposure by the absorption factor (percentage of absorption). Expressing the exposure in terms of absorbed or internal dose is particularly important when a route-specific toxicological threshold or cancer slope is not available. In this case, risk assessment must rely on toxicity data extrapolated from other routes. For example, it may be necessary to use the oral toxicity NOEL for assessing the risk of inhalation exposures.

Human exposures can also be estimated on the basis of biological monitoring data. This is a useful alternative to the dosimetric approach using data from environmental monitoring and exposure parameters. Anwar [21] provided a list of references for pesticide biomarkers in blood, urine, breast milk, and adipose tissue. Important to a creditable use of biomonitoring data is the understanding of the pharmacokinetics and the timing of monitoring [22]. Although the absorbed dose can be directly measured in the blood, it can only be estimated through the metabolites detected in the urine. Urinary measurements may also reflect the composite exposures of more than one pesticide that have a common metabolite. Whereas the dosimetric method calculates the exposure from a single route or pathway of exposure, biomonitoring data represent the total exposure aggregated over all pathways during a given time period.

3.3.1 Dietary Exposure

To date, dietary exposure is the best-developed area in all pathways of exposure, owing in large part to the availability of computer programs and quality data on both food consumption and pesticide residue. Following the basic exposure equation, dietary exposure is calculated as the residue in food multiplied by the food consumption rate. For example, the exposure of a 60 kg woman consuming a medium-sized apple (300 g) containing 1 part per million (ppm; mg/kg) of a pesticide active ingredient (AI) is

$$\text{Exposure} = \frac{1\ \text{mg}_{\text{AI}}}{\text{kg}_{\text{Apple}}} \times 0.3\ \text{kg}_{\text{Apple}} \div 60\ \text{kg}_{\text{BWt}} = 0.005\ \text{mg/kg}$$

This illustrative calculation for a single individual receiving exposure through a single food form is manageable by hand or with a calculator. However, the calculation quickly becomes complicated in real life. Consider that apples in the diet consist of not only fresh apples but also apples in the form of pie, juice, and sauce. Consider further that many other agricultural commodities in a person's diet besides apples may also contain the pesticide of interest. Then consider the wide range of individual consumption rates for each of these commodities, all varying with age, gender, physiological status (e.g., pregnancy), geographic locations, and season. Realistically, the iteration of exposure calculation for a population becomes impossible without the aid of computer programs. These programs generally are capable of computing exposure profiles for a specified population subgroup. A typical analysis consists of more than 20 population subgroups by age (infants $\geq$ 1 year old, children 1–6 and 6–12 years old, teens, adults above 20 years old), gender, physiological status (nursing, pregnant, within childbearing age), ethnicity (Hispanic, non-Hispanic white, black, others), seasons, and geographic regions in the United States.

Consumption Data. Traditionally, the lack of consumption data to reflect current eating patterns is one major source of uncertainties in the dietary exposure analysis. This dilemma is greatly eased by the sizable survey data from the Continuing Survey of Food Intake by Individuals (CSFII). They are currently the most used food consumption data in pesticide risk assessment for the U.S. population. During 1989–1990, 1990–1991, and 1991–1992 (CSFII 1989–1991), the three consecutive day surveys had a combined total sample size of approximately 11,500 individuals [23–25]. During 1994–1995, 1995–1996, and 1996–1997 (CSFII 1994–1996), the two nonconsecutive day surveys had a combined total of approximately 16,000 individuals [26]. A Supplemental Children's Survey (SCS) of approximately 5000 children up to age 9, when available, will add to the CSFII 1994–1996 data for better characterizing the consumption profiles of infants and children. For pesticide dietary exposure assessment, the reported foods that are consumed are coded into food forms and commodities for which tolerances exist. This is accomplished on the basis of ingredient and nutrient labels, recipes, and estimated serving sizes.

Residue Data. The consumption database is standardized by data availability, but the selection of residue data for dietary exposure analysis has greater flexibility. In the order of increasing representativeness of what people eat, the choices of residue data are tolerance, data from field studies conducted for supporting tolerance determination, data from enforcement monitoring and surveillance, and data from monitoring studies designed for risk assessment. Field studies are typically conducted at a maximum application rate and with the shortest preharvest interval. Enforcement programs aim at timely detection of residues exceeding tolerance. They tend to have higher detection limits (e.g., within 10% of the tolerance) and are biased toward finding violations. Monitoring programs for risk assessment are statistically designed to represent what people eat. Among these programs, the yearly Pesticide Data Program (PDP) of the USDA beginning in 1991 provides the most extensive and useful data [27]. Samples are collected from distribution centers, prepared (washed, peeled, cored), and analyzed for multiple pesticides with minimum detection limits (MDLs) generally lower than those in the enforcement programs.

Exposure Analysis. The use of data from the various residue programs is dependent on data quality and availability as well as the tier level of dietary exposure analysis. A general description of current practices by the USEPA [28] and the California Department of Pesticide Regulation [29] is briefly described.

A screening analysis may assume that residues for all commodities are all at the highest levels, such as the tolerance or the high end of field trial data. No adjustment is made for residue reduction due to food preparation and processing. The analysis is deterministic, resulting in point estimate exposure. In the next tier, some consideration is given to the use of residue monitoring data instead of tolerances. One consideration is to adjust for residue reduction due to food preparation (washing, peeling, removing nonedible portions) and processing (cooking, canning). Another consideration is to use monitoring data from composite sampling (many units of food, e.g., 10 apples, in a sample) for foods that are commonly "mixed" (e.g., juice, grain, oil). Alternatively, monitoring data may be used for both mixed and nonmixed commodities. The rationale is that, compared to the extreme value of the tolerance, residue levels detected in monitoring programs are more representative of the food people eat. The next refining tiers of analysis may advance from a point estimate to a probabilistic analysis. The entire range of residue data is used to produce a distribution of exposure. Instead of assuming that the entire supply of a commodity has been treated with a pesticide, data on percentage of crop treatment can be factored into the distribution of residues.

3.3.2 Drinking Water Exposure

The focus of regulation with respect to ground and surface water contamination by pesticides has been on prevention. Based on the leaching potential of pesticides determined from their physical and chemical properties and their fate (e.g., degradation, dissipation) in the environment, buffer zones are established for protecting against groundwater contamination. Measures to reduce pesticide runoff and discharge to surface waters were also implemented, largely to address ecological concerns. Accordingly, the existing survey data generated for contamination prevention are largely inadequate for assessing the exposure in humans. These data are mostly sporadic, small in sample size, and limited to a relatively few geographic locations. Moreover, most of the groundwater surveys are taken from a single water source (e.g., a water well), which does not necessarily represent the residue level in municipal tap water that has gone through treatments and mixing from other water sources.

With the mandate of the 1996 FQPA, the shortage of reliable residue data became the main impediment in realistically including the drinking water pathway in the total (aggregate) pesticide exposure. An alternative to the use of monitoring data for HRA is model simulation. However, a basic weakness in drinking water simulation is that the existing models have not been through sufficient validation for use in estimating human exposures. As such, models are used with caution and in conjunction with all pertinent monitoring data. When a pesticide is judged to have the potential to contaminate drinking water sources, the "worst case" simulation may be used for the screening tier analysis. This may include simulations of a maximum runoff to a minimum size body of water or a shallow and vulnerable well. Understandably, these scenarios are unlikely to represent the sole source for drinking water. When this screen level risk exceeds the level of concern, the subsequent refining assessment may use models that allow more realistic considerations while making sure that the model output would not underestimate the potential exposure [30].

3.3.3 Residential Exposure

Residential exposure has long been recognized as a potential source of significant pesticide exposure [31]. As a part of the FQPA mandate, attention is now given to formulating a process for exposure assessment. Residential uses of pesticides may include foggers, crack and crevice treatments, structural fumigations, paint and wood treatments, and treatments of home gardens and trees, swimming pools, and pets [32]. These diverse applications can result in all routes of exposure. Inhalation exposure occurs from breathing the air containing residues from drift, volatilization, and dust resuspension. Dermal exposure occurs from contact with carpets, countertops, surfaces of household structures, toys, and pets. Oral exposure occurs from deposition on foods and home-grown produce and conveyance by hand from contaminated surfaces. Much emphasis has been placed on the exposure of infants and children because of their longer residential time, lawn and floor activities, and mouthing behavior [33]. Concerns have also been raised regarding children of farm workers [34] and those who live in and around a farm [35]. Residential contamination could occur through air, dust, and work clothes and shoes.

Unique to the residential exposure assessment is the inordinate number of scenarios and their possible combinations. They vary geographically and temporally and are dependent on individual lifestyles and activities. The sparsity of residue data at the present time further limits the extent to which the exposure can be reliably estimated. Although a systematic approach to exposure estimation is yet to be established, it is clear that the focus should be on a manageable number of scenarios with the most significant exposures. This can be achieved by first identifying all possible pathways for the population of interest, then screening to eliminate those pathways that would not result in significant exposures.

3.3.4 Occupational Exposure

Pesticide workers include those who mix, load, and apply the pesticides to agricultural fields and parks, structures, livestock and pets, and areas for vector control. They also include flaggers, harvesters, and agricultural scouts who inspect the fields for pests after pesticide application. Occupational exposures can be through oral (e.g., residues on hands and foods) or inhalation (e.g., breathing the volatilized form, dust, aerosol) pathways. More often, they occur through dermal contacts with the formulation, application solutions, or foliage coated with residues. Dermal exposures from direct contact with the pesticide are most often estimated dosimetrically by multiplying the amount of pesticide on the skin by the area of contact. Dermal exposure from contact with foliage has been roughly estimated by multiplying the dislodgeable foliar residue (DFR; $\mu g/cm^2$) by a transfer factor (TR). TR (usually in cm^2/hr) is an empirically determined ratio of dermal exposure ($\mu g/hr$) and the DFR. Default values are established for groups of chemicals based on their canopy stance and harvesting practices [36,37]. The absorbed dose is then calculated by multiplying the estimated exposure by the absorption factor.

When chemical-specific data are not available for estimating the exposure of handlers, data from the Pesticide Handlers Exposure Database (PHED) are often used. PHED (version 1.1) is a database compiled jointly by a task force representing Health Canada, the USEPA, and the American Crop Protection Association [38]. It is used as a generic database with the assumption that handlers' exposure is more a function of the physical parameters of pesticide handling than of the chemical properties. The reduction of exposure from label-specified personal protective equipment (PPE) is taken into account. The PPE may include long-sleeved shirts, long pants, socks, overalls, chemical-resistant suits, face shields, respirators, gloves, and self-contained breathing apparatuses.

Several factors are included in estimating the overall exposure of a worker. Both the occupational exposure at work and exposure as a member of the general public while away from the work site are taken into account. The latter could include not only dietary exposure but also drinking water and residential exposures, albeit with less residential time than a home-bound individual. Therefore, it is prudent to separate workers as a population subgroup in an exposure assessment. This would also facilitate the identification of areas for mitigation when the overall risk exceeds the acceptable level.

3.4 Risk Characterization

Risk characterization describes the anticipated risk to humans based on information from the other three HRA components. Following the two basic models of the dose–response relationship regarding thresholds, the nononcogenic and oncogenic risks are characterized differently. Nononcogenic risk is expressed as a ratio of the toxicity threshold (e.g., NOEL) and the exposure. Oncogenic risk is expressed as a probability.

3.4.1 Nononcogenic Effects—Margin of Exposure

The risk of adverse effects other than cancer or tumors is commonly characterized as the margin of exposure (MOE):

$$\text{MOE} = \frac{\text{NOEL}}{\text{exposure (or dose)}}$$

Given an existing threshold below which no adverse effect is likely, an "adequate MOE" is established. Although there is some flexibility for including value judgments, the required MOE is established mainly on the basis of toxicological considerations. Uncertainty factors (UFs), usually ranging from 1 to 10, are used to address these considerations [39]. Unless sufficient evidence indicates otherwise, the current default assumption is that, on a dose per body weight basis, humans can be tenfold more sensitive than laboratory animals (interspecies variation) and that there may be a tenfold variation of sensitivity among humans (intraspecies or interindividual variation). Accordingly, the basic default requirement for an acceptable MOE is 100 when it is calculated based on a NOEL determined in animals, and 10 when calculated based on a NOEL determined in humans. Other UFs have also been used to further ensure the adequacy of the MOE for the protection of human health. For example, a UF of 10 may be used when it is necessary to estimate a chronic NOEL from a subchronic NOEL. Another UF of up to 10 may be used when it is necessary to estimate a NOEL from a LOEL. Yet another UF may be used when the toxicity database is deficient. For example, an additional FQPA uncertainty factor of up to 10 is used when the current toxicity database is inadequate to ensure the safety of infants and children.

3.4.2 Oncogenic Effects—Risk

When a chemical shows sufficient oncogenic weight of evidence, the oncogenic risk is calculated as

$$\text{Risk} = \text{exposure (or dose)} \times \text{potency}$$

Oncogenic risk assessment has sometimes been referred to as "quantitative risk assessment," because "risk" is a quantitative expression. Unlike the acceptable MOE determination, the level of risk a society is willing to accept is a value judgment that takes into account not only the health risk but also socioeconomic considerations and the balance between risk and benefit.

3.4.3 Standards of Exposure

Besides characterizing the risk, a risk assessment is also used to establish the standards of exposure.

Reference Dose. A reference dose (RfD) is an estimated daily oral dose for the human population that is likely to be without an appreciable risk of deleterious (nononcogenic) effects in a lifetime [40]. A comparable term is acceptable daily intake (ADI). RfD is calculated as

$$\text{RfD} = \frac{\text{NOEL}}{\text{uncertainty factors}}$$

The UFs take into account the interspecies and interindividual sensitivity and any other additional factors as previously discussed. The total UF should be the same as the MOE judged necessary for an acceptable risk. According to the traditional definition, RfD is a standard for long-term exposures. Using the same approach, an acute RfD can also be calculated based on an acute NOEL.

Reference Concentration. Reference concentration (RfC) is a term defined similarly to RfD except that it is a concentration term for inhalation exposures. Unlike the RfD, the calculation of RfC takes into account the patterns of human exposure (e.g., breathing rate, body weight) and is usually established for a 24 hr exposure duration.

3.4.4 FQPA Emphasis

Although the specific requirements under the FQPA pertain to the safety evaluation and setting of tolerances, they require a more thorough consideration for risk assessment that have wider implications beyond pesticide food safety. These considerations have become integral parts of risk assessment presentation for pesticides.

Safety to Infants and Children. It has long been recognized that infants and children differ from adults in their response to chemicals [6,41]. Depending on the chemical, the young are not always more sensitive. A weight-of-evidence evaluation of age-specific sensitivity is conducted under hazard identification and dose–response assessment. A particular concern is the lifetime effect from in utero exposure. Unfortunately, this has generally not been thoroughly tested for most pesticides, especially with respect to developmental neurotoxicity, immunotoxicity, and oncogenicity [6,42]. In addition, infants and children generally have higher exposure than adults because of their greater intake rates (e.g., food consumption, breathing rate) on a body weight basis. This can be accounted for in the exposure assessment, provided that reliable data are available to ensure that the exposure for the individual pathway and the aggregate are not underestimated. When the overall database is insufficient to ensure adequate protection of infants and children including their exposure in utero, the FQPA additional tenfold UF is applied both for determining the adequacy of MOE and for establishing the RfD.

Aggregate Exposure. Route-specific aggregate exposure can be calculated as the sum of exposure from all pathways (e.g., total oral exposure from the dietary, drinking water, and residential exposures). Summing high end values (worst case, 95th percentile) from all pathways would most likely overstate the exposure. A more realistic approach may be to add the high end of exposure from the major pathway(s) to the average exposure from the remaining pathway(s). Risk is then calculated based on the aggregate exposure. Finally, the route-specific risks are combined into a total risk or "risk cup." This can be achieved by summing the hazard quotients (ratio of RfD to exposure; RfD/exposure) from all pathways or summing the inverse values of MOEs. An alternative to the point estimate approach is a probabilistic approach that allows some more realistic combinations of exposure pathways based on pesticide use patterns and the spatial and temporal interrelations among these pathways.

Cumulative Risk. With the widespread use of pesticides, it is reasonable to expect that humans will come into contact with more than one pesticide on a daily basis. Of particular concern is the cumulative effect of those chemicals that have a common mechanism of toxicity (e.g., organophosphates). The first set of organophosphate cumulative risk assessment is scheduled for completion in 2002.

Critical issues exist in both the toxicity and the exposure sides of the risk assessment equation [43]. On the one hand, it is essential to properly determine the relative toxicity of chemicals that have the same mechanism of toxicity so that their contribution to the collective risk can be equitably estimated. On the other hand, the pattern of coexisting exposure to more than one pesticide would have to be established so that the estimated exposure is not an exaggerated level from stacking together many "conservative" parameters that has little or no real-life implications. Understandably, not all pesticides with the same mechanism of toxicity are used on a crop at a given time. Pesticide use records can be used to characterize the probable pattern of multiple chemical exposure. Some statewide records are available. For example, under a full use reporting requirement, comprehensive use data on all agricultural applications of pesticides have been published since 1991 by the California Department of Pesticide Regulation, which publishes yearly pesticide use reports [44]. Residue monitoring data showing multiple pesticides in a given sample could also be used for realistically narrowing down the number of pesticides in one commodity.

Endocrine Disruption. An endocrine disruptor may interfere with the role of natural hormones in several ways [45]. The current focuses include not only interference with the normal actions of estrogen, which alter the natural development of the reproductive tract and sexual differentiation of the brain (e.g., precocious puberty, disrupted cycling), but also the non–sex steroid–based functions, such as the thyroid. Although serious effects could occur over all life stages, exposure during the developing stage when the organism may be more vulnerable and the effects longer lasting has been of particular concern [45].

The complexity of the issues with respect to the many potential mechanisms and target organs cannot be overstated. Presently, data to address these effects are largely lacking. Under the auspices of the USEPA Risk Assessment Forum, in 1998 the Endocrine Disruptor Screening and Testing Advisory Committee (EDSTAC) completed a report outlining the tier approach to testing chemicals for endocrine disruption potential [46]. In light of the lack of sufficient information as well as consensus in the scientific community, the USEPA's current science policy is that "the Agency does not consider endocrine disruption to be an adverse endpoint per se, but rather to be a mode or mechanism of action potentially leading to other outcomes" [45].

Testing guidelines and criteria for hazard identification are needed for a clear evaluation of the endocrine disruption potential.

3.4.5 Descriptive Presentation

Up to this point, risk characterization has been presented in a numerical fashion. Obviously, the quality of the risk assessment and the certainty of the estimated risk are dependent on the quality of information from each of the other three components. Greater certainty about the safety of pesticides can be achieved only with a better understanding of the inherent toxicities of a chemical, a more accurate description of the dose–response relationship, and a more realistic estimation of human exposure.

Risk assessment is an important component in environmental policy decisions. As such, it is also a medium that communicates risk to policy makers (risk managers) and the public. Therefore, it is essential that risk assessment be presented in a clear and transparent fashion [47,48]. The presentation will not be complete unless the numerical results (e.g., MOE, risk) are accompanied by a concise description of all the associated key uncertainties and variabilities. This may include discussions of the approaches and their sufficiencies to account for interspecies (animal to human) and intraspecies (interindividual) variations. It may also include discussions of strengths and weaknesses of the existing data for toxicity evaluation and exposure assessment and for extrapolating the slope to a low-response range. The point estimate of risk is presented in the context of its expected range (e.g., worst case, high end, 95th percentile, a statistical bound, central tendency). The overall strengths and limitations of the assessment are clearly articulated, separating scientific justification from policy judgments.

4 ECOLOGICAL RISK ASSESSMENT

Ecological risk assessment (ERA) is an evaluation of the likelihood that adverse ecological effects may occur or are occurring as a result of exposure to one or more stressors [49]. It is a part of the pesticide registration process to ensure that the use of pesticides will not pose an unreasonable risk to nontarget species, wildlife, and the environment. The Endangered Species Act (ESA) of 1973 also requires the protection of endangered or threatened species against any harm.

4.1 General Framework

The previously rather situational processes to address local and specific concerns (e.g., bird, fish, and bee kills; exceeding environmental criteria) are being formalized into a framework described in the ERA guidelines [49]. The process of ERA is similar to the HRA with respect to the use of scientific information to identify hazards, assess the dose–response relationship and the exposure, and characterize the risk. However, the scope and dynamic interrelationships in the biological organization (e.g., individuals, populations, communities, ecosystems) beyond the individual and population levels present a far more complex task than HRA. Whereas the goal and scope for HRA are commonly acknowledged, the initial process of planning is crucial for an ERA. It defines the goal and scope of the assessment, the resource availability, and the type of decisions the ERA is to support.

The ERA process consists of three phases: problem formulation, analysis, and risk characterization. The initial phase of problem formulation determines the assessment endpoints and generates preliminary hypotheses. The exposure and the stressor–response relationship are described in the subsequent analysis phase. In the final phase of risk characterization, data on the exposure and effects are integrated to evaluate the likelihood of adverse effects.

4.2 Pesticide Ecological Risk Assessment

The basic tests for ecological effects of pesticides include oral (acute) and dietary toxicities in avian species and acute toxicities in aquatic species (40 CFR, Part 158). These studies focus mainly on lethality as endpoints for the determination of the LD_{50} and LC_{50}. Reports in the literature on incidents of environmental perturbation also contribute to the database for assessment. Nevertheless, there is an enormous gap between the data available and the data needed to adequately address the ecological concerns, especially beyond species and populations. Data for exposure parameters for mammalian, bird, amphibian, and reptile species across North America are available in *Wildlife Exposure Factors Handbook* [50].

In the past few years, there has been substantial progress in formalizing the process and methods of pesticide ERA within the framework of FIFRA. Several recommendations for improvement are made by the FIFRA Scientific Advisory Panel (SAP). The general direction is to use endpoints other than lethality and to advance into using a probabilistic rather than a deterministic approach. Field tests are needed to generate site-specific data, both for monitoring pesticide uses and for validating the formulated hypothesis on ecological interactions. Resource limitations would necessitate prioritization to focus on direct effects of pesticides and on high risk species.

5 FUTURE DIRECTIONS

The ultimate goal of a pesticide risk assessment is to realistically characterize the present and/or anticipated risk to ensure that pesticide uses will not result in unreasonable adverse outcomes to humans and the environment. Compared to other environmental contaminants, pesticides have a rather extensive database useful for risk assessment. However, several critical issues remain. Toxicity tests are needed to better assess the sensitivity of infants and children, including in utero stages. A common approach and tools are being developed for a more realistic assessment of the aggregate exposure. Data are being generated and collected for reducing the uncertainties in the parameters for population exposure pathways. A reasonable approach is being formulated for addressing cumulative risks of pesticides with a common mechanism of toxicity.

There are also many overarching risk assessment issues to be investigated.

One issue is the risk of exposure to inert ingredients and solvents in formulations. This is particularly important for workers' protection. A related need is better surveillance and treatment of workers' illnesses. Another issue is the need to better characterize the toxicities of degradation products. When data are not available, toxicity equal to that of the parent compound on a molecular basis is often assumed. However, this may not be true in some cases. For example, toxicity data support a much higher toxicity equivalence factor for methyl paraoxon that is found in the air after methyl parathion application [51]. A related need is a comprehensive monitoring of toxicologically significant degradation products. Yet another remaining issue is the adequacy of the UF to address interindividual sensitivity. Patterns of genetic polymorphism identified in human populations for several key metabolic enzymes such as cytochrome P450, glutathion-S-transferase, and paraoxonase [52,53] suggest that variations of susceptibility to environmental contaminants may be greater than tenfold. Finally, it is essential that the approach and practices for evaluating pesticide safety be globally harmonized to ensure that governmental boundaries will not create unnecessary barriers that hinder commercial trade across borders.

Risk assessment is never a closed book. As toxicological research and understanding continue to advance, the toxicity database may reveal additional concerns that need to be assessed. Pesticide use patterns also change over time. New pesticides and formulations are being introduced. Continuing monitoring of use patterns and residue levels in foods, water, and environmental media is necessary for identifying major changes that warrant a reassessment of risk.

REFERENCES

1. PH Nicholson. 1963. Contamination of water with pesticides: Pesticide pollution studies in the southeastern states. In: New Developments and Problems in the Use of Pesticides. Food Protection Committee, Food and Nutrition Board. Publ 1082. Washington DC: Natl Acad Sci—Nat Res Council, 1963, pp 65–71.
2. WB Deichmann. Pesticides and the Environment: A Continuing Controversy. New York: Intercontinental Med Book Corp, 1973.
3. R Cremlyn. Pesticides: Preparation and Mode of Action. New York: Wiley, 1978, pp 210–221.
4. WM Upholt. Pesticide highlights—The seventies new legislation. In: WB Deichmann, ed. Pesticides and the Environment: A Continuing Controversy. New York: Intercontinental Med Book Corp, 1973, pp 19–22.
5. National Academy of Science (NAS). Risk Assessment in the Federal Government: Managing the Process. Committee on the Institutional Means for Assessment of Risks to Public Health, Commission on Life Sciences, and National Research Council. Washington, DC: Natl Acad Press, 1983.
6. National Academy of Science (NAS). Pesticides in the Diets of Infants and Children. Washington, DC: Natl Acad Press, 1993.

7. National Center for Environmental Assessment (NCEA). Office of Research and Development, U.S. Environmental Protection Agency. [Online, June 1999: http://www.epa.gov/ncea/]
8. Office of Pesticide Programs (OPP) of U.S. Environmental Protection Agency (USEPA). FQPA science policies and guidance documents. [Online, August 2000: http://www.epa.gov/pesticides/trac/science/]
9. U.S. Environmental Protection Agency (USEPA). Chapter 8: Precautionary Labeling. In: Label Review Manual. USEPA Office of Pesticide Programs. Updated June, 2001. [Online, February 2002: http://www.epa.gov/oppfead1/labeling/lrm/chap-08.htm]
10. CA Kimmel. Quantitative approaches to human risk assessment for noncancer health effects. Neuro Toxicol 11:19–198, 1990.
11. DG Barnes, GP Daston, JS Evans, AM Jarabek, RJ Kavlock, CA Jimmel, C Park, HL Spitzer. Benchmark dose workshop: Criteria for use of a benchmark dose to estimate a reference dose. Regul Toxicol Pharmacol 21:296–306, 1995.
12. RC Gupta, GT Patterson, W-D Dettbarn. Comparison of cholinergic and neuromuscular toxicity following acute exposure to Sarin and VX in rat. Fundam Appl Toxicol 16:449–458, 1991.
13. S Brimijoin. Enzymology and biology of cholinesterases. In: Proceedings of the USEPA Workshop on Cholinesterase Methodology, Dec 4–5, 1991. US Environ Protection Agency, 1992.
14. U.S. Environmental Protection Agency (USEPA). Science Policy on The Use of Data on Cholinesterase Inhibition for Risk Assessments of Organophosphate and Carbamate Pesticides. Prepared by William F. Sette, PhD, for the Office of Pesticide Programs, U.S. Environmental Protection Agency, Washington, DC, Aug 18, 2000. [Online, February 2001: http://www.epa.gov/pesticides/trac/science/cholin.pdf]
15. LL Carlock, WL Chen, EB Gordon, JC Killeen, A Manley, LS Meyer, LS Mullin, KJ Pendino, A Percy, DE Sargent, LR Seaman, NK Svanborg, RH Stanton, CI Tellone, DL Van Goethem. Regulating and assessing risks of cholinesterase-inhibiting pesticides: divergent approaches and interpretations. J Toxicol Environ Health B Crit Rev 2(2):105–160, 1999.
16. KS Crump. An improved procedure for low-dose carcinogenic risk assessment from animal data. J Environ Pathol Toxicol 5:675–684, 1980.
17. D Krewski, KS Crump, J Farmer, DW Gaylor, R Howe, C Portier, D Salsburg, RL Sielken, J Van Ryzin. A comparison of statistical methods for low dose extrapolation utilizing time-to-tumor data. Fundam Appl Toxicol 3:140–160, 1983.
18. US Environmental Protection Agency (USEPA). Draft Report: a cross-species scaling factor for carcinogen risk assessment based on equivalence of $mg/kg^{3/4}/day$; Notice. Fed Reg 57(109):24152–24173, 1992.
19. US Environmental Protection Agency (USEPA). Exposure Factor Handbook. Office of Research and Development, National Center for Environmental Assessment. U.S. Environmental Protection Agency, Washington, DC. EPA/600/P-95/002Fa. Natl Tech Infor Service (NTIS) PB98-124217, 1998.
20. U.S. Environmental Protection Agency (USEPA). Policy for Use of Probabilistic Analysis in Risk Assessment and the Guiding Principles for Monte Carlo Analysis. EPA/630/R-97/001. Washington, DC: US Environ Protect Agency, 1997.

21. WA Anwar. Biomarkers of human exposure to pesticides. Environ Health Perspect 105(suppl 4):801–806, 1997.
22. DB Barr, JR Barr, WJ Driskell, RH Hill Jr, DL Ashley, LL Needham, SL Head, EJ Sampson. Strategies for biological monitoring of exposure for contemporary-used pesticides. Toxicol Ind Health 15:168–179, 1999.
23. US Department of Agriculture (USDA), Agricultural Research Service. Data tables: Results from USDA 1989 Continuing Survey of Food Intakes by Individuals (CSFII) and 1989 Diet and Health Knowledge Survey. US Dept Commerce, Natl Tech Info Service, PB93-500411, 1993.
24. US Department of Agriculture (USDA), Agricultural Research Service. Data tables: Results from USDA 1990 Continuing Survey of Food Intakes by Individuals (CSFII) and 1990 Diet and Health Knowledge Survey. US Dept Commerce, Natl Tech Info Service, PB93-504843, 1993.
25. US Department of Agriculture (USDA), Agricultural Research Service. Data tables: Results from USDA 1991 Continuing Survey of Food Intakes by Individuals (CSFII) and 1991 Diet and Health Knowledge Survey. US Dept Commerce, Natl Tech Info Service, PB94-500063, 1994.
26. US Department of Agriculture (USDA), Agricultural Research Service. Data tables: Results from USDA 1994–96 Continuing Survey of Food Intakes by Individuals (CSFII) and 1994–96 Diet and Health Knowledge Survey. ARS Food Surveys Res Group, 1997.
27. US Department of Agriculture (USDA), Agricultural Marketing Service. Pesticide Data Program, Annual summary calendar year 1997. Pub 1998.
28. US Environmental Protection Agency (USEPA). 1996 Acute Dietary Exposure Assessment; Office Policy. Office of Pesticide Programs. [Online, June 1999: http://www.epa.gov/opphed01/acutesop.htm]
29. L Lim, NR Reed, W Carr Jr. Regulation of pesticide residue levels in food: Approaches and issues in the USA. Pesticide Outlook 8(4):6–11, 1997.
30. US Environmental Protection Agency (USEPA). Interim guidance for conducting drinking water exposure and risk assessments (Sept 16, 1998). Health Effects Division, Office of Prevention, Pesticides and Toxic Substances. Washington, DC: USEPA, 1998.
31. US Environmental Protection Agency (USEPA). Nonoccupational pesticide exposure study (NOPES). U.S. Environmental Protection Agency, Research Triangle Park, NC. EPA 600/3-90/003. Springfield, VA: Natl Tech Infor Service, PB 90-152-224/AS, 1990.
32. US Environmental Protection Agency (USEPA). Standard Operating Procedures (SOPs) for Residential Exposure Assessments, Draft. Prepared by the Residential Exposure Assessment Work Group, includes Office of Pesticide Programs, Health Effects Division, Versar, Inc., January 1999.
33. S Gurunathan, M Robson, N Freeman, B Buckley, A Roy, R Meyer, J Bukowski, PJ Lioy. Accumulation of chlorpyrifos on residential surfaces and toys accessible to children. Environ Health Perspect, 106:9–16, 1998.
34. C Loewenherz, RA Fenske, NJ Simcox, G Bellamy, D Kalman. Biological monitoring of organophosphorus pesticide exposure among children of agricultural workers in central Washington state. Environ Health Perspect 105:1344–1353, 1997.

35. BC Gladen, DP Sandler, SH Zahm, F Kamel, AS Rowland, MCR Alavanja. Exposure opportunities of families of farmer pesticide applicators. Am J Ind Med 34: 581–587, 1998.
36. G Zweig, JT Leffingwell, W Popendorf. The relationship between dermal pesticide exposure by fruit harvesters and dislodgeable foliar residues. J Environ Sci Health B20:27–59, 1985.
37. R Brouwer, DH Brouwer, SCHA Tijssen, and JJ van Hemmen. Pesticides in the cultivation of carnations in greenhouses: Part II. Relationship between foliar residues and exposures. Am Ind Hyg Assoc J 53:582–587, 1992.
38. Pesticide Handlers Exposure Database (PHED). Version 1.1. Developed by the Task Force representing Health Canada, the U.S. Environmental Protection Agency, and the American Crop Protection Association, 1995.
39. M Dourson, JF Stara. Regulatory history and experimental support of uncertainty (safety) factors. Regul Toxicol Pharmacol 3:224–238, 1983.
40. DG Barnes, M Dourson. Reference dose (RfD): Description and use in health risk assessments. Regul Toxicol Pharmacol 8:471–486, 1988.
41. International Life Sciences Institute (ILSI). Similarities and Differences Between Children and Adults: Implications for Risk Assessment. PS Guzelian, CJ Henry, SS Olin, eds. Washington, DC: ILSI Press, 1992.
42. NR Reed. Assessing risks to infants and children. In: D Tennant, ed. Food Chemical Risk Analysis. London, UK: Blackie Academic and Professional, Chapman & Hall, 1997, pp 219–239.
43. US Environmental Protection Agency (USEPA). Guidance on cumulative risk assessment of pesticide chemicals that have a common mechanism of toxicity. Office of Pesticide Programs, January 14, 2002.
44. California Department of Pesticide Regulation (CDPR). Annual Pesticide Use Reports, 1991–1998. Calif Environ Protect Agency, Sacramento, CA. [Online, August 2000: http://www.cdpr.ca.gov/docs/pur/purmain.htm]
45. US Environmental Protection Agency (USEPA). Special report on environmental endocrine disruption: An effects assessment and analysis. EPA/630/R-96/012, 1997.
46. Endocrine Disruptor Screening and Testing Advisory Committee (EDSTAC). Final Report, USEPA, 1998.
47. National Academy of Science (NAS). Science and Judgment in Risk Assessment. Committee on Risk Assessment of Hazardous Air Pollutants, Board on Environmental Studies and Toxicology, Commission on Life Sciences. Washington, DC: Natl Acad Press, 1994.
48. US Environmental Protection Agency (USEPA). Elements to consider when drafting EPA risk characterizations. March 1995. Updated Sept 22, 1997. [Online, June 1999: http://www.epa.gov/ORD/spc/rcelemen.htm].
49. US Environmental Protection Agency (USEPA). Guidelines for Ecological Risk Assessment; Notice. Fed Reg 63(93):26846–26924, 1998.
50. US Environmental Protection Agency (USEPA). Wildlife Exposure Factors Handbook. Office of Health and Environmental Assessment, Office of Research and Development. EPA/600/R-93/187. Washington, DC: US Environ Protect Agency, 1993.

51. NR Reed. Evaluation of Methyl Parathion as a Toxic Air Contaminant—Health Effects Evaluation. Dept Pesticide Regul, Calif Environ Protect Agency, 1999.
52. MS Miller, DG McCarver, DA Bell, DL Eaton, JA Goldstein. Genetic polymorphisms in human drug metabolic enzymes. Fundam Appl Toxicol 40:1–14, 1997.
53. JE Hulla, MS Miller, JA Taylor, DW Hein, CE Furlong, CJ Omiecinski, TA Kunkel. Symposium overview: the role of genetic polymorphism and repair deficiencies in environmental disease. Toxicol Sci 47:135–143, 1999.

5

Environmental Fate of Pesticides

James N. Seiber
Western Regional Research Center
Agricultural Research Service
U.S. Department of Agriculture
Albany, California, U.S.A.

1 RATIONALE

Assessing the transport and fate of pesticides in the environment is complicated. There are a myriad of transport and fate pathways at the local, regional, and global levels. Pesticides themselves represent a diverse group of chemicals of widely varying properties and use patterns. And the environment is, of course, diverse in makeup and ever-changing, from one location to another and from one time to another.

Environmental sciences have evolved as a means of understanding and dealing with the complexities in nature by sorting out and defining underlying principles. These can serve as starting points or steps in the assessment of chemical processing important to the health of the environment, humans, and wildlife.

In the past, particularly from roughly the 1940s to 1970, knowledge of how pesticides and other chemicals behaved in the environment was obtained by retrospective analysis for these chemicals after they had been used for many years. By analyzing soil, water, sediment, air, plants, and animals, environmental scientists were able to piece together profiles of behavior. Dibromochloropropane (DBCP), ethylene dibromide (EDB), and chemicals with similar uses as soil nematicides and similar properties came to be recognized as threats to groundwater in general use areas. DDT and other chlorinated insecticides and organic compounds of similar low polarity, low water solubility, and exceptional stability threatened some aquatic and terrestrial animals because of their potential for undergoing bioaccumulation and their chronic toxicities. The chlorofluorocarbons (CFCs) and methyl bromide were found to be exceptionally stable in the atmosphere and able to diffuse to the stratosphere, where they entered into reactions that result in destruction of the ozone layer.

But as large a testimony as these examples and others were to the skill of environmental analytical chemists, environmental toxicologists, ecologists, and other environmental scientists in detecting small concentrations and subtle effects of chemicals, the retrospective approach is fraught with difficulty.

1. Adverse chemical behavior might be discovered too late, after considerable environmental damage (e.g., decline of raptorial bird species in the case of DDT/DDE, or contamination of significant groundwater reserves in the case of EDB and DBCP) was already done.
2. By analyzing for the wrong chemical, or the wrong target media, the problem may be misdefined or completely overlooked. For example, parent pesticides such as aldicarb and aldrin yield products in the environment (aldicarb sulfoxide and sulfone; dieldrin and, eventually, photodieldrin) which may be the primary offenders. Initial analyses may miss this, by targeting only the parents rather than the products.

The trend from roughly the 1970s to the present has thus focused on ways to predict environmental behavior before the chemical is released. For economic materials (pesticides, industrial chemicals in general), premarket testing of environmental fate and effects is now built into the regulatory processes leading to regulatory approval. The Environmental Fate Guidelines of the U.S. Environmental Protection Agency (USEPA) [1,2], for example, specify the tests and acceptable behavior required for registration of candidate pesticides in the United States. Europe [3] Canada [4], Australia [5], and other nations and economic organizations produce similar guidelines and test protocols to screen for potential adverse environmental behavior characteristics.

Another stimulus for developing both better analytical and better predictive tests was the onset of risk assessment as a formal methodology for evaluating risks of chemicals in the environment. Risk assessment and risk science in general are relatively new fields, dating from the late 1970s and early 1980s for human health risk assesment [6] and even later for ecological risk assessment [7]. In both the hazard identification component, which includes measuring and/or estimating emissions to the environment, and the exposure assessment component of risk assessment, which involves measuring or modeling exposures via food, water, air, etc., predictive tools (models) are undergoing rapid development for use in regulatory actions, both for premarket screening and for decisions on continuing use. Many pesticides, as well as hazardous air pollutants [8] and other substances of environmental concern, have undergone or are now in the process of risk assessment review [9].

Although regulatory agencies might be seen as primarily responsible for stimulating predictive methods, industry has also played early and continuing roles. It is clearly in the best interests of companies to screen out potential environmental problems early in the development process and to focus resources on chemicals that have the potential for long-term environmental compatibility. For example, environmental scientists at Dow Chemical in the early 1970s developed a "benchmark approach" to evaluating environmental characteristics of candidate pesticides [10]. The benchmark approach and other early developments in screening or predicting environmental behavior, including modeling, became formalized in the new field called environmental chemodynamics, which may be generally defined as [11,12]

> The subject dealing with the transport of chemicals (intra and interphase) in the environment, the relationship of their physical-chemical properties to transport, their persistence in the biosphere, their partitioning in the biota, and toxicological and epidemiological forecasting based on physicochemical properties.

Another factor in developing a predictive capability for environmental behavior and fate is the rapidly changing nature of pesticide chemicals. The highly stable lipophilic organochlorines, organophosphates of high mammalian toxicity, and environmentally persistent triazine and phenoxy herbicides that dominated pesticide chemistry until the 1970s are either gone entirely from the pesticide markets or are undergoing replacement. In their place are synthetic pyrethroids, sulfonylureas, aminophosphonic acid derivatives, biopesticides, and many other classes and types whose environmental fate and ecotoxicological effects are less straightforward and in need of detailed evaluation. Some of the new pesticides are attractive because they degrade relatively rapidly and extensively in the environment. However, this can multiply the number of discrete chemicals that need to be evaluated in terms of mobility, fate, and nontarget effects. Relying solely on experimentation in the environment could significantly slow regulatory approval, arguing again for the use of predictive screening assessment tools as an integral component of premarket testing.

Increasing pressure is being exerted on environmental scientists to define tests for subtle environmental effects that go beyond the leaching, bioaccumulation, and acute/chronic toxicity testing so prominent in environmental fate tests of the past. A current example is provided by concerns over environmental endocrine disruption caused by trace levels of chemicals and chemical mixtures [13,14]. Ideally, environmental chemists would be able to detect interactions of endocrine-disrupting chemicals (EDCs) with mammalian tissues and ecosystems by biobased testing for the chemicals themselves or biomarkers indicating that exposure to EDCs had occurred. The methods and approaches to screening for EDCs, under intense development from the stimulus of the Food Quality Protection Act [15], have the potential for adding complexity to the already complicated business of "environmental chemodynamics."

Much of our current capability in environmental sciences for determining the transport and fate of pesticides and other chemicals may be traced directly to the tremendous developments in analytical chemistry of the past quarter century or so. Detection limits of low parts per billion (ppb) and even parts per trillion (ppt) are now achievable by better methods of extracting, preparing, and, particularly, determining residues of pesticides and breakdown products in a variety of matrices (e.g., Fong et al. [16]). Developments in gas and liquid chromatography, mass spectrometry, and immunoassay have been among those most useful to environmental scientists, but computer data-handling capabilities have also enabled the routine use of these sophisticated techniques in industry, academic, agency, and commercial laboratories.

2 PRINCIPLES

2.1 The Dissipation Process

Once a substrate (agriculture commodity, body of water, wildlife, soil, etc.) has been exposed to a chemical, dissipation processes begin immediately. The initial residue dissipates at an overall rate that is a composite of the rates of individual processes (volatilization, washing off, leaching, hydrolysis, microbial degradation, etc.) [17]. When low-level exposure results in the accumulation of residues over time, as in the case of bioconcentration of residues from water by aquatic organisms, the overall environmental process includes both the accumulation and dissipation phases. However, for simple dissipation, such as occurs in the application of pesticides and resulting exposure from residues in food or water or air, the typical result is that concentrations of overall residue (parent plus products) decrease with time after end of exposure or treatment (Fig. 1).

Because most individual dissipation processes follow apparent first-order kinetics, overall dissipation or decline is also observed to be first-order. This has important ramifications. Because first-order decline processes are logarithmic, that is, a plot of remaining residue concentration versus time is asymptotic to the time axis, residues will approach zero with time but never cease to exist entirely (Fig. 1a). That is, all environmental exposures lead to residues that have, theoretically, unlimited lifetimes. However, our ability to detect remaining residues is limited by the detectability inherent in the methods of gas chromatography, high performance liquid chromatography, mass spectrometry, immunoassay,

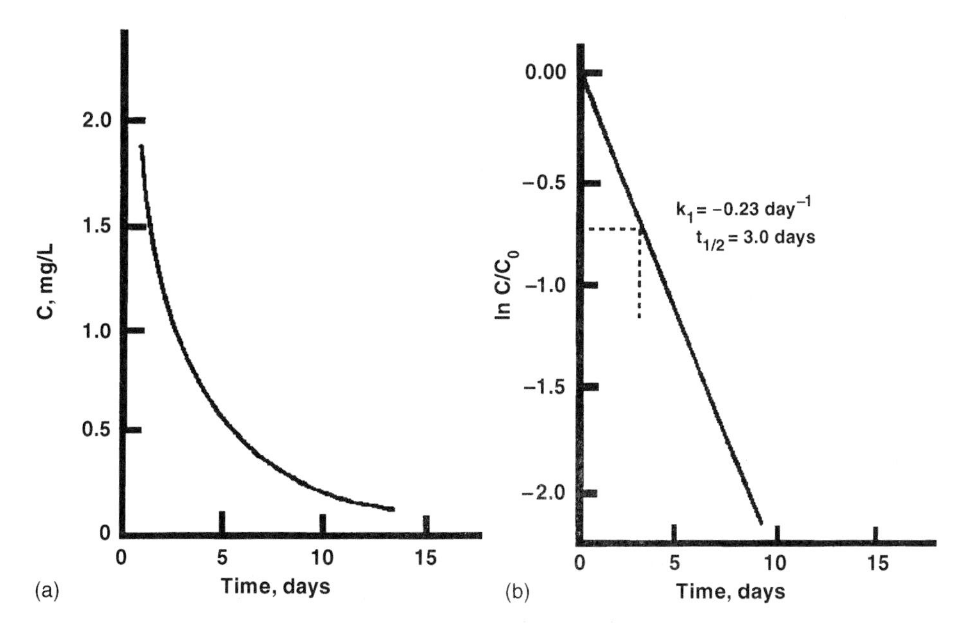

FIGURE 1 Dissipation rate of molinate from a rice field at 26°C (a) as a dissipation curve and (b) as a first-order plot. C_0 is the initial concentration and C the concentration of time *t*. (From Ref. 26. See Ref. 86 for original data.)

and other analytical approaches. The trick is to have sufficient detectability to be able to follow, or track, residues to the point where they are well below any plausible potential for adverse biological effects. This presents an inherent dilemma, because biological significance is subject to frequent reevaluation (e.g., with endocrine-disrupting chemicals). Thus, more sensitive analytical techniques are in constant demand so that dissipation processes can be followed longer, to lower concentration levels, and in more chemical product detail, anticipating reevaluation of environmental effects.

2.2 Environmental Compartments

Once a pesticide gains entry to the environment by purposeful application, accidental release, or waste disposal, it may enter one or more compartments, illustrated in Figure 2. The initial compartment contacted by the bulk of the pesticide will be governed largely by the process of use or release. In time, however, residues will tend to redistribute and favor one or more compartments or media over others, in accordance with the chemicals' physical properties, chemical reactivity, and stability characteristics and the availability and quality of compartments in

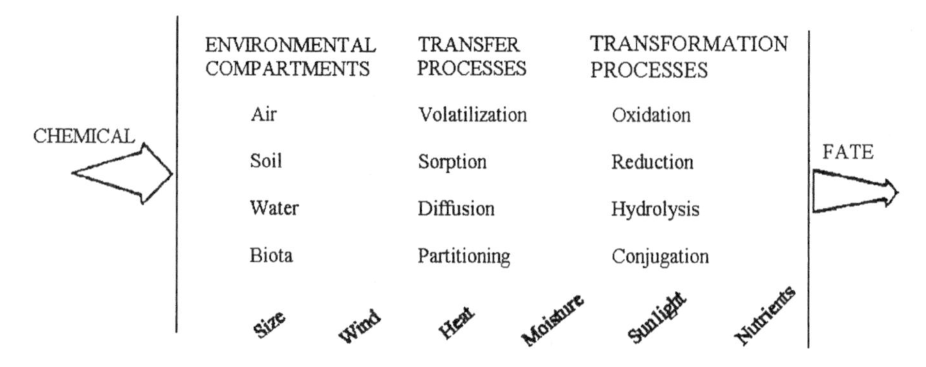

FIGURE 2 A schematic of the components of the fate of a chemical in the environment (From Ref. 17.)

the environmental setting where the use or release has occurred. Figure 2 tabulates the compartments, the transfer/transformation process, and the environmental characteristics that are involved in transport and fate in a very general way. Clearly, the nature of the chemical of interest will dictate what pathways are to be favored, so that environmental dissipation and fate must be evaluated on a chemical-by-chemical basis as well as on an environment-specific basis. This is illustrated in Figure 3 for chemical behavior in a pond environment, for which the properties of the chemical of interest must be taken into account along with,

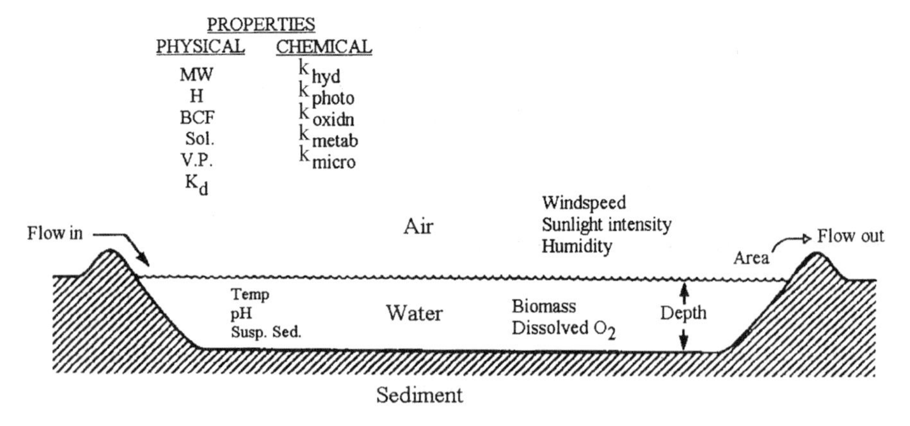

FIGURE 3 Intrinsic and extrinsic properties governing the distribution and fate of a chemical in a pond environment. (From Ref. 49.)

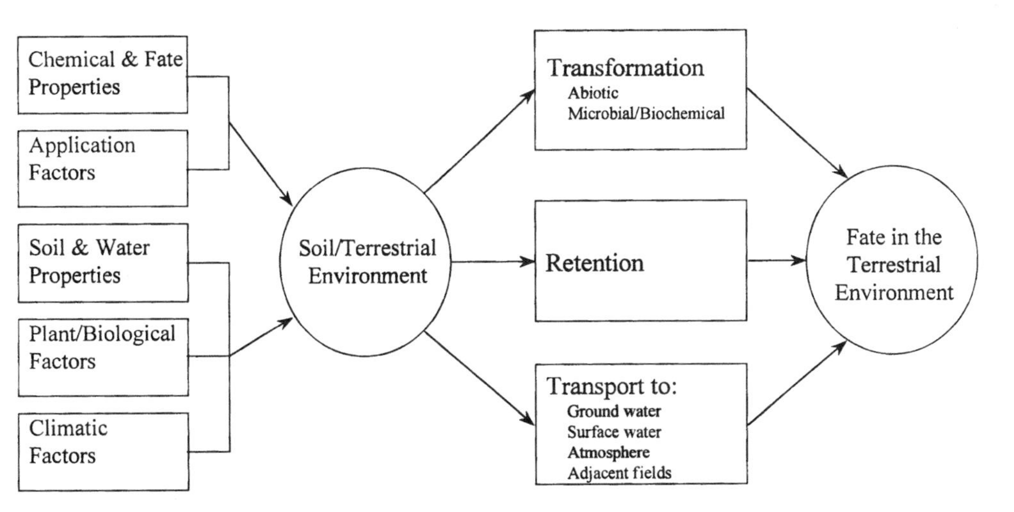

FIGURE 4 Conceptual model of the factors affecting the field dissipation of a chemical. (Adapted from Ref. 18.)

and as influenced by, the properties of the pond environment. Cheng [18] constructed an analogous schematic for chemical behavior in a soil environment (Fig. 4).

Some chemicals inherently favor water and thus will migrate to it when the opportunity arises. These are primarily chemicals of high water solubility and high stability in water, such as salts of carboxylic acid herbicides (2,4-D, MCPA, TCA). Others favor the soil or sediment compartment because they are preferentially sorbed to soil and they may lack other characteristics (volatility, water solubility) that lead to removal from soil. Examples include paraquat, which is strongly sorbed to the clay mineral fraction of soil, and highly halogenated pesticides such as DDT, toxaphene, and the cyclodienes, which sorb to and are stabilized in soil organic matter. Others, such as the fat-soluble organochlorines, favor storage in fatty animal tissue when the opportunity arises. Volatile chemicals such as methylbromide and telone (1,3-dichloropropene) migrate to the air compartment. The elements of predicting environmental behavior, based on properties of the chemical of interest, become apparent through these well-established "benchmark" chemicals.

2.3 Structure

The key to how a chemical will behave is contained in its structure. The development of the field of structure–activity relationships in pesticide chemistry has followed the development of those in drug chemistry and, more generally, pharmacology and toxicology.

An example of the importance of even small structural changes is provided by contrasting the biological activity and behavior of the two closely related chemicals DDT and dicofol (Table 1).

The subtle structural change due to the substitution of the OH of dicofol for the H of DDT at the central carbon has major ramifications. Biological activity is significantly altered. DDT is a broad-spectrum insecticide, whereas dicofol is a poor insecticide but a good acaricide and miticide. DDT has moderately high acute mamalian toxicity and is a tumorigen and carcinogen in rodents. Dicofol is of relatively low acute mammalian toxicity and has not exhibited carcinogenicity or tumorigenicity. DDT degrades slowly in the environment, and its primary breakdown products, DDE and DDD, are also very stable. Dicofol degrades rather rapidly in the environment, and its principal breakdown product, dichlorobenzophenone (DCBP), is also degraded further rather rapidly. DDT and DDE/DDD are highly lipophilic, showing strong tendencies to bioconcentrate in aquatic organisms and also, through accumulation in the food chain, in terrestrial animals and humans. Dicofol has much lower lipophilicity because of the presence of the polar OH group and a greater tendency to break down, and it does not significantly bioconcentrate or bioaccumulate. Its primary breakdown products do not exhibit these negative characteristics either. Even though there has been much experience with both DDT and dicofol, new information continues to surface.

Because of these differences in toxicity and environmental behavior, DDT was banned in the United States for most uses in 1972, whereas dicofol is still registered for use. Thus the answer to the question "Does structure matter?" is clearly yes, for closely related structures such as DDT and dicofol and certainly so for more structurally diverse chemicals. As has been pointed out, if methylchlor and methiochlor had been included in the synthetic program of Paul Müller, the Swiss chemist who discovered DDT, we might still be using "DDT-like" insecticides in agriculture. Methylchlor and methiochlor are good insecticides and biodegrade in the environment [19].

2.4 Activation–Deactivation

Most environmental transformations lead to products that are less of a threat to biota and the environment in general. The products may be less toxic than the parent or of lower mobility and persistence relative to the parent. They may, in short, be simply transient intermediates on the path to complete breakdown, that is, mineralization of the parent chemical. Thus, 2,4-D may degrade to oxalic acid and 2,4-dichlorophenol. The latter is of some concern, but it lacks the herbicidal toxicity of 2,4-D and appears to be further degraded in most environments by sunlight, microbes, etc. Organophosphates can be hydrolyzed in the environment

TABLE 1 Influence of Structure on Biological Activity, Environmental Behavior, and Regulatory Status of DDT and Diocofol

Property	DDT	Dicofol (Kelthane®)
Activity as pesticide	Insecticide	Acaricide
Mammalian toxicity		
Acute	High (LD_{50}, mg/kg)	Low (LD_{50}, g/kg)
Chronic	Causes tumors in rodents	Noncarcinogen/tumorigen
Environmental reactivity	Stable. Breakdown products (DDE and DDD) also stable	Breaks down. Primary breakdown product (DCBP) also stable
Bioconcentration potential	High, aquatic and food chain	Low
Regulatory status (U.S.)	Banned	Still registered

to phosphoric or thiophosphoric acid derivatives and a substituted phenol or alcohol. These products, in the case of most organophosphates, are not serious threats to humans or the environment.

Environmental activation represents the relative minority of transformations that lead to products with one or more of the following characteristics:

Enhanced toxicity to target and/or nontarget organisms
Enhanced stability, leading to greater persistence
Enhanced mobility, leading to contamination of groundwater or other sensitive environmental media
Enhanced lipophilicity, leading to bioconcentration and bioaccumulation

Notable examples of activations [20,21] include the (1) formation of DDE, which is apparently the agent responsible for causing thin eggshells in birds that have bioaccumulated DDT or DDE from their prey, and DDD, which can persist for years in some soil and water systems; (2) formation of dieldrin and eventually photodieldrin from aldrin, as noted previously; (3) oxidation of organophosphate thions to the more toxic "oxon" form; (4) oxidation of aldicarb (and some other *N*-methylcarbamates) to the more water-soluble and, in some cases, more persistent (and equally toxic relative to the parent) sulfoxide and sulfone forms; (5) formation of the volatile fumigant methyl isothiocyanate (MITC) from metam sodium, the commercial precursor of MITC, when the parent is applied to moist soil; and (6) formation of the carcinogen ethylenethiourea (ETU) from ethylenebisdithiocarbamate (EBDC) fungicides.

In part because of the concern over environmental activation, the USEPA requires extensive information on the occurrence and toxicity of environmental and metabolic transformation products of pesticides submitted for registration [2]. The tests include products of hydrolysis, photolysis, oxidation, and microbial metabolism in both laboratory and field tests. But, increasingly, regulations are also geared to products that might be formed during illegal use or during fires, explosions, spills, disinfection, and other situations that expose chemicals to conditions for which they were not intended [22]. Unfortunately, not all such situations can be anticipated, requiring continual vigilance by the registrant and regulatory agencies as a part of product stewardship and environmental protection.

3 TOOLS FOR PREDICTION

3.1 Physicochemical Properties

Important physical properties that determine transport, partitioning, and fate of pesticides are illustrated in Figure 5. Major advances were made in the last quarter of the twentieth century in defining, measuring, and using behavior and fate characteristics, both in the environment and in human and animal systems. The defi-

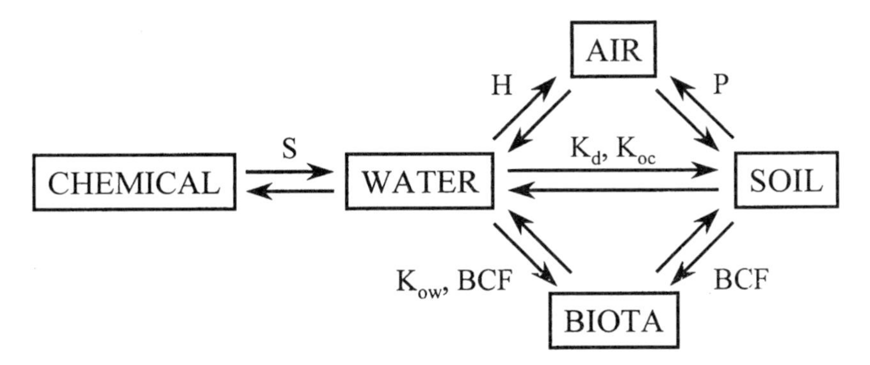

FIGURE 5 Key physical properties and distributions affecting transfer of chemicals in the environment. S = Saturated water solubility; P = vapor pressure; K_{ow} = octanol-water partition coefficient; BCF = bioconcentration factor; H = Henry's law coefficient; K_d = soil sorption coefficient; K_{oc} = soil sorption coefficient expressed on an organic carbon basis.

nitions, and means of measuring properties, have been summarized in a number of works [17,23–27] and will not be repeated in detail here. Notable developments have been made, leading to means for estimating properties from structures or chromatographic behavior, correlations between properties that are also useful for estimation, and particularly the use of properties to gauge some aspect of environmental behavior.

The estimation of properties from structures has been best developed for the octanol–water partition coefficient (K_{ow}), which is a useful estimate of a chemical's polarity, water solubility (S), and bioconcentration factor (BCF). Log K_{ow} may be estimated by summarizing contributions from atoms and groups of atoms and from bonds and other structural features. As long as a chemical's structure can be written, log K_{ow} can be calculated, usually in very good agreement with experimental values. A computer program is now available that can help to minimize uncertainty when several pathways exist for calculating log K_{ow} from the same structure [23]. Compilations of experimental log K_{ow} values are given by Leo et al. [28] and Hansch and Leo [29] for comparisons with calculated values. Compilations of experimental log K_{ow} values for pesticides and other environmentally relevant chemicals can also be found in several references and compendia (e.g., Mackay et al. [30], Shiu et al. [31], and Suntio et al. [32], in the computer database PestChem, and in database files for other computerized environmental fate programs such as CalTox.

The concept of correlation of properties is illustrated in the examples of water solubility, octanol–water partition coefficient, and bioconcentration factor in Table 2. Correlation equations, sometimes included in linear free energy relations (LFERs), have been defined for the following:

Property 1 (y)	Property 2 (x)	Slope
Log K_{ow}	log S	Negative
log K_{ow}	log BCF	Positive
log S	log BCF	Negative
log K_{ow}	log K_{oc}	Positive
log S	log K_{oc}	Negative

The equations of each correlation will vary depending on the database of chemicals included. One can find tight correlations when chemicals of the same general type (polycyclic aromatic hydrocarbons, chlorinated benzenes, etc.) are correlated, and fairly loose correlations when chemicals of diverse structures (all pesticide types, as in sample listing for K_{ow} vs. S in Table 2) are correlated. One needs to choose the published correlation that best fits the chemical(s) of interest or even to construct tailored ones by selecting data from the appropriate analogs, homologs, or class members that most resemble the chemical(s) of interest (see examples in Schwartzenbach et al. [25] and Lyman et al. [23]).

There is also a structure–activity relationship (SAR) for calculating boiling point [23] and from it the vapor pressure based upon structure. These methods are most applicable to the simpler structures of molecular weight less than 400.

The experimental database for vapor pressures for complex, higher molecular weight chemicals including many pesticides is spotty at best, and many errors exist and have been propagated in secondary compilations. A particularly good resource for pesticides is that of Suntio et al. [32] who list all available vapor pressures for listed chemicals along with an indication of the most reliable one when several exist. Other sources that include primarily or solely pesticides include Mackay et al. [30], the PestChem computer database, and Montgomery [33].

In order to determine whether a given value of a physical property is reasonable or not, two types of quality checks may be run. For condensed phase properties, such as S, K_{ow}, and K_{oc}, Johnson et al. [34] used an outlier test for the reasonableness of (S, K_{oc}) pairs compared against a correlation constructed from 109 data pairs [35] of pesticides, aromatic hydrocarbons, halogenated biphenyls, and biphenyl oxides and a second correlation from 123 different pesticides, some of which had multiple entries for either or both S and K_{oc}. The two correlations were

$$\log K_{oc} = 3.64 - 0.55 \log S \qquad \text{(Ref. 35)}$$

TABLE 2 Linear Energy Relationships Between Octanol–Water Partition Constants and (Liquid) Saturated Aqueous Solubilities for Various Sets of Compounds

			$\log K_{ow} = a \log C_w^{sat} (l,L) + b$	
Set of compounds	n	R^2	$a(\pm\sigma)$	$b(\pm\sigma)$
Polycyclic aromatic hydrocarbons	8	0.99	0.87(±0.03)	0.68(±0.16)
Substituted benzenes				
Only nonpolar substituents	23	0.98	0.86(±0.03)	0.75(±0.09)
Including polar substituents	32	0.86	0.75(±0.05)	1.18(±0.16)
Miscellaneous pesticides	14	0.81	0.84(±0.12)	0.12(±0.49)

Source: Ref. 25.

and

$$\log K_{oc} = 3.08 - 0.277 \log S \qquad \text{(Ref. 34)}$$

Errors due to coding mistakes, miscalculations, and incorrect chemical identification codes for outlier (S, K_{oc}) pairs were about twice those of pairs that conformed to the regression equation.

A second check, which involves straightforward experimentation, can be based on chromatographic data. There are good correlations between log K_{ow} (and thus also log K_{oc} and log S) and HPLC reversed-phase retention times [25] and between vapor pressure and gas-liquid chromatography (GLC) retention data [36]. In the latter case, one selects a reference standard of similar structure and/or polarity for which the vapor pressure is known accurately at several temperatures and then extrapolates data from GLC temperatures to ambient temperatures. This results in the vapor pressure of the subcooled liquid of the chemical of interest [$P_0(L)$] if it is normally a solid at ambient temperature, which may then be corrected to the vapor pressure of the solid [$P_0(S)$] using the melting point (T_m) correction [25]

$$\ln \frac{P_0(S)}{P_0(L)} \cong -[6.8 + 1.26(n - 5)] \frac{T_m}{T} - 1$$

For Henry's constant, Mackay et al. [37] published an experimental method based upon the rate of stripping of the compound from water purged with air or nitrogen and, later, a summary of all available experimental and estimation methods [32].

Generalizing, use should be made of the popular estimation method

$$H = P/S$$

where P = vapor pressure and S = water solubility, when reliable values are available for P and S at the same temperature. This equation is most useful for compounds of moderate to low water solubility, which include the majority of pesticides.

Estimation methods have also been derived for some of the nonstandard distributions, such as the air/leaf wax [38] and air/soil organic matter distributions. The washout ratio is a useful distribution for calculating the tendency of chemicals to be scrubbed from air by rain or fog droplets. The washout ratio (WR) is simply the reciprocal of H', the dimensionless Henry's constant, where [39,40]

$$H' = C_a/C_w$$

$$H' = H/RT$$

and

$$\mathrm{WR} = \frac{1}{H'} = \frac{C_w}{C_a}$$

where C_w and C_a have concentration units of moles per liter.

3.2 Leaching

Leaching is a physical process whereby chemicals are moved from the surface layers of soil, where pesticides will initially reside after a typical application, to (and through) the soil vadose zone and eventually to groundwater. It is a mass transport process carried by the downward movement of water following rain or irrigation. The most important physical properties are the chemical's water solubility and sorption coefficient. However, the rate of breakdown is important too, because if a chemical is unstable in soil it will not have sufficient residence time for the process of leaching, which is generally slow (order of weeks to months). Similarly, volatilization is a counteracting process because if a chemical is very volatile it will evaporate and not remain in soil sufficiently long for leaching. Using this kind of reasoning, a "leaching index" may be described as [23]

$$\text{Leaching index} = \frac{St_{1/2}}{PK_d}$$

where S = water solubility, $t_{1/2}$ = degradation half-life in soil, P = vapor pressure, and K_d = soil sorption coefficient.

California's Department of Pesticide Regulation used this index as a starting point for classifying chemicals according to their leaching tendencies [41,42]. Chemicals with the characteristics

$$t_{1/2}\ (\text{hydrolysis}) > 14\ \text{days}$$

or

$$t_{1/2}\ (\text{aerobic metabolism}) > 610\ \text{days}$$

and

$$S > 3\ \text{ppm} \qquad \text{or} \qquad K_{oc} < 1900$$

were classified as potential leachers, for which registration in California would not be granted until the registrant provided field test results indicating that under conditions of proposed use the chemical would not leach significantly. In the original dataset [41] for 26 pesticides found by monitoring to occur in at least one instance in groundwater, the California guidelines predicted that 19 should be "leachers." Four were predicted not to be leachers even though they were found at least once in groundwater, and three had insufficient information to classify.

Of 27 chemicals never before reported in groundwater in the United States, 14 were expected to be leachers using the California guidelines, while 13 were classified as nonleachers. Clearly, a shortcoming of this analysis is the experimental criteria used for denoting true leachers as chemicals found at least once in groundwater; a positive finding may not be indicative of leaching but rather of an incorrect analytical result or entry to groundwater by some process other than leaching (i.e., improper disposal of a residual tank mix or formulation by pouring into a well or onto the ground next to a well casing). Also, of those chemicals never found in groundwater but whose properties suggested a potential for leaching, low or infrequent usage, insufficient analytical detectability, or registered uses in cropping situations where the depth to groundwater was large or groundwater recharge rate was low could result in improper classification. The specific numerical values are constantly refined as new data are presented [42].

Woodrow et al. [43] described a correlation for predicting the initial rate of volatilization of chemicals from soil, water, and plant foliage. They compiled volatilization rates measured in the field and lab chamber and regressed these against selected properties as follows:

Application surface	Property
Foliage	P
Soil	$P/K_{oc}S_w$
Water	P/S_w

The resulting correlation equations are

Foliage:

$$\ln \text{Flux}\ [\mu g/(m^2 \cdot hr)] = 11.779 + 0.85543 \ln P$$

Soil:

$$\ln \text{Flux}\ [\mu g/(m^2 \cdot hr)] = 28.355 + 1.6158 \ln [P/K_{oc}S_w]$$

Water:

$$\ln (\text{Flux}/[mg/L]) = 13.643 + 0.8687 \ln (P/S_w)$$

where [mg/L] = water concentration.

Vapor pressure (P) is expressed in pascals. These ln–ln correlations were used to estimate the flux for pesticides with known physiochemical properties (P, K_{oc}, S_w). The estimated flux values were used as source strengths in an atmospheric dispersion model (e.g., USEPA's SCREEN-3) to calculate downwind concentrations near treated fields for short time periods following application. Calculated downwind concentrations compared reasonably well (within a few percent to within a factor of 2) with concentrations measured near treated fields for at least 10 different pesticides and application situations. This approach is useful for prioritizing pesticides that pose potential health hazards and for which monitoring should be considered.

3.3 Other Properties

Information on the degree of ionization, bioavailability, chemical and microbial degradation pathways, and rates of both physical and chemical processes are needed for complete assessment of environmental fate pathways. With the exception of ionization potentials [25], quantitative information, including rate constants, is often difficult to come by or to estimate. Clearly these are important processes that occur simultaneously with simple phase partitioning and transfers represented by physical properties discussed in the preceding section.

3.4 Rate Constants for Physical Fate Processes

Distribution coefficients tell the expected direction of a transfer but not the rate at which the transfer process occurs. The influence of local conditions (wind speed, temperature, soil moisture, relative size and proximity of compartments) is important in rates of volatilization, adsorption, bioconcentration, and the like. Ideally, one might wish to have available methods that allow calculation of rates given the chemical's physiochemical properties and local environmental conditions.

An example is provided by rates of volatilization from water and other surfaces. There exists a good correlation between H, the Henry's law constant, and the rate of volatilization from water. Lyman et al. [23] summarized the available data and pointed out that the environmental conditions most likely to influence rates (see Fig. 2) were wind speed, water depth, water mixing depth and rate, and temperature. The model of volatilization includes contributions from diffusion of solute to the air/water surface, transfer across the surface, and diffusion of volatilized solute away from the surface. All of these processes can be described mathematically and related to diffusion coefficients, Henry's constants, and the like [44].

For compounds of very low water solubility, such as chlorinated insecticides, PCBs, and polynuclear aromatic hydrocarbons, the rate of volatilization from water cannot be simply related to the rate of cleansing because of two additional factors. Much or most of the residue of these materials in a body of water such as a lake or river is likely to be bound in the sediment or suspended particulate matter rather than dissolved in the water. In that case, the bulk of the material is not "available" for volatilization or other waterborne fate processes. Rates of volatilization alone will not suffice to describe these residues, which have an additional rate process of desorption that must be accounted for. This can often be the rate-limiting step in the cleansing of a body of water by volatilization.

Another, often overlooked, factor is the competition between volatilization and deposition. For compounds with significant concentrations in ambient air, such as the ubiquitous organochlorine chemicals, loss by volatilization from water may be counteracted by deposition of fresh residue [45]. The net flux may be positive or negative for large water bodies such as oceans and the Great Lakes. Examples have been provided for toxaphene, PCBs, and other chemicals in these systems [46]. Methyl bromide provides still another dimension, because it can be produced in the oceans (from metabolism of seawater bromide), so that there is some uncertainty as to whether the oceans are a net source or net sink for methyl bromide (see references cited in Ref. 47).

Another approach, mentioned previously, to estimating rates of volatilization of pesticides is to subjectively correlate rates determined from actual experiments to physical properties [48]. The rate expressions that include only physical properties of the solute require modification by water depth, wind speed, temperature, etc., in order to be applicable to a specific field condition.

Estimates of rates of other processes, such as the rate of adsorption and desorption from soil and the rate of uptake and elimination from fish and other aquatic life, are not easily obtained either from experimentation or from estimation [23,25,49].

3.5 Bioavailability

The concepts of "bound residues" and bioavailability have been defined in some detail in recent years. The difficulty in extracting all of a pesticide residue from soil or crops by organic solvent extraction gave initial evidence for the presence of a physically sorbed or covalently bound phase so tightly held in the matrix that it could not easily be mobilized. For soil, an example is provided by paraquat, which is so strongly sorbed to the clay mineral fraction that it can be removed only by treatment with a strong acid. The concept was advanced that residues so tightly bound to soil were of no biological significance because of their lack of bioavailability [50].

Another example is provided by chlorinated dibenzodioxin residues in soil and sediments that are tightly sorbed to organic matter and, because of binding and low solubility, essentially immobile. Although of inherently high toxicity, the chemicals such as dioxins in soil may have little significance in most situations and so, some have argued, should not always command Draconian measures for remediation [51].

3.6 Ionization

Covalent acids and bases will display markedly different environmental partitioning behavior depending on whether they exist in the un-ionized or ionized form [25]. Simple calculations show radically different K_{ow} values, for example, for 2,4-dichlorophenoxyacetic acid (2,4-D) and its salt forms. Methods for calculating ionization constants (pK_a, pK_b) for organic acids and bases for environmentally relevant chemicals are straightforward extensions of Hammett sigma-rho constants from physical organic chemistry. The "Hammett correlation" is perhaps the best known of the linear free energy relationships (LFERs). Unfortunately, it holds strictly only for substituted benzoic, phenylacetic, and a few other types of carboxylic acids [25]. Whether a compound is ionized or not at the pH that characterizes its aqueous environment will influence its

Extractability during analysis
Uptake and elimination by aquatic organisms
Sorption to sediment
Interfacial concentration
Volatilization rate
Bioavailability

In most cases, it is not an all-or-nothing situation. The pK_a effect on the ionization of most organic acids extends over two or three pH units as one goes from 100% ionized (at pH > 7 for pentachlorophenol) to 0% ionized (pH < 4 for pentachlorophenol). Even a small amount that is un-ionized or ionized may be enough to facilitate a specific uptake or other fate process that depends on the availability of the solute, even in a relatively small percentage.

3.7 Chemical Reactions

By far the greatest complication in fully defining, or predicting, environmental fate processes arises with chemical degradation of the parent chemical into an array of degradation products. Abiotic reactions include hydrolyses and oxidations that occur in air, in water, and at the surface of soils, with or without light activation, but without intervention by microorganisms, plants, or animals. Biotic reactions are under enzymatic control, but both kinetics and degree of degradation vary considerably depending on whether plants, animals, or microorganisms are involved and, for microorganisms, the population density (cells per milliliter or gram). The pathways of biotic and abiotic degradation are often the same, so that analysis for product profiles does not always help in detecting which type of process operates or predominates in a given setting. However, there are experimental techniques for differentiating biotic and abiotic reactions, just as there are for separately determining the operation of chemical and physical dissipaton processes and the type of process.

Any attempt at in-depth coverage of reaction pathways for pesticides here would be superficial and incomplete because of the variety of pesticides and, consequently, reaction products. A few generalizations, however, will be offered, followed by a discussion of reaction rates emphasizing microbial degradation (by far the most important from an overall environmental perspective) and citation of relatively recent references to the subject.

Environmental reactions fall into just a relatively few reaction types, each summarized with a few generic examples in Table 3. Some are favored over others depending on the medium of occurrence: oxidations in air and on surfaces

TABLE 3 Environmental Reactions: Types, Reagents, and Examples

Type	Exogenous reagents	Example
Oxidation	O_2, O_3, ·OH, H_2O_2, Cl_2, Fe^{3+} Aerobic microorganisms Mixed function oxidase	Ar—CH_3 → Ar—COOH R—O—CH_3 → R—OH + CO_2 —P=S → —P=O R—S—R' → R—S(=O)—R' → R—S(=O)(=O)—R' C=C → C—C (epoxide, O) Ar—NR_2 → Ar—NHR R—S—H → R—S—S—R
Hydrolysis	H_2O, OH^-, H^+ Microorganisms, plants, animals	RCOOR' → RCOOH R—Br → R—OH P(=S)—OR → P(=S)—OH R—O—R' → ROH + R'OH
Reduction	Fe^{2+} and its complexes Anaerobic microorganisms	Ar—Cl → Ar—H R—NO_2 → R—NH_2 + intermediates R—CHO → R—CH_2OH
Conjugation	Sulfate, glucose, glucuronic acid, amino acids	Ar—OH → Ar—O—SO_3^- Ar—OH → Ar—O—GLU
Isomerization	OH^-, H^+, $h\nu$	P(=S)—OR → P(=O)—SR H(Ar)C=C(H)(Ar) → Ar(H)C=C(H)(Ar)
Elimination	OH^-, $h\nu$	Ar—CH_2—CCl_3 → Ar—CH=CCl_2

exposed to air, reductions in anaerobic sediments and soils, hydrolyses in water and moist soil, and conjugations in plants and animals. In some cases, several types of reactions take place in environmental degradation pathways, as is illustrated for DDT, which may be oxidized (to dicofol, dichlorobenzophenone, and *p*-chlorobenzoic acid), reduced (to DDD and dichlorodiphenylacetic acid), and subject to elimination of HCl (to DDE) in the same field or body of water [26]. Parathion can undergo oxidation (to paraoxon), reduction (to aminoparathion), and hydrolysis (to *p*-nitrophenol and diethylphosphorothioic acid) in the same, or similar, environments.

3.8 Microbial Degradation (Biodegradation)

The important role played by microorganisms in degrading pesticides has been studied in great detail during the past 25 years or so. It is believed that degradation by microbes (bacteria, fungi, algae) accounts for over 90% of all degradation reactions in the environment and is the nearly exclusive breakdown pathway in most surface soils, near plant root zones (micorrhyzae), and in nutrient-rich waters including sewage ponds and sewage treatment systems [52,53].

The proficiency with which microorganisms carry out chemical transformations is due to their simplicity in absorbing chemicals from exogenous sources and excreting transformation products, and their enzymatic content. Bacteria predominant among microorganisms, representing single-cell organisms existing in great numbers (up to 10^5 or more cells per gram of soil) with a facility to adapt to different environments and to different chemicals as food sources.

There are three types of bacterial chemical degradation possibilities, differentiated by the kinetics of breakdown of the chemical substrate [23,27].

Type a. Substrate degradation begins immediately upon contact. This indicates that the substrate can be used immediately as an energy source by the bacterial community, resulting in consumption of the substrate and a population increase among the degraders. Substrates that most resemble natural energy sources for bacteria—sugars and other simple carbohydrates, amino acids and simple proteins, aliphatic alcohols and acids, etc.—are the favored substrates in this class. Pesticides such as methomyl, glyphosate, sulfonylureas, and some prethyroids are included in this group because of their similarity to natural substrates in physicochemical properties and/or ability to act as nutrient sources.

Type b. Substrate degradation occurs after a lag period of bacterial acclimation. Group b includes substrate–bacteria combinations that require adaptation or acclimatization of the bacteria before the substrate can be used as an energy source. Adaptation may involve an induction of latent enzymes in the microbial community or a population shift to favor degrading species in a mixed microbial population, or some combination of the two. Once adaptation has occurred, the degradation rate increases until the substrate source is depleted, the same as in Type a systems. Type b systems predominate for most pesticide–microbe combinations, perhaps because most pesticides have structures different enough from (i.e., foreign to) natural food sources that enzyme systems are not immediately present in the natural microflora for deriving energy from them. After several exposures, at a constant level of exposure, adaptation occurs and degradation can become immediate and occur at ever-increasing rates. Although attractive from an environmental cleansing viewpoint, adaptation of microbes in agricultural field soils or water can result in loss of efficacy of, e.g., soil-applied insecticides and herbicides, or pesticides applied to rice paddy water. Loss of efficacy of aldicarb to control insect pests of rice in the tropics is one example, and the loss of efficacy of various herbicides in fields in the midwest United States is another [53]. Soils that possess a high population of degraders have been termed "aggressive" or "problem" soils. Management options include pesticide rotation or simply elimination of use in areas or crops where the problem occurs.

Adaptation can also be a benefit, in reducing pesticide residue carryover from one season or crop to another and in decontaminating environments with problematic residue buildup. Researchers have developed microbial cocktails enriched in adapted organisms to decompose pesticides that are improperly disposed of or spilled (see, e.g., several chapters in Bourke et al. [22]). The possibility of using bioengineered microorganisms has not yet been taken advantage of commercially for pesticide cleanup in the environment, although it shows promise for the future [54]. A low-tech approach at biodegradation of pesticide wastes involves the use of naturally enriched sources, such as horse manure, added to a "reactor" through which pesticide-contaminated wastewater can be circulated. A commercial system is available that is based on this principle [55].

Type c. The substrate is not significantly usable as an energy source by microbial populations. If a chemical cannot be used as an energy source, even after prolonged periods of adaptation and addition of nutrients, water, air, etc., it is regarded as "recalcitrant" to microbial degradation. Several chlorinated hydrocarbon insecticides, chlorodibenzodioxins, and polychlorinated biphenyls (PCBs) fall into this group, along with synthetic polymers and certain other organic chemicals. Recalcitrant chemicals can be transformed by microbes, but the transformation is incidental to the normal metabolism of acceptable substrates by the microorganisms ("cometabolism"). The slow microbial conversion of DDT to DDE or DDD in soil is an example, as is the anaerobic dechlorination of highly

Table 4 Contrasts in Metabolism of Pesticides by Animals, Plants, and Microorganisms

	Animal	Plant	Microorganism
Reaction pathway	Discrete steps, one at a time	Discrete steps, one at a time	Complete mineralization
Elimination pathway	In urine or feces, as polar metabolites of reduced toxicity	Stored in vacuoles, as polar metabolites of reduced toxicity	Evolution of CO_2; diffusion-elimination of ions
Storage of stable products	Fat	Lipid layers or vacuoles	No storage
Rate of degradation (half-lives)	Hours–days	Hours–days	Minutes–days (if adapted)

chlorinated aliphatic insecticide mixtures such as toxaphene, a process that can be used effectively to decontaminate soil because the lower chlorinated products are more volatile and more water-soluble than the parents [56].

Recalcitrant molecules generally possess low water solubility and a high degree of halogenation. One could surmise that the electron-rich surface of a polyhalogenated hydrocarbon may hinder microbes from extracting carbon from the compound, and absorption is limited as well because microbial absorption favors the substrate in aqueous solution. The Kelthane–DDT example (Table 1) is applicable here, because the OH substitution increases aqueous solubility and also provides a “handle” for more facile enzymatic conversion of the parent structure.

Plants and animals can affect biodegradation of pesticides, but there are interesting contrasts relative to microorganisms (Table 4). Plants and animals degrade enzymatically but generally to intermediate products by just one or a few discrete reactions, and the products are then either eliminated (animals) or stored in vacuoles (plants). Formation of more polar transformation products, including conjugates (Table 3) favors elimination (animals) or storage (plant vacuoles). Unlike microorganisms, for which “mineralization” (formation of simple elements and compounds naturally present in the biosphere: CO_2, Cl^-, PO_4^{3-}, NO_3^-, SO_4^{2-}, etc.) is the rule in biodegradation, plant and animal metabolism of xenobiotics usually stops partway through the process and any further degradation of the terminal metabolites may well occur by microbial action.

4 TOOLS FOR PREDICTION: MODELS

Because of the cost and complexity of environmental experimentation and the need to be able to manipulate variables, various approaches to modeling environmental transport and fate have been developed. They range from the use of field plots (specified in the USEPA registration requirements; see Ref. 2 and more recent EPA updates) to laboratory or greenhouse chambers to virtual (computer) models. The latter allows developers of new candidate pesticides to screen for potential adverse environmental behavior very early in the development process, in some cases before the candidate chemical is even synthesized for the first time.

4.1 Physical Models

A major development of the 1970s was the introduction of various microcosm approaches to environmental fate testing. In these chambers, often just modified aquaria, simple elements of the ecosystem could be simulated and a test chemical added and monitored. The early chambers included:

Model ecosystem or "farm pond microcosm" [57]
Terrestrial model ecosystem [58]
Agroecosystem chambers [59]

These early chambers were useful for comparing or ranking chemicals in terms of their abilities to biodegrade, bioconcentrate, bioaccumulate, volatilize, etc., but they did not generate information that could be immediately transferred to field conditions, probably because of their high degree of artificiality and elimination or minimal accommodation of key features (e.g., wind or precipitation) that play major roles in field dissipation processes. More sophisticated chambers, including lysimeters and wind tunnels, have been described more recently (Chaps. 2–5 in Ref. 46; references cited in Refs. 48 and 60).

4.2 Mathematical Models

Schwarzenbach et al. (Chap. 15 of Ref. 25) summarized the use of models for estimating the loading and partitioning of chemicals in lakes. Of particular interest are organochlorine pesticides, such as DDT/DDE and toxaphene, and PCBs. McCall et al. [61] described an equilibrium distribution model, based upon box model principles, that allowed for estimating environmental partitioning of organic chemicals in model aquatic ecosystems. For a water–sediment–air–fish system of defined dimensions, one could calculate compartmental distributions for chemicals whose physical properties (K_{ow}, K_d, BCF, vapor pressure) were known or could be estimated. This was an excellent starting point for estimating concentrations expected for various media given a specified loading of chemical, to compare with monitoring data and to predict exposures and potential effects of aquatic life. Figure 6 shows the calculated percentages and concentrations for chlorpyrifos in this model system.

The equilibrium distribution or partitioning model can be used only to calculate expected compartmental contents at equilibrium in the absence of degradative pathways. This, of course, is only part of the information needed. To predict the dissipation of chemicals from each component and from the entire system, rate constants or half-lives have to be added in. A tandem partitioning–dissipation computer model flowchart is given in Figure 7 that illustrates the steps and outputs.

The Exposure Assessment Modeling System (EXAMS) has proven useful for estimating all fate pathways for contaminants in streams and other surface waters [62]. Applications have also been made to pesticides in rice paddies [63] and to predicting loss from waste ponds and other impoundments [64]. Given an input of key parameters of the water environment, physicochemical properties of the chemical of interest, and the loading of chemical into the system, EXAMS provides output data in the following table:

Output	Based upon
1. Distribution at equilibrium, in sediment, suspended sediment, water, fish, air	Physical properties, distribution coefficients
2. Rate of dissipation, from each medium in the above distribution or from the system	Rate constants for, e.g., volatilization, microbial degradation, sorption, bioaccumulation
3. Relative importance of each dissipation process, i.e., percent lost by volatilization, hydrolysis, biodegradation, etc.	Rate constants and distribution coefficients

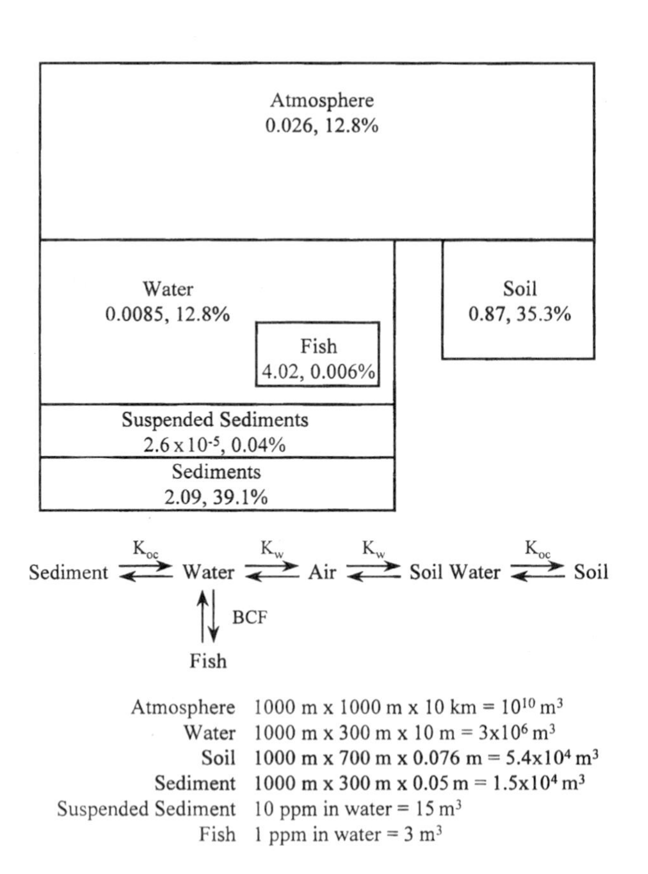

FIGURE 6 Calculated distribution of chlorpyrifos in a model ecosystem using the partitioning model of McCall et al. [61]. Concentrations are in parts per million for all media except atmosphere ($\mu m/m^3$) and suspended sediment (ppm waste basis). The specified load was 200 kg to the system.

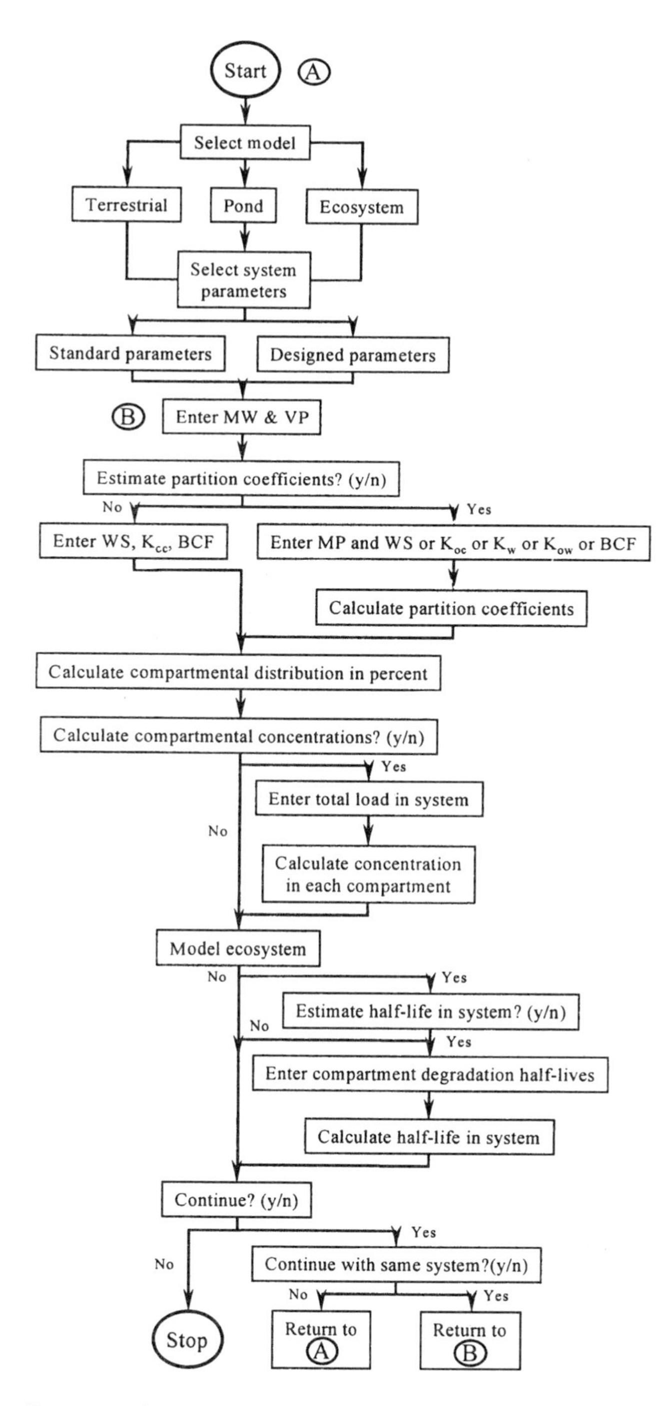

FIGURE 7 Steps in operating an environmental fate computer model that calculates compartmental distributions, compartmental concentrations, and half-lives.

EXAMS is an excellent tool for gauging the rate of decontamination of a body of water, for guiding sampling schedules (what medium to sample, and how frequently), and for contrasting behavior of individual chemicals in a series of chemicals. All of this helps in making good choices, early on, of chemicals that can be used safely in or around water bodies and what actions(s) to take when an accidental release or unexpected contamination occurs.

Matthies [65] and Clendening et al. [66] summarized models applicable to pesticide movement and persistence in the soil and vadose zone. Mackay [67] proposed and developed fugacity approaches to modeling and pointed out the advantages over compartmental distribution and partitioning models. The fugacity approach has been incorporated into CalTox and other regulatory models with regulatory uses in predicting exposures associated with toxic waste sites and other chemical exposure sources. Laskowski [68] described probabilistic modeling in environmental fate and Solomon [69] described the overall framework of probabilistic ecotoxicological risk assessment. A current frontier is faced in marrying environmental fate models of the type described briefly in this section with landscape-scale processes and landscape-scale models, which have arisen somewhat independently in the domain of landscape ecology [70]. The integration of models so that areawide, regional, or global environmental processing of pesticides can be better understood, integrated with exposure and toxicity data, and used to manage chemicals represents a challenge for development in the twenty-first century.

5 FRONTIER AREAS IN ENVIRONMENTAL FATE

5.1 Air

Concerns over chemical contamination of the air have historically focused on the "criteria" air pollutants—NO_x, SO_x, ozone, particulate matter, CO, and hydrocarbons. But increasingly society has become concerned with other chemicals—solvents such as chloroform, benzene, and methyl *t*-butyl ether, polynuclear aromatic hydrocarbons, and pesticides and other "economic" materials. At the U.S. federal level, these may be referred to as "hazardous air pollutants" (HAPs) resulting from designation in the Clean Air Act amendments [71]. In California, the term "toxic air contaminant" (TAC) refers to HAPs and other airborne chemicals that have undergone appropriate risk assessment and designation guidelines in the state's Toxic Air Contaminant Act [72]. Pesticides are among the HAPs and TACs, and, indeed, pesticides and pesticide transformation products are prominent on the CAAA-90 list of 189 HAPs and California's growing list of TACs [73].

Partly because of these two broadbrush pieces of air quality legislation and also because of the need to safeguard the health of residents downwind of pesticide applications, methods have evolved for measuring or estimating the downwind drift of residues emanating from treated areas by spray drift or postapplication volatilization. The experimental methods are summarized by Seiber and Woodrow [74]. Estimation of volatilization flux based upon a chemical's physiocochemical properties and the type of surface to which it has been applied [48] was described above. Armed with this estimate of the emission source term, one may then proceed to estimate downwind exposure concentrations using an air dispersion model, such as Industrial Source Complex—Short Term (ISC-ST), CalPuff, or ALOHA.

Airborne transport and fate of pesticides represents an intriguing area of environmental science in need of further definition and study. If one assumed that 25% of all applied pesticides enter the air by drift during application or postapplication volatilization—not an unreasonable estimate according to experimental information summarized in Glotfelty et al. [75] and Taylor and Glotfelty [76]—it is striking how little of the airborne residue can be accounted for by potential dissipation processes. These include deposition to downwind foliage [38,77] and water [45] by both wet and dry deposition processes and chemical degradation. The latter aspect is in particular need of new experimental data from chamber and field experiments [78].

Interaction of airborne pesticides (both particles and vapors) with atmospheric moisture is another area of much current interest. Seiber [79] reviewed research on pesticides in fogwater, for which the phenomenon of "enrichment" in fogwater [80] over that predicted from Henry's law-based partitioning of pesticide vapor in the atmosphere into suspended droplets of fogwater has still not been adequately explained. The fate of pesticides suspended on dust, particularly the fine dusts ($PM_{2.5}$ and below) of current interest, is virtually unexplored.

5.2 Water

Current interest is focused on understanding the routes of entry of pesticides into surface water [81], following several years of studies focused on aquatic fate, exposure, and risk assessment for humans and wildlife. The surface runoff of pesticides from treated fields and orchards, although usually just a fraction of the total applied, represents an unacceptable off-target threat in the immediately adjacent areas as well as to the quality of lakes, rivers, and ecosystems. Current interest is in understanding the runoff process itself and how it might be modified or controlled by use of e.g., vegetative filter strips at the field edge that can sorb residue and prevent further runoff from the field environment [82]. Formulation and application technology can help in minimizing runoff losses, as can better integration with weather forecasting to guide timing of applications and, of course, the choice of which chemical to apply in a situation where runoff represents a distinct possibility.

Barbash and Resek [83] provide a comprehensive summary of pesticides in groundwater, along with perspectives on distribution, concentrations, trends, and mitigation. All of their summarized areas represent opportunities for research aimed at understanding and preventing contamination in the future.

5.3 Soil

Challenges exist also with respect to understanding the interaction of soil with pesticides—at the molecular, microscopic, and macroscopic levels—and how it affects mass transport, diffusion, bioavailability, reactivity, leaching, volatilization, etc. [18]. Renewed interest in the soil component of global cycling of carbon should provide new experimental approaches and models with applicability to pesticides and other organic chemicals in the soil environment.

5.4 Biota

Tremendous challenges exist in understanding how exposures occur; pathways of adsorption, distribution, metabolism, and elimination; intereaction at the organ tissue, cellular, and enzymatic levels; and, of course, resulting effects at the single organism, community, and population levels [7]. The physiocochemical properties, distribution, and reactions involved in environmental behavior and fate also operate within organisms and in communities and ecosystems. Concerns over endocrine disruption and decline of whole genera of amphibians and other wildlife are raising new interest in this subject area. Rather than study these aspects separately, environmental chemistry, environmental toxicology, exposure, risk assessment, and risk management will almost certainly be integrated in multidisciplinary approaches to environmental science in the future [84]. The advent of genetically modified plants and food animals, including those modified to combat insects and disease and those modified to accelerate metabolism of pesticides, will almost certainly pose new challenges for assessing environmental and human health safety in the environmental sciences.

6 SUMMARY

The 1970–1990 era began with the banning of pesticides that were problematic from an environmental viewpoint—DDT and other organochlorines, DBCP, EDB, and others—and the creation of regulatory measures (creation of the USEPA and passage of the Toxic Substances Control Act; Amended Federal Insecticide, Fungicide, and Rodenticide Act; and Food Quality Protection Act in the United States, matched by similar measures in European and Asian countries) designed to ensure that pesticide use and the pest control agents of the future would not pose these kinds of environmental risks. The development of risk sciences dealing with both human health and ecological concerns provided a framework in which environmental chemistry, environmental toxicology, environmental modeling, and related scientific disciplines had the opportunity to make important contributions.

As a result of these activities, pesticides and pest control practices at the beginning of the twenty-first century are safer than pre-1970 for those employing them, their neighbors, consumers of treated commodities, and wildlife and other segments of the environment [85]. Challenges remain in integrating environmental exposure data with environmental effects assessment and keeping pesticides confined to their intended targets without off-target movement in surface and ground waters and air. These efforts are important as society learns more about, and experiences continuing concern over, potential long-term impacts of low levels of pesticide residues toward people and wildlife. Multidisciplinary research, with local, regional, and global purviews, will increasingly command the attention of pesticide environmental scientists in this century.

REFERENCES

1. EPA. Pesticide Assessment Guidelines. Subdivision N Chemistry: Environmental Fate. EPA-540/9-82-021. Washington, DC: US Environ Protect Agency, Office of Pesticides and Toxic Substances, 1982.
2. MF Kovacs Jr. EPA guidelines on environmental fate. Residue Rev 85:3–16, 1983.
3. B Thomas. Pesticide registration in Europe. In: Regulation of Agrochemicals. Washington, DC: Am Chem Soc, GJ Marco, RM Hollingworth, JR Plimmer, 1991, pp 73–79.
4. Agriculture Canada, Environment Canada, Department of Fisheries and Oceans. Environmental Chemistry and Fate Guidelines of Pesticides in Canada. Ottawa, July 15, 1987.
5. J Holland. Environmental fate: A Down Under perspective. In: GT Brooks, TR Roberts, eds. Pesticide Chemistry and Bioscience. The Food-Environment Challenge. Cambridge, UK: Roy Soc Chem, 1999.
6. Nat Res Council. Risk assessment in the Federal Government: Managing the Process. Washington, DC: Nat Acad Press, 1983.
7. GW Suter, LW Barnthouse, SM Bartell, T Mill, D Mackay, S Patterson. Ecological Risk Assessment. Boca Raton, FL: Lewis, 1993.
8. Nat Res Council. Science and Judgment in Risk Assessment. Washington, DC: Nat Acad Press, 1994.
9. Nat Res Council. Pesticides in the Diets of Infants and Children. Washington, DC: Nat Acad Press, 1993.
10. CI Goring, JN Hamaker. Organic Chemicals in the Soil Environment, Vols 1 and 2. New York: Marcel Dekker, 1972.
11. R Haque, VH Freed, eds. Environmental Dynamics of Pesticides. New York: Plenum Press, 1975.
12. LJ Thibodeaux. Chemodynamics: Environmental Movement of Chemicals in Air, Water, and Soil. New York: Wiley-Interscience, 1979.

13. S Jobling, T Reynolds, R White, MG Parker, JP Sumpter. A variety of environmentally persistent chemicals, including some phthalate plasticizers, are weakly estrogenic. Environ Health Perspect 103:582–587, 1995.
14. Inter-Organization Programme for the Sound Management of Chemicals. International Workshop on Endocrine Disruptors. Report. Geneva, Switzerland: UNEP Chemicals. 1997.
15. Food Quality Protection Act. US Congress, Washington, DC, 1996.
16. WG Fong, HA Moye, JN Seiber, JP Toth. Pesticide Residues in Foods: Methods, Techniques, and Regulations. New York: Wiley, 1999.
17. JN Seiber. General principles governing the fate of chemicals in the environment. In: JL Hilton, ed. Agricultural Chemicals of the Future. Beltsville Symp Agric Res No. 8, Totowa, NJ: Rowan and Allanheld, 1985.
18. HH Cheng. Pesticides in the Soil Environment: Processes, Impacts, and Modeling. Soil Sci Soc Am Book Ser No. 2. Madison, WI: Soil Sci Soc Am, 1990.
19. RL Metcalf. A century of DDT. J Agric Food Chem 21:511–519, 1973.
20. JR Coats. Pesticide degradation mechanisms and environmental activation. In: L Somasundaram, JR Coats, eds. Pesticide Transformation Products: Fate and Significance in the Environment. ACS Symp Ser 459. Washington, DC: Am Chem Soc 1991, pp 10–31.
21. MF Wolfe, JN Seiber. Environmental activation of pesticides. In: DJ Shusterman, JE Peterson, eds. De Novo Toxicants: Combustion Toxicology, Mixing Incompatibles, and Environmental Activation of Toxic Agents. Occup Med: State of the Art Rev, Vol. 8. Philadelphia: Hanley and Belfus, 1993, pp 561–573.
22. JB Bourke, AS Felsot, TJ Gilding, JK Jensen, JN Seiber, eds. Pesticide Waste Management: Technology and Regulation. ACS Symp Ser 510. Washington, D.C.: Am Chem Soc, 1992.
23. WJ Lyman, WF Reehl, DH Rosenblatt. Handbook of Chemical Property Estimation Methods. New York: McGraw-Hill, 1982.
24. JW Biggar, JN Seiber, eds. and tech coordinators. Fate of Pesticides in the Environment. Proc Tech Seminar. Publ 3320. Berkeley, CA: Univ. California, Div Agric Nat Resources, 1987.
25. RP Schwarzenbach, PM Gschwend, DM Imboden. Environmental Organic Chemistry. New York: Wiley, 1993.
26. DG Crosby. Environmental Toxicology and Chemistry. New York; Oxford Univ Press, 1998.
27. RS Boethling, D Mackay. Handbook of Property Estimation Methods for Chemicals. Boca Raton, FL: Lewis, 2000.
28. A Leo, C Hansch, D Elkins. Partition coefficients and their uses. Chem Rev 71: 525–651, 1971.
29. C Hansch, AJ Leo. Substituent Constants for Correlation Analysis in Chemistry and Biology. New York: Wiley, 1979.
30. D Mackay, WY Shiu, KC Ma. Illustrated Handbook of Physical-Chemical Properties and Environment Fate for Organic Chemicals, Vols I–V. Boca Raton, FL: Lewis, 1992.
31. WY Shiu, KC Ma, D Mackay, JN Seiber, RD Wauchope. Solubilities of pesticide chemicals in water. Part I: Environmental physical chemistry. Rev Environ Contam Toxicol 116:1–13, 1990.

32. LR Suntio, WY Shiu, D. Mackay, JN Seiber, D Glotfelty. Critical review of Henry's law constants for pesticides. Rev Environ Contam Toxicol 103:1–59, 1988.
33. JH Montgomery. Agrochemicals Desk Reference: Environmental Data. Boca Raton, FL: Lewis, 1993.
34. B Johnson, C Johnson, JN Seiber. The use of regression equations for quality control in a pesticide physical property database. Environ Manage 19:127–134, 1995.
35. EE Kenaga, CI Goring. Relationship between water solubility, soil, sorption, octanol-water partitioning, and bioconcentration of chemicals in biota. In: JG Eaton, PR Parrish, AC Hendricks, eds. Aquatic Toxicology: Proceedings of the Third Annual Symposium on Aquatic Toxicology. ASTM Spec Tech Pub 707. Philadelphia, PA: ASTM, 1980, pp 78–113.
36. Y-H Kim, JE Woodrow, JN Seiber. Evaluation of a gas chromatographic method for calculating vapor pressures with organophosphorus pesticides. J Chromatogr 314:37–53, 1984.
37. D Mackay, WY Shiu, PR Sutherland. Determination of air-water Henry's law constants for hydrophobic pollutants. Environ Sci Technol 13:333–337, 1979.
38. C Gaggi, D Calimari, E Bacci. Bioconcentration of non polar xenobiotics in terrestrial plant biomass. In: D Calimari, ed. Chemical Exposure Predictions. Boca Raton, FL: Lewis, 1993, pp 147–160.
39. JN Galloway, SJ Eisenreich, BC Scott, eds. Toxic Substances in Atmospheric Deposition: A Review and Assessment. Rep NC-141, Nat Atmos Deposition Program EPA 560/5-80-001. July 1980.
40. MS Majewski, PD Capel. Pesicides in the Atmosphere: Distribution, Trends, and Governing Factors. Chelsea, MI: Ann Arbor Press, 1995.
41. MR Wilkerson, KD Kim. The Pesticide Contamination Prevention Act: Setting Specific Numerical Values. Environ Hazards Assessment Program. Sacramento, CA: Calif Dept Food Agric, State of California, 1986.
42. B Johnson. Setting revised specific numerical values. EH-916. Sacramento, CA: Calif Dept Food Agric 1991.
43. JE Woodrow, MM McChesney, JN Seiber. Modeling the volatilization of pesticides and their distribution in the atmosphere. In: D Kurtz, ed. Long-Range Transport of Pesticides. Chelsea, MI: Lewis, 1990, pp 61–81.
44. P Isnard. Volatilization of chemicals from bodies of water. In: D Calamari, ed. Boca Raton FL: Lewis, 1993, Chemical Exposure Predictions. pp 63–83.
45. WMJ Strachan, SJ Eisenreich. Mass balance accounting of chemicals in the Great Lakes. In: D Kurtz, ed. Long-Range Transport of Pesticides. Chelsea, MI: Lewis, 1990, pp 291–301.
46. DA Kurtz ed. Long Range Transport of Pesticides. Chelsea, MI: Lewis, 1990.
47. PS Honaganahalli, JN Seiber. Health and environmental concerns over the use of fumigants in agriculture: The case of methyl bromide. In: JN Seiber, JA Knuteson, JE Woodrow, NL Wolfe, MV Yates, SR Yates, eds. Fumigants: Environmental Fate, Exposure and Analysis. ACS Symp 652. Washington, DC: Am Chem Soc, 1997, pp 1–13.
48. JE Woodrow, JN Seiber. Correlation techniques for estimating pesticide volatilization flux and downwind concentrations. Environ Sci Technol 31:523–529, 1997.
49. JN Seiber. Principles governing environmental mobility and fate. In: NN Ragsdale,

RJ Kuhn, eds. Pesticides: Minimizing the Risks. ACS Symp Ser 336. Washington, DC: Am Chem Soc, 1987, pp 88–105.
50. JL Hamelink, PF Landrum, HL Bergman, WH Benson. Bioavailability: Physical, Chemical, and Biological Interactions. Boca Raton, FL: Lewis, 1994.
51. M Alexander. How toxic are toxic chemicals in soil? Environ Sci Technol 29:2713–2717, 1995.
52. M Alexander. Biodegradation of chemicals of environmental concern. Science 211: 132, 1981.
53. PC Kearney, JS Karns. Microbial metabolism. In: JW Biggar, JN Seiber, eds. Fate of Pesticides in the Environment. Berkeley, CA: Univ. Calif Div Agric Nat Resources, Pub 3328. 1987, pp 93–101.
54. JS Karns. Biotechnology in bioremediation of pesticide contaminated sites. In: JB Bourke, AS Felsot, TJ Gilding, JK Jensen, JN Seiber, eds. Pesticide Waste Management: Technology and Regulation. ACS Symp Ser 510. Washington, DC: Am Chem Soc, 1992, pp 148–156.
55. JE Woodrow, LS Aston, T Shibamoto, JN Seiber. The assessment of a biological system for biodegradation and recycling of pesticide wastes. In: DT Teddler, FG Pohland, eds. Emerging Technologies for Hazardous Waste Management VI. 1996, pp 43–59.
56. SG Mirsatari, MM McChesney, AC Craigmill, WL Winterlin, JN Seiber. Anaerobic microbial dechlorination: An approach to on-site treatment of toxaphene-contaminated soil. J Environ Sci Health B22:663–690, 1987.
57. R Metcalf. Model ecosystem studies of bioconcentration and biodegradation of pesticides. In: MAQ Khan, ed. Pesticides in Aquatic Environments. New York: Plenum Press, 1977, pp 127–144.
58. PH Pritchard. Model ecosystems. In: RA Conway, ed. Environmental Risk Analysis for Chemicals. New York: Van Nostrand, 1982, Chap 8.
59. RG Nash, ML Beall Jr, WG Harris. Toxaphene and 1,1,1-trichloro-2,2-bis(p-chlorophenyl)ethane (DDT) losses from cotton in an agroecosystem chamber. J Agric Food Chem 25:336–341, 1977.
60. V Walter. Pesticide volatilization: A comparison of methods for measuring and approaches to fuzzy logic modeling. Dissertation. Humboldt Univ, Berlin, 1998.
61. PJ McCall, DA Laskowski, RL Swann, HJ Dishburger. Estimation of environmental partitioning of organic chemicals in model ecosystems. Residue Rev 85:231–244, 1983.
62. LA Burns, SM Cline, RR Lassiter. Exposure Analysis Modeling System (EXAMS): User Manual and System Documentation. Athens, GA: US Environ Protect Agency, Environ Res Lab, 1981.
63. JE Woodrow, MM McChesney, JN Seiber. In: DA Kurtz, ed. Long Range Transport of Pesticides. Chelsea, MI: Lewis, 1990, pp 61–81.
64. PF Sanders, JN Seiber. Organophosphorus pesticide volatilization: Model soil pits and evaporation ponds. In: RF Krueger, JN Seiber, eds. Treatment and Disposal of Pesticide Wastes. ACS Symp Ser 259. Washington, DC: Am Chem Soc, 1984, pp 279–295.
65. M Matthies. Transport and behavior in soil. In: D Calamari, ed. Chemical Exposure Predictions. Boca Raton, FL; Lewis, 1993, pp 103–113.

66. LD Clendening, WA Jury, FF Ernst. In: DA Kurtz, ed. Long Range Transport of Pesticides. Chelsea, MI: Lewis, 1990, pp 47–60.
67. D Mackay. Finding fugacity feasible. Environ Sci Technol 13:1218–1223, 1979.
68. DA Laskowski. Landscape-scale environmental modeling. In: GT Brooks, TR Roberts, eds. Pesticide Chemistry and Bioscience. Cambridge, UK: Roy Soc Chem, 1999, pp 302–312.
69. KR Solomon, Integrating environmental fate and effects information: The keys to ecotoxicological assessment of pesticides. In: GT Brooks, TR Roberts, eds. Pesticide Chemistry and Bioscience. Cambridge UK: Roy Soc Chem, 1999, pp 313–326.
70. EPA. Mid-Atlantic Landscape Indicators Project Plan. EPA 620/R-95/003. Washington, DC: U.S. Environ Protect Agency, Office Res Develop, June 1995.
71. US Congress, Washington, DC. Clean Air Act Amendments 1990.
72. JN Seiber. Toxic air contaminants in urban atmospheres: Experience in California. Atmos Environ 5:751–756, 1996.
73. LW Baker, DL Fitzell, JN Seiber, TR Parker, T Shabamoto, MW Poore, KE Longley, RP Tomlin, R Propper, DW Duncan. Ambient air concentrations of pesticides in California. Environ Sci Technol 30:1365–1368, 1996.
74. JN Seiber, JE Woodrow. Origin and fate of pesticides in air. In: NN Ragsdale, PC Kearney, JR Plimmer, eds. Washington, DC: Am Chem Soc, 1995, ACS Conf Proc Ser. Eighth International Congress of Pesticide Chemistry: Options 2000. pp 157–172.
75. DE Glotfelty, AW Taylor, BC Turner, WH Zoller. Volatilization of surface-applied pesticides from fallow soil. J Agric Food Chem 32:638–643, 1984.
76. AW Taylor, DE Glotfelty. Evaporation from soils and crops. In: R Grover, ed. Boca Raton, FL: Environmental Chemistry of Herbicides, Vol 1. pp 89–129, 1988.
77. LS Aston, JN Seiber. Fate of summertime airborne organophosphate pesticide residues in the Sierra Nevada mountains. J Environ Qual 26:1483–1492, 1997.
78. R Atkinson, R Guicherit, RA Hites, W-U Palm, JN Seiber, P DeVoogt. Transformations of pesticides in the atmosphere. A state of the art. Water, Air, Soil Pollut 115: 219–243, 1999.
79. JN Seiber. Transport and fate of pesticides in fog in California's Central Valley. In: TR Steinheimer, LJ Ross, TD Spittler, eds. Fate and Movement: Perspective and Scale of Study. ACS Symp Ser 751. Washington, DC: Am Chem Soc, 1999, pp 323–346.
80. DE Glotfelty, JN Seiber, LA Liljedahl. Pesticides in fog. Nature 325:602–605, 1987.
81. SJ Larson, PD Capel, MS Majewski. Pesticides in Surface Waters. Chelsea, MI: Ann Arbor Press, 1997.
82. TR Steinheimer, LJ Ross, TD Spittler, eds. Fate and Movement: Perspective and Scale of Study. ACS Symp Ser 751. Washington, DC: Am Chem Soc, 1999.
83. JE Barbash, EA Resek. Pesticides in Groundwater. Chelsea, MI: Ann Arbor Press, 1996.
84. JJ Cech, BW Wilson, DG Crosby, eds. Multiple Stresses in Ecosystems. New York: Lewis, 1998.
85. National Research Council. The Future Role of Pesticides in U.S. Agriculture. Washington, DC: Nat Acad Press, 2000.
86. CJ Soderquist, JB Bowers, DG Crosby. Dissipation of molinate in a rice field. J Agric Food Chem 25:940–946, 1977.

6

Pesticide Residue Procedures for Raw Agricultural Commodities: An International View

S. Mark Lee and Sylvia J. Richman
Center for Analytical Chemistry
California Department of Food and Agriculture
Sacramento, California, U.S.A.

1 INTRODUCTION

Pesticides are modern-day miracles. These chemicals have helped us to grow food in abundance and eliminate pests. Unfortunately, many pesticides can also have negative effects both on the environment and on humans. The use of pesticides must consequently be carefully controlled and closely monitored to maximize their benefits and minimize harmful effects. To support good stewardship of pesticide uses, many analytical methods have been developed to measure levels of specific pesticide residues in food [1] and in the environment [2].

There are a large number of analytical methods for the analysis of specific pesticides on individual matrices. Analytical methods for a pesticide may vary depending on the sample type and the purpose of the analysis. In the United

This chapter was not prepared on behalf of the California Department of Food and Agriculture and therefore does not represent any official policy of that department.

States, over 700 pesticides are registered for use in food production, and many analytical methods for pesticides are described in the literature.

The number of pesticides that must be monitored to safeguard the public interest is substantial even in the case of a single commodity. Farmers can choose from many different pesticides to control the multitude of insect pests, fungi, and weeds that attack their crops. Rotations of different pesticides on a crop are recommended to reduce the buildup of resistance by pests, potentially further increasing the number of residues that may be found on a commodity. Finally, mixtures of pesticides are often used for more effective control of pests. A greater variety of pesticides are used in growing fruits and vegetables than for any other food items [3]. Because it is not possible to know which pesticide residues you might find on a given crop, samples need to be screened for all possible residues.

The purpose of this chapter is to describe the analytical process and to present the regulatory methods that are used internationally for analysis of food.

2 SINGLE-RESIDUE METHODS VS. MULTIPLE-RESIDUE METHODS: PAM II AND PAM I

2.1 Single-Residue Methods

The U.S. Federal Insecticide, Fungicide and Rodenticide Act [4] and the Federal Food, Drug and Cosmetic Act [5] state that a pesticide registrant must submit to the United States Environmental Protection Agency (USEPA) a valid analytical method for the pesticide (and its pharmacologically significant metabolites) as a tool for tolerance enforcement in food and feed. These single-residue methods (SRMs) describe analysis of a single pesticide (or a group of related compounds derived from it) in a specific crop because they have been developed to register particular pesticides for particular applications or crops. As part of the registration process the USEPA Registration Laboratory in Fort Meade, MD, validates each method by reproducing the results independently. Once a method has been reviewed, validated, and accepted by the USEPA, it is included in Volume II the *Pesticide Analytical Manual* (PAM II), which is maintained by the U.S. Food and Drug Administration (FDA) [6]. Because the method was developed for a specific pesticide–matrix combination and independently validated, it is useful as a second method for confirmation of positive findings. Because of the length of time required to register a pesticide and validate the method, the method will often undergo revision or updating to include more recent developments in technology or instrumentation before it is published in PAM II.

For these reasons regulatory laboratories often adopt multiresidue methods (MRMs)—methods that can be used for assaying a wide range of pesticides in many different types of samples. To reach the broadest application of pesticide residue analysis, this review focuses on MRMs for screening a wide range of pesticides on a wide variety of matrices such as fresh fruits and vegetables. By focusing on the methods used by regulatory laboratories, an extra dimension of complexity is added: unknown pesticide application history. Regulatory multiresidue methods represent the best of modern pesticide residue analyses. This chapter also summarizes several countries' most current regulatory MRMs for monitoring and surveillance of fresh fruits and vegetables.

2.2 Multiresidue Methods

Not one but many different pesticides are used during food production [7], and many of these pesticides exert known harmful effects on humans. Thus, pesticide residue levels in foods must be monitored, and pesticide regulatory levels established for the intentional or unintentional presence of pesticides must be enforced.

It is outside this chapter's scope to discuss whether or not the regulatory limits established for pesticides are adequate to protect the public from harmful effects. The fact is that the public is concerned about potential exposure to pesticides through residues remaining in the foods they eat. Due to differences in quantities required to control target pests, pesticides can be legally present in food at different levels (a tolerance is the maximum residue level that may be present) in different crops and even in different parts of a single crop [2]. In addition, it is not uncommon to find more than one pesticide residue in a single crop. When foods containing several food components (e.g., pizza) are examined it is likely that several widely used pesticides will be present. Recent Pesticide Data Program monitoring studies [8] indicated that multiple pesticide residues exist in a food sample such as "spinach with red pepper," "mushroom salad," or "banana smoothie." Potential combinations of multiple pesticides in many different crops make MRMs the analytical methods of choice and SRMs far less practical.

Fortunately many pesticides have similar physical and chemical properties. This is true not only for pesticides of the same chemical families but also for pesticides of different families having similar functional groups, solubility, adsorption characteristics, vapor pressure, etc. These similarities allow the analysis of relatively large groups of pesticides with the use of a single analytical method. Most commercial pesticides are marketed as formulations designed to disperse in water, but the active ingredient is often more soluble in organic solvents than in water. This characteristic allows water-miscible solvents such as acetonitrile and acetone to be used effectively for extracting pesticide residues from all types of matrices. Once extracted into organic solvents, pesticide residues with similar chemical properties can be concentrated and purified using the same procedure. Individual pesticides are separated using chromatography, often gas-liquid chromatography (GLC), and detected based on the presence of certain common heteroatoms or functional groups. Thus, the SRMs of organophosphate [9], chlorinated hydrocarbon [10], phenylurea [11], and carbamate [12] pesticides can also be assayed quite effectively using MRMs. Some new classes of pesticides such as sulfonylurea and imidazolinone pesticides can also be assayed efficiently with MRMs owing to similarities in their physical and chemical properties.

Volume I of the *Pesticide Analytical Manual* (PAM I) [13] describes five different MRMs used not only in the United States but also by many countries worldwide. For this chapter, we compiled 12 different MRMs used around the world: PAM-I (Luke and Storres methods); European standards I, II, and III [14,15]; those of Sweden [16], the Netherlands [17], United Kingdom [18], and Canada [19]; the modified Luke method [20]; the California Department of Food and Agriculture method [21]; and those of Japan [22,23], Australia [24], and South Korea [25]. The methods presented here represent a small percentage of the more widely known methods. These methods are often used with in-house method validation and verification procedures.

2.2.1 Regulatory Method of Choice

The presence of residues in fruits and vegetables makes pesticide residue testing a real challenge. Regulatory samples arrive at the laboratory with only minimal sample information—typically what the matrix is and when and where it was collected. The analyst will generally not know the history of what pesticides were applied to the crop, how recently they were applied, or what application rates were used. Consequently, regulatory fresh fruit and vegetable samples range from those that contain no pesticide residues to those that contain several residues at varying levels. Customarily, regulatory laboratories receive several different types of samples on any given day, depending on the season, location, and availability of fruits and vegetables for sale. It is not uncommon for them to test five or six different fruit and vegetable samples at the same time. Furthermore, rapid analysis is essential for assaying perishable samples such as lettuce, strawberries, and cucumbers. It is challenging for any chemist and for any method to analyze for unknown pesticides in a variety of matrices in a short time. For many regulatory laboratories, it is often the goal to complete the analysis the same day the samples are received. Even though MRMs may sometimes provide less method sensitivity or analytical precision than SRMs, they are the methods of choice for regulatory pesticide residue analysis because of their ability to detect a large number of pesticides, their applicability to a wide range of matrices, and the relative ease and speed of sample analysis. The following section describes the components of the analytical process.

2.2.2 Techniques Involved in Multiresidue Methods

Like other chemical analyses, MRMs in general consist of the same five fundamental steps as trace chemical analysis:

1. *Sample processing.* A process to generate a homogeneous laboratory sample from the sample submitted
2. *Extraction.* A procedure in which analytes in a sample are dissolved and transferred into a suitable organic solvent or a mixture of solvents
3. *Purification (cleanup).* A series of steps that reduce sample matrix components and enrich target analytes in the sample extract
4. *Separation and detection.* A technique employed to separate analytes into individual identifiable components and quantify them
5. *Confirmation.* A measurement or process that provides the same analytical results by alternative physical or chemical means

Table 1 summarizes the steps for the 12 MRMs used in selected countries throughout the world. The steps shown in the table correspond to separate procedures for the chemist, and correlate in a general way to the following sequence.

Sample Processing. Sampling is not discussed because it is often not considered part of the laboratory analytical method although it is an important factor influencing the final results of analyses. Samples submitted to laboratories may consist of several individual fruits or vegetables. The exact numbers and sizes of samples vary depending on each nation's regulations. In general, the samples range from five to 20 individual fruits or plants or from 10 to 20 kg in total weight depending on the particular commodity. Some sample manipulation, such as the removal of outer layers of leafy vegetables, removal of cores of fruits, and washing, may be required by regulations. In the United States, unless otherwise indicated in the Code of Federal Regulations (CFR-40), regulatory samples cannot be manipulated through brushing, washing, peeling, removing outer leaves, or any other procedure that could affect the magnitude of pesticide residues.

Samples may require further preparation for analysis such as cutting and chopping coarsely prior to extraction. Most laboratories chop and homogenize entire samples unless the applicable government regulation requires the preservation of an unaltered portion of the submitted sample. Samples are often homogenized by using common commercial food processors (size of processor may vary depending on sample type but could be as large as 30 kg capacity), providing both maceration and mixing at the same time. A subsample, usually in the range of 25–200 g, is taken for extraction.

Extraction. Water-miscible solvents such as methanol, acetonitrile, and acetone are the most common extracting solvents, along with ethyl acetate, which also extracts significant amounts of water. Much of the weight of fruits and vegetables—80–95% [26]—is due to water, and this water derived from the commodity mixed with the solvent becomes an efficient pesticide extraction medium [27]. For example, a 50 g apple sample (80% moisture content) combined with 100 mL of acetonitrile yields an extracting solvent that is ~70% acetonitrile in water.

TABLE 1 Summary of Multiresidue Methods for Nonfatty High-Moisture Foods

Method	Step 1	Step 2	Step 3	Step 4	Step 5	Compounds	Detector
European Std L [14] Acetone extraction	*Xtrct smpl* Blend 100 g smpl, 200 mL acetone, 30 s. (Celite optional). Filter. Rinse all w 50 mL acetone.	*Liq–liq part'n* Dil 50 mL xtrct (1/5 total) w 250 mL water, add 25 g NaCl. Xtrct 2× w 50 mL DCM. Dry DCM w 30 g Na_2SO_4. Conc to 2 mL, add DCM to 10 mL.	*Chrom: Silica cartridge* Load conc smpl on (20 g sil + 1 g activ charcoal) col; collect. Elute w 140 mL 5/5/1 DCM/Tol/Ace, collecting.	*Solv xchg* Evap joined xtrcts to 2 mL; adj to 5 mL w hexane.		CH pesticides OP pesticides N pesticides GC-able pesticides	HECD/ECD FPD/NPD NPD MSD
European Std M [14] Acetone extraction	*Xtrct smpl* Blend 100 g smpl, 200 mL acetone. Note volume	*Liq/liq part'n* Xtrct 80 mL Ace xtrct w 100 mL DCM, 100 mL PE (3 min). Dry org w/3 g Na_2SO_4. Add 7 g NaCl to aq phase, xtrct 2× w 100 mL DCM (30 s). Join. DCM xtrcts.	*Solv xchg* Conc org to 2 mL. Add 100 mL PE, conc to 2 mL and repeat. Dissolve in 2 mL Ace (no cleanup) or 1 mL Ace then dilute to 10 mL w PE (Florisil cleanups).	*No cleanup*		CH pesticides	HECD
						N and P pesticides	FPD/NPD
				Chrom: Florisil 1 Load on 20 g Florisil col, collecting. Elute w 200 mL Eth/PE 6/94 = frac 1; elute w 200 mL Eth/PE 15/85 = frac 2; elute w 200 mL Eth/PE 50/50 = frac 3 OR *Chrom: Florisil 2* As above, but elute w 200 mL DCM/PE 2:8 = frac 1; elute w 200 mL DCM/PE/Acn 50:49.65/0.35 = frac 2; elute w 200 mL DCM/PE/Acn 50/48.5/1.5 = frac 3.	*Adj. Volume* Adjust each fraction to a suitable known volume.	CH pesticides	ECD

European Std N [14] Acetone extraction	*Xtrct smpl* Blend 100 g smpl, 100 x g H_2O (x is g H_2O in matrix), 200 mL acetone, 3 min. Add 10 g Celite, blend 10 s. Filter.	*Liq/liq part'n* Xtract 200 mL Ace xtrct + 20 g NaCl w 100 mL DCM for 2 min. Collect org and dry 30 min w 25 g Na_2SO_4. Filter, conc to just dry.	*GPC: SX-3* Diss in 7.5 mL EtOAc, add 7.5 mL Chex, load on 50 g SX-3 col. Elute w EtOAc: Chex 1:1 eluent at 5 mL/min. Collect pest frac, conc to 1 mL, adj to 5 mL w EtOAc.	*SPE: silica* Add 5 mL isooct to 2.5 mL xtrct, evap to 1 mL. Load on 1 g deact sil col, elute w 2 + 6 mL Hex:Tol 65:35 = frac 1 (adj to 10 mL), elute w 2 + 6 mL Tol = frac 2 (to 10 mL). Repeat w Tol:Ace 95:5, Tol (frac 3) Ace 8:2 (frac 4) and Ace (frac 5).	*Adj. volume* Adjust each fraction to 10 mL with the addition of the solvent used to elute it.	CH pesticides	ECD/HECD
						N & P pesticides	NPD
						All GC-able pesticides	MSD
				No cleanup		OP pesticides	FPD
European Std O [14] Acetonitrile extraction	*Xtrct smpl* Combine 100 g smpl, 200 mL Acn, 10 g Celite. If 5–25 g sugar in smpl add 50 mL H_2O. Blend 2 min, filter; meas vol.	*Liq/liq part'n* Xtrct Acn xtrct w 100 mL PE 2 min. Xtrct w 600 mL H_2O and 10 mL sat NaCl 15 s, discard aq soln. Wash org 2× w 100 mL H_2O, meas vol, dry w Na_2SO_4 (15 g), conc to 5–10 mL.	*Chrom: Florisil* Load on 10 cm × 22 mm activ Florisil col and wash w PE, collecting. Elute w 200 mL Eth/PE 6/94 = frac 1; elute w 200 mL Eth/PE 15/85 = frac 2; elute w 200 mL Eth/PE 50/50 = frac 3.	*Concentrate* Evap each fraction to suitable known volume.		CH pesticides	ECD/HECD
						N & P pesticides	NPD/FPD
						All GC-able pesticides	MSD

TABLE 1 Continued

Method	Step 1	Step 2	Step 3	Step 4	Step 5	Compounds	Detector
European Std P [14] Ethyl acetate extraction	*Xtrct smpl* Blend 50 g smpl, 100 mL EtOAc, 50 g Na_2SO_4, 2–3 min. Filter, rinse 2× w 25 mL EtOAc. Measure vol and evap 1/4 to 5 mL w EtOAc.	*GPC: SX-3* To EtOAc xtrct add 5 mL Chex, load on 50 g SX-3 col and elute w EtOAc: Chex 1:1 eluent at 5 mL/min. Collect pest frac, conc to 1 mL, adj to 5 mL w EtOAc.	*Concentrate* Evap xtrct to ~1 mL, adj to 5 mL w EtOAc.			P pesticides	NPD/FPD
		No cleanup				P pesticides	NPD/FPD
"New" Luke [20] Acetone extraction	*Xtrct smpl* Blend 100 g smpl, 200 mL Ace 2 min. Filter.	*SPE: C_{18}* Push 40 mL xtrct thru 0.45 μm filter/0.5 g C_{18} combo, follow w 10 mL 30% H_2O in Ace. Collect.	*Salt out water* Add 10 g fructose to xtrct, shake 15 s; add 10 g $MgSO_4$, shake 15 s; add 20 g NaCl, shake 3–4 min.	*Conc/Solv xchg* Trnsfr 20–25 mL xtrct to KD, add 50 mL Ace, 100 mL PE, evap to ~1 mL. Add 10 mL Ace, 50 mL PE, evap to <2 mL. Adj to 5 mL w Ace.	*SPE: SAX/PSA* Xtrct + 10 mL PE, load on 0.5 g SAX/PSA comb, elute w 1:2 Ace: PE 2 × 10 mL + 40 mL. Collect in KD, evap <2 mL, add 10 mL Ace, evap to 2 mL.	P and S pesticides N and P pesticides CH pesticides	FPD NPD HECD

Sweden [16] EtOAc extraction	*Xtrct/Dry smpl* Blend 75 g smpl, 200 mL EtOAc, 40 g Na_2SO_4, 3 min. Filter thru 20 g Na_2SO_4, add 10 g more.	*Concentration/ solv. xchg.* Conc 100 mL to 5 mL final vol in EtOAc: Chex 1:1.	*GPC: SX-3 10 × 400* EtOAc: Chex 1:1 eluent. Inject 1 mL (7.5 g). Collect pest frac, conc to 3 mL 95:5 Chex: EtOAc. (2.5 g/mL).	*No cleanup (most commod)*	*Dilute* Xtrct to 1.5 g/mL w EtOAc.	P and S pesticides	NPD/FPD
						GC-able pesticides	MSD
					Dil xtrct to 0.3 g/mL w Chex.	CH pesticides	ECD
				SPE: silica (some commod) 0.6 mL xtrct dissolved in 20 mL Chex, evap to 1 mL. Repeat. Load on 1 g cart, elute w 15 mL Tol: Chex. 15:85.	*Concentrate* Evap to 1 mL, adj to 3 mL w Chex.	CH pesticides	ECD
				SPE: Silica (some commod.) 2 mL xtrct dissolved in 20 mL Chex, evap to 1 mL. Repeat. Load on 1 g cart, elute w 25 mL Tol: Chex: Ace 6:3:1.	*Concentrate* Evap to 1 mL, adj to 2 mL w Chex.	P and S pesticides	NPD/FPD
				Carbamate cleanup			HPLC post-col
				Concentrate to 5–6 g/mL.		Various	HPLC DAD/ FLD
					Partition into pH 2.2 buffer.	H_2O-soluble	HPLC DAD/ FLD
					SPE: Silica Load 1 mL xtrct, elute w 4 mL Chex, 6 mL EtOAc: Hex 1:3 = frac 1. Dry, elute w 15 mL 0.04% TEA, pH 2.2, Buff = frac 2.	Imazalil, carbendazim, thiophanate methyl	HPLC DAD/ FLD

Table 1 Continued

Method	Step 1	Step 2	Step 3	Step 4	Step 5	Compounds	Detector
Netherlands [17] EtOAc extraction	*Xtrct smpl /Adsorb H_2O* Blend 50 g smpl, 100 mL EtOAc, 50 g Na_2SO_4, 2–3 min. Filter.	*Concentration/ Solv. xchg* Evap 25 mL xtract at 65°C, diss in 5 mL Isooct: Tol 9:1.				P and S pesticides All GC-able pesticides	FPD ITD
Acetone partition (miniaturized)	*Xtrct/Part'n smpl* Homog 15 g smpl, 30 mL Ace 30 s. Add 30 mL DCM, 30 mL PE, (Int Std optional). Homog 30 s. Centrifuge at 4000 rpm 5 min, collect upper phase. (If early OPs in sample, repeat xtrction w addn of 7.5 g Na_2SO_4.)	*Concn/ Solv xchg* Evap 25 mL xtract at 65°C, diss in 5 mL Isooct: Tol 9:1.				P and S, N and P, and all GC-able pesticides	FPD, NPD, ITD
		Concn/solv xchg Evap 200 μl xtract, dissolve in 1 mL isooct: Tol 9:1.				CH's, pyrethroids	ECD
		SPE: Aminopropyl Dry 2 mL xtrct, diss in 1 mL DCM, load on 100 mg Aminoprop cartridge, el w 0.5 mL DCM, 1 mL DCM: MeOH 99:1. Collect.	*Conc/Solv xchg* Evap xtract to nr dry at 50°C. Diss in 1 mL of 0.05 mg/mL trimethacarb in Acn: H_2O 20:80.			Carbamates (1) Phenylureas (2)	HPLC postcol hydrolysis (1) or photolysis (2)
		SPE: Aminopropyl Dry 6 mL xtrct, diss in 2 mL DCM, load 1 mL on 500 mg Aminoprop cartridge, el w 2 mL DCM, 2 × 2 mL DCM. Collect.	*Conc/Solv xchg* Evap xtrct to nr dry at 50°C. Diss in 1 mL of 0.05 mg/mL trimethacarb in Acn: H_2O 20:80.			Benzoylphenylureas	HPLC DAD
		SPE: Diol					

	Evap 3 mL xtrct nr dry, diss in 2 mL MeOH w Int Std, load on 500 mg diol cartridge, wash in 1 mL MeOH, elute w 2 mL MeOH: 0.1 MH_3PO_4, 1:1; add 0.1 mL 1 M NaOH.			Benzimidazoles	HPLC-FLD
	SPE: Conazoles			Conazoles	HPLC
Either of above two extraction methods	*Concn/Solv xchg* Rotovap dry at 40°C, diss in 5 mL PE, dry, diss in PE to 2 g/mL.	*Alumina chrom* Add 1 mL xtrct to 1 g $AgNO_3$-coated alumina, elute w 9 mL, collecting.		CH pesticides	ECD
	Concn/Solv xchg Rotovap dry at 40°C, diss to 5 g/mL with DCM.	*Chrom: Silica (triaz)* Add 1 mL xtrct to 1 g dry silica, wash w 15 mL DCM: Ace 99.5: 0.5, elute w 10 mL DCM: Ace 85:15.	*Concentrate* Evap triazine xtrct to 1 mL.	Triazines	ITD
	Concn/Solv xchg Rotovap dry at 40°C, diss to 0.5 g/mL w hexane.	*Chrom: Silica (pyreth)* Add 1 mL xtract to 1 g dry silica, wash w 20 mL 99.8:0.2 Hex: EtOAc, elute w 35 mL 9:1 Hex: EtOAc.	*Concentrate* Evap pyrethroid xtrct dry, diss in 1 mL decane.	Pyrethroids	ECD

TABLE 1 Continued

Method	Step 1	Step 2	Step 3	Step 4	Step 5	Compounds	Detector
Canada [19] Acetonitrile extraction	*Xtrct smpl/Salt out water* Combine 50 g smpl, 100 mL Acn, blend 5 min. Add 10 g (8 mL) NaCl, blend 5 min.	*SPE: C_{18}* Pass 13 mL Acn xtrct thru, (prewash SPE w 2 mL smpl). Collect. Add 2 cm^3 Na_2SO_4, shake, remove 10 mL, conc to 0.5 mL w N_2.	*SPE: Carb, NH_2 prop* Load on 6 mL Envicarb–aminoprop combo, wash 2 × 1 mL and elute w 23 mL Acn: Tol 3:1.	*Conc/Solv xchg* Evap to <2 mL, add 10 mL Ace, evap and repeat, add 50 μL Int Std and adj to 2.5 mL. Use 0.5 mL for MSD. Reserve 2 mL for carbamates.	*Conc/Solv Xchg* Evap 2 mL left to <0.2 mL, add 1 mL MeOH, evap to <0.2 mL, add carb Int Std, adj to 0.8 mL w pH 3 H_2O.	Carbamates	HPLC post-col
					No changes	All GC-able pesticides	MSD
PAM-I 302-a/b [13] Acetone extraction	*Xtrct Smpl* 100 g smpl, blend 2 min w 200 mL acetone, filter.	*Three Choices* *Part'n w DCM, PE.* Ext 80 mL Ace xtrct w 200 mL PE and 200 mL DCM. Add 7 g NaCl to aq phase, xtrct w 2× 100 mL DCM. *Hydramatrix Col.* Load 40 mL Ace xtrct on 40 g col, elute w 2× 50 +200 mL DCM. *Sm. Hydramatrix.* Load 40 mL Ace xtrct on 25 g col, elute w 2× 25 + 150 mL DCM. *Always dry all xtrcts w 1.5" Na_2SO_4 in funnel.*	*Conc: KD* Evap on KD to 2 mL, add 100 mL PE and evap to 2 mL, add 50 mL PE and evap to 2 mL. Add 20 mL Ace, evap to 2 mL. Make up to 7 mL for part'n extract.	*No cleanup*	*Conc: KD* Evap separate fractions on KD to known vol.	P and S pesticides N and P pesticides CH pesticides	FPD NPD HECD
				Chrom: Florisil C5 Dil xtrct to 10 mL w Ace, dil to 100 mL w PE, load on 4" × 22 mm Florisil col, elute at 5 mL/min w 200 mL Eth:PE 15:85 = fraction 1. El w 200 mL Eth:PE 1:1 = fraction 2.	*Conc: KD* Evap on KD, e.g., to 1.0 mL.	Biphenyl o-Ph phenol	FID
			Conc: KD for Florisil C1 Evaporate to <5 mL, add 50 mL Ace and evap again, e.g., to 2 mL or 7 mL.	*Chrom: Florisil C1* Dil 1 mL xtrct to 10 mL w Hex, load on 4 g Florisil col, rinse w Hex, elute at 5 mL/min w 50 mL DCM: Acn: Hex 50:1.5:48.5.		Pyrethroids CH pesticides S pesticides	ECD

			Conc: KD for C1$_{18}$ Conc xtrct on KD to 2.0 mL, then to near dry (0.1 mL) w N_2.	*SPE: C_{18} C4* Diss resid w 2 mL MeOH, load on 2.8 mL C_{18} cartridge (collect). Elute w MeOH until nr 5 mL. Adj to 5 mL.		Carbamates	HPLC post-col
			Conc: Rotovap Evap: xtrct at 35°C to just dry, diss in 10 mL DCM.	*Chrom: Car/sil Celite C3* Load at 5 mL/min on 0.5 g silanized Celite + 5 g charcoal/Celite 1:4 col. Collect. Elute w 10 mL DCM, 25 + 100 mL Tol:Acn 25:75.	*Conc/Solv xchg* Rotovap solvent to just dry, replace w 5 mL MeOH.	Carbamates	HPLC post-col
USDA PDP [8] *(Calif, incl MSD screen)* Acetonitrile extraction	Xtrct smpl Blend 50 g smpl, 100 mL Acn, 2 mL 2 M PO_4 buff, pH 7 for 3 min. Filter.	*SPE: C_{18}* Add 2 mL sat NaCl to xtrct and push thru 1 or 2 g cart. Coll, add 40 g NaCl and let sit 1 h. Take 5 mL ea for CH, OP, Carb, 10 mL for MSD.	*Conc/Solx xchg* Evap at 45°C w air, diss in 1 mL hexane.	*SPE: Florisil* Load smpl on 1 g cart, transfer w 2 × 5 mL Hex:Ace 9:1. Collect all.	*Concentrate* Evap at 45°C to nr dry, diss in 5 mL hexane. Filter.	CH pesticides	HECD/ECD
			Conc/Solv xchg Evap at 45°C w air, diss in 1 mL acetone.			OP pesticides	FPD
			Conc/Solv xchg Evap at 45°C w air, diss in 1 mL MeOH: DCM 1:99.	*SPE: NH_2 prop* Load smpl, transfer w 2 × 4 mL 1% MeOH/DCM. Collect all.	*Conc/Filter* Evap at 45°C to nr dry, diss in 1 mL MeOH. Filter (0.2 μm).	Carbamates	HPLC/post-col
			SPE: NH_2 prop Load smpl, transfer w 2 × 1 mL Acn. Collect all.	*Conc/Filter* Evap at 45°C to <2 mL, rinse w 1 mL Ace, then nr dry, add 0.5 mL Int Std		MS screen	MSD

TABLE 1 Continued

Method	Step 1	Step 2	Step 3	Step 4	Step 5	Compounds	Detector
California [10] *and Korea* Acetonitrile extraction	*Xtrct smpl/Salt out water* 50 g smpl + 100 mL Acn, blend 2 min, filter into cylinder w 10 g NaCl, shake 1 min, let separate. Pipet three 10 mL aliquots and evap to near dryness in a beaker at 40–70°C.	*Solv xchg: Hexane* Resuspend in 2 mL hexane (for CHs).	*SPE: Florisil* Load smpl on 1 g col, transfer w 2 × 5 mL Hex: Ace 9:1. Collect all.	*Concentrate* Evap at 40°C w air, diss in 5 mL hexane. Filter.		CH pest, pyrethroids	ECD
		Solv xchg: Acetone Resuspend in 5 mL acetone. Filter w 0.2 μm nylon filter.				P and S pesticides	FPD
		Solv xchg: MeOH/DCM Resuspend in 2 mL 1% MeOH/DCM.	*SPE: NH_2 Prop* Load smpl transfer w 2 × 2 mL 1% MeOH/DCM. Collect all.	*Concentrate* Evap at 40°C w air, diss in 2.5 mL MeOH. Filter.		Carbamates	HPLC/post-col
Japan [22] *(official)*	*Extraction* Blend 20 g smpl w 150 mL acetone for 3 min. Filter and conc to 30 mL.	*Hydramatrix col.* Pass sample thru Chem-Elut (discard eluent), then elute w 150 mL EtOAc.	*GPC: SX-3* Load pesticide, elute w EtOAc: Chex 1:1. Collect pest frac. (Take aliquot for carbamate analysis.)	*SPE: Silica* Load pesticides, wash w Ace: Hex 1:1, elute w 20 mL Ace:Hex 1:1.	*SPE: Florisil* 18 mL 15% Eth/Hex (Frac 1), 15 mL 15% Ace/Hex (Frac 2).	CH pesticides, pyrethroids	ECD
					No cleanup	P and S pesticides N and P pesticides	FPD NPD
				No cleanup		Carbamates	HPLC/post-col
Japan [23] *(MSD)*	*Extraction* Blend 50 g smpl, 100 mL Acn for 3 min.	*SPE: C_{18}* Filter sample thru 1 g C_{18} to trap nonpolars. Collect.	*Salt out water/pH* Add 10 mL 2 M PO_4 buff pH 7, 15 g NaCl. Shake 3 min, keep 60 mL Acn layer.	*Conc/Solv xchg* Dry w Na_2SO_4, add 0.3 mL Int Std, conc and adj to 3 mL Hex: Ace 1:1.	*SPE: PSA* Load on 500 mg cart., elute w 3 × 3 mL Hex:Ace 1:1. Collect and adj to 2 mL w Hex: Ace 4:1.	Al GC-able pesticides	MSD

UK [18] EtOAc extraction	*Xtrct smpl/Adsorb H_2O* Homogenize 30 g smpl, 60 mL EtOAc, 35–40 g Na_2SO_4 (+ 5–6 g Na_2CO_3 or 1 mL 5% H_2SO_4 for basic or acidic xtrctn) at 27–33°C for ≥30s. Filter.	*Concentrate* Evap 5 mL to <1 mL. Adj to 1 mL w EtOAc. Do same w blank xtrct for stds.				Early OP pesticides All GC-able pesticides	FPD/NPD ITD/MSD
		Chrom: Alumina Pass 4 mL xtrct (6 mL for EI and CI anal) thru 0.4 g deact Al (30 mL H_2O per 200 g) pipet col until 2.5 mL collected. Take 2 mL of cln xtrct, add 25 μL TDE int std.	*Concentrate* Evap xtrct w N_2 at 27 ± 5°C to ~0.2 mL. (For EI and CI anal, 0.4 mL is prepd from 4 mL cln xtrct, plus 50 μL TPE.				
		Oxidation of S pest Evap 10 mL xtrct+ 6 drops propylene glycol:Ace 1:1 to keeper. Add 5 mL tBuOH and shake w 25 mL 0.2% $KMnO_4$ 1 min. Let stand 10 min.	*Part'n/Recovery* Add 25 mL 5% Na_2SO_4 and 50 mL DCM, shake. Collect DCM layer, dry w Na_2SO_4. Repeat 2× w 50 mL DCM. Rinse w 25 mL DCM. Evap org at 30°C to nr dry.	*Solv xchg* Evap joined xtrcts at 30°C to nr dry. Add 2 mL EtOAc, evap to nr dry, and diss in final vol of 2 mL EtOAc.		Oxidizable S pesticides (e.g., Demeton-S-Methyl)	NPD/FPD
Acetone extraction	*Xtrct smpl* Homog 35 g smpl w 105 mL Ace for 3 min. Adj to pH ~7.	*Partition* Xtract w 2× 105 mL DCM:Chex. Dry and coll org. Xtrct ag w 2× 70 mL DCM. Dry orgs and join. Evap to nr dry, diss in 5 mL EtOAc: Chex 1:1.	*HPGPC: Envirosep* Inj 1 mL xtrct on 2 col in series (tot L = 41–45 cm, ID = 19–21 mm) at 5 mL/min EtOAc:Chex. Collect ~25 mL pest fraction.	*Concentrate* Evap to 0.2 mL. Adj to 2 or 5 mL w EtOAc.		All GC-able pesticides OP pesticides	ITD/MSD FPD

Abbreviations: ~, approximately; Ace, acetone; Acn, acetonitrile; act, activated; adj, adjust; AmPr, aminopropyl; Aminoprop, aminopropyl; NH_2 prop, aminopropyl; Aq, aqueous (phase); buff, buffer; car, carbon; carb, carbamate; cart, cartridge; cc, cubic centimeter; cent, centrifuge; CH, chlorinated hydrocarbon; Chex, cyclohexane; chrom, chromatography; cln, clean; col, column; coll, collect; combo, comb, combination; commod, commodities; conc, concentration; DAD, diode array detector; DCM, dichloromethane; deact, deactivated; dil, dilute; disc, discard; diss, dissolve; ECD, electron capture detector; el, elute; Eth, ethyl ether; EtOAc, ethyl acetate; evap, evaporate; filt, filter; fin, final; FLD, fluorescence detector; FPD, flame photometric detector; Fr, frac, fraction; g, grams; GC, gas chromatography; GPC, gel permeation chromatography; HECD, (Hall) electrolytic conductivity detector; Hex, hexanes; homog, homogenize; HPLC, high pressure liquid chromatography; intl, int, internal; isooct, IsoOct, isooctane; KD, Kuderna-Danish concentrator; liq, liquid; lyr, layer; meas, measure; MeOH, methanol; min, minutes; MSD, mass spectral detection; N, nitrogen (containing); nr, near; NSD, nitrogen phosphorus detection; OP, organophosphate; org, organic (phase); P, phosphorus (containing); part'n, partition; PE, petroleum ether; pest, pesticide(s); PSA, phenylsulfonic acid; resid, residue; rotovap, carry out rotary evaporation; rpm, rotations per minute; s, seconds; sat, saturated; SAX, strong anion exchange; Sil, silica; solv, solvent; SPE, solid phase extraction; spl, smpl, sample; std, standard; tBuOH, tert-Butanol; thru, through; Tol, toluene; trfr, tfer, trnsfr, transfer; vol, volume; w, with; x, times; xchg, exchange; xtrctn, extraction.

Aqueous organic solvents with a similar solvent/water ratio have been reported to be the best possible extracting solvents [28]. It is not possible (nor is it practical or necessary) to achieve exactly the same solvent/water ratio for every sample, because each sample type may have a different moisture content or state of hydration (e.g., wilted lettuce vs. fresh lettuce). If necessary, additional water can be added to compensate for the low moisture status of some samples. For example, 10–20 mL of water is often added to low-moisture samples (e.g., wheat, rice, soybeans) to increase the aqueous proportion of the extraction solvent.

Extraction of pesticides into organic solvent is often enhanced by further blending and shearing of the homogenate. Several types of blenders are used. Most common extraction devices have a rotating blade mounted at either the top or bottom of the vessel (Omni Mixer and Waring Blender, respectively). Two to five minutes of blending at a moderate speed (2000–5000 rpm) normally suffices for the extraction of pesticides. To accomplish a more thorough extraction, MRMs can specify a device that disrupts samples through the generation of cell-rupturing ultrasound (e.g., Polytron Tissumizer) in addition to mechanical mixing and shearing. Repeated extractions to ensure complete recovery of residues are often omitted from MRMs to save time and effort. Immediately after extraction, solvent is separated from nonextractable plant material. This procedure is optimized for speed and efficiency. Different MRMs may accomplish this step in different ways depending on the circumstances of the laboratory. Simple filtering to remove plant material may be accomplished by using Sharkskin™ filter paper, which is designed for quick filtration. Centrifugation is also used to separate insoluble materials from soluble extracts.

Centrifugation of several samples at the same time reduces processing time. In MRMs not every step needs to be quantitatively precise. For example, the filtration of aqueous/organic solvent away from plant material does not require complete removal of the solvent. The methods require only that a sufficient volume of the solvent be collected for further cleanup. The ratio of the sample weight (e.g., 50 g) to the volume of the extraction solvent (e.g., 100 mL) is used to determine the final concentration of residues (e.g., 2 mL/g sample). To conserve time and cost, most MRMs do not attempt to recover all solvent from the homogenate, just a representative portion. For the same reason, a superior extraction technique (e.g., Soxhlet extraction) is time-consuming and is therefore not used in routine regulatory MRMs for fresh fruit and vegetables requiring quick turnaround time.

It is a common practice, but not a part of the method, to discard the remaining homogenized samples except for a small portion, which is often stored at 4°C for subsequent analysis. The storage period of the homogenate varies depending on the organization's internal protocol and the status of the final results. In the case of negative findings, stored samples are discarded shortly after the validation of results to make space for the large numbers of samples that must pass through the laboratory. For positive findings, especially commodities containing pesticide residues for chemicals that are not approved for the specific commodity or for residues in excess of the legal limits, homogenates are stored under suitable long-term storage conditions (e.g., −20°C).

Purification (Cleanup). The filtrate resulting from sample extraction is a complex mixture that contains organic solvent(s), water, biochemicals (lipids, sugars, amino acids, and proteins), and secondary metabolites (organic acids, alkaloids, terpenoids, etc.) at high concentrations with very minute amounts of the pesticide residues of interest. It is a challenging task to isolate and detect pesticide residues of interest in the presence of high levels of background chemicals, often called matrix interferences. Most crude extracts require some purification prior to analysis.

Purification involves the removal of water, evaporation of excess organic solvent, and selective trapping to separate pesticides from interferences. Most MRMs utilize one or more techniques for this purification process. The greater the number of cleanup steps included in a method, the greater the losses of analytes and the longer it takes to carry out the analysis. Most of the water must be removed from the extract to further concentrate the desired analytes. Much water is quickly removed by partitioning the organic solvent with sodium chloride–saturated water (see Table 1). Other approaches accomplish the same result by adsorbing water [29]. These techniques remove large amounts of the water, but the remaining traces of water must be removed by filtering or adding dehydrated hygroscopic salts (e.g., Na_2SO_4) to the organic phase of the extract. Recoveries of extremely water-soluble pesticides such as acephate and methamidophos can vary depending on the concentration of other solutes and the mechanism of water removal. Adsorbing the water present in organic solvents yields greater and more consistent recoveries of extremely water-soluble pesticides than does the partitioning process.

Even after the removal of many water-soluble coextractives, extracts still contain large amounts of interfering compounds and only trace levels of pesticide residues. Buffering the aqueous phase close to a pH of 7 prior to removing water causes more ionic and polar biochemicals to partition into the water and results in removing large quantities of organic acids (i.e., phenolic acids, citric acids, oxalic acids, and tannic acids) from the organic phase, which contains neutral and nonpolar chemicals, including pesticide residues [30]. It is also possible to remove large amounts of nonpolar plant constituents (lipids, waxes, some pigments, and secondary metabolites) before removing the water by filtering the aqueous/organic extract through reversed-phase solid-phase extraction (SPE) material [20,30] or an activated carbon sorbent [31]. This is an efficient way to remove highly nonpolar chemicals, because a ~30%/70% aqueous–organic combination solvent elutes most pesticide residues very effectively from the sorbent but leaves these nonpolar interfering chemicals behind. Removing these nonpolar chemicals in the early steps of the cleanup process allows the sample to be manipulated more easily. For example, some nonpolar chemicals precipitate during sample concentration, affecting the recovery of residues. Nonpolar chemicals that remain in extracts often interfere with chromatographic separation and detection. Solvent partitioning of the aqueous/organic extract with a nonpolar solvent (i.e., hexane or petroleum ether) is also used in some MRMs to remove nonpolar interfering chemicals [32].

Because extracts must be concentrated 100-fold or more, a requirement for trace residue analyses, rapid and efficient concentration techniques are preferred. Various solvent concentration techniques are used in MRMs. Rotary evaporators and Kuderna-Danish sample concentrators are good for concentrating thermally labile and highly volatile pesticides [15,20]. Heating the extract in an open beaker with a stream of gas (air or nitrogen) is also an efficient and inexpensive way to achieve concentration [21]. The sample concentration step often varies for different MRMs, seemingly depending on the laboratory's preference rather than on performance. Any of these techniques carefully applied yields similar results. Concentrated extracts, even after being subjected to the purification process, often contain quantities of interference chemicals that can easily interfere with analysis by overwhelming a chromatographic system and/or saturating a detector. Additional cleanup of extracts for MRMs maximizes the difference between physical and chemical properties of pesticide residues and those of interference chemicals. Two common techniques for cleanup are solid-phase extraction (SPE) [20,30] and size-exclusion chromatography (gel permeation chromatography) [14,16]. Differences in cleanup techniques among MRMs reflect some method performance differences but are mostly the result of a laboratory's experience, availability of supplies, and programmatic and regulatory needs of the parent agencies rather than technical or performance criteria.

Chromatography and Detection. Multiresidue methods rely on chromatographic techniques to separate pesticide residues, to determine an analyte's identity on the basis of elution time (retention time), and to quantify responses from a specific detector. To this end, two chromatographic techniques are most common among MRMs: gas chromatography (GC) and high performance liquid chromatography (HPLC) [33].

GAS CHROMATOGRAPHY. Gas chromatography is perhaps the single most important advancement in analytical chemistry in making trace pesticide residue testing possible. Many review articles address GC techniques and GC applications for pesticide residue analysis [34]. The gas chromatograph has become the primary analytical instrument for pesticide residue screening because the physical and chemical properties of many common pesticides ($\sim$560) are ideally suited to the GC technique [35]. These pesticides are semivolatile with different vapor pressures, relatively stable to high temperature, and soluble in organic solvents, and they contain elements distinguishable from background interferences. Most current MRMs (Table 1) have chosen open tubular columns over the packed colums used in earlier MRMs (PAM-I). The use of wide-bore columns allows the introduction of larger amounts of sample into the gas chromatograph and enables trace pesticide residues to be detected more easily.

GC DETECTION SYSTEM. The flame photometric detector (FPD), the electron capture detector (ECD), the alkaline flame ionization detector (NPD), and the electrolytic liquid conductivity detector (ELCD) are relatively insensitive to interfering substances and exhibit selective sensitivity to many pesticide classes. In fact, many MRMs can be characterized on the basis of the detection modes used. Reliance on selective and specific detectors reduces the number of false positive findings. Without selective detection systems, GC responses would be difficult to interpret and offer too many possibilities. For this reason, MRMs relying on universal detection systems, such as full-scan electron impact (EI) mass spectrometers, the flame ionization detector (FID), and the thermionic detector are less useful for identification. The sensitivity of the electron capture detector compensates for its lack of specificity, and the selectivity of the ELCD compensates for its lack of sensitivity. Perhaps the most important factor in the usefulness of detectors is ruggedness. All of the above detectors have proved over the years that they are durable and easy to maintain with heavy daily use.

HIGH PERFORMANCE LIQUID CHROMATOGRAPHY. As is the case for GC, much is written about HPLC techniques [36]; this chapter does not review HPLC techniques and their application in detail. Not many MRMs use HPLC despite the fact that more pesticides are suited to HPLC analysis than to GC analysis. The one reason for the low utilization of HPLC in pesticide screening might be the lack of detection systems comparable to those available for GC. HPLC still does not have a detection system that is selective, sensitive, and definitive in identifying pesticides. With the exception of the fluorescence detector and mass spectrometer, HPLC detectors (i.e., UV/Vis, refractive index, and electrochemical) are not selective and sensitive enough to perform trace residue analysis. The postcolumn reaction technique coupled with fluorescence detection made possible the analysis of *N*-methyl carbamate pesticides [37] and phenylurea herbicides [11] as MRMs. Conditions for these methods are listed in Table 1. Liquid chromatography coupled with mass spectrometry (LC/MS) is a promising technique for trace residue analysis, but as yet no MRM based on LC/MS has been reported for routine regulatory testing [38].

Another reason for the low utilization of the HPLC technique for MRMs is the limited resolution of solvent gradient systems. Reversed-phase HPLC separates chemicals by varying the concentration of organic modifiers (methanol, acetonitrile, and mixtures of water-miscible solvents) with water through a column containing a hydrophobic liquid phase bonded to a solid-phase stationary material. The separation efficiency (theoretical plates) of HPLC does not provide sufficient resolution within a practical time period to resolve many analytes in a reproducible manner. Furthermore, the solvent gradient system cannot be varied as easily as the temperature gradient technique used in GC.

Identification and Quantification. There are comprehensive reviews and books that describe identification and quantification techniques using GC and HPLC for trace pesticide residue analysis [39,40]. Identification of pesticides using chromatography is based on the characteristic retention time of the pesticide on a particular chromatographic column under a given elution condition used in the separation. Retention times (R_t) of pesticides are often listed as a part of MRMs. Two factors are major influences of R_t: types of columns used (liquid phases) and separation conditions (column oven temperature in GC and eluting solvent composition in HPLC). Gas chromatographic MRMs rely on multiple temperature gradient programs to enhance separation of pesticides and to reduce overall chromatographic time. Three different liquid phases are commonly used in MRMs: methyl silicone, 5% phenyl methyl silicone, and 50% phenyl methyl silicone. Each liquid phase gives a slightly different elution pattern for some pesticides.

Most HPLC MRMs rely on reversed-phase separation because of its reliability and cost-effectiveness. There are many different bonded liquid phases with different carbon loads and end capping that give different performance characteristics. The most commonly used HPLC column phases are octyl (C_8) and octadecyl (C_{18}) bonded to silica stationary phase. Methanol and acetonitrile are the two organic modifiers most commonly used with water in MRMs. As is the situation with the GC technique, HPLC MRMs rely on mobile-phase gradient schemes to vary the composition of the organic modifier to achieve the same goal of speed and cost efficiency. Although an R_t value is commonly used for "identification" of a compound, chromatographic behavior does not provide unequivocal information regarding identity. True identification of pesticide requires structural information for the specific compound. The most common approach used in modern analytical chemistry is mass spectrometry as discussed in the following subsection.

Quantification of pesticide levels is as important as identification of pesticides for the regulatory laboratory, because the regulation of pesticide use is based on the maximum residue level (MRL or tolerance) that may be present. Thus, the correct estimation of pesticide residue concentration in a given matrix is critical, because levels that exceed the MRL are illegal. Most MRMs rely on external calibration techniques to quantify residues. Three to five concentrations of given pesticides are used to generate the GC, HPLC, or other calibration curve. The concentration of an incurred residue is quantified by comparison to the concentration–response curve.

There are several difficulties in correctly estimating or quantifying residue concentration. First, it is impractical to generate daily calibration curves for all analytes of interest. There are over 200 pesticides of interest in GC MRMs. Second, external calibration curves are often generated with standard pesticides in neat solvent (acetone or hexane) and not in a matrix blank. The so-called matrix effect on quantification of analytes is well known to analytical chemists. Sample matrix components (or coextractives) significantly influence the response of analytical instruments to pesticide residues. It would be ideal to use external calibration standards made in a matrix blank.

Different laboratories and organizations use different procedures for ensuring the best estimate of residue levels. The following is an example used in the California Department of Food and Agriculture laboratory. MRMs are validated initially by using a handful of representative pesticides. The external calibration curves for these pesticides are created by using standards in solvent on a daily basis. Over long periods of time, laboratories establish external calibration curves for all pesticides of interest and demonstrate the range of detection and linearity of detector response to the concentration of pesticides. When a pesticide residue is detected in a sample during a routine screening process, the estimation is made by using the external calibration curve. A pair of bracketing concentrations of the specific pesticide are chosen, and new external calibrations are then made using the same pesticide in a previously saved matrix blank. These calibration solutions are used to determine the residue concentration. In some cases only a single level of calibration might be used to reduce the time of analysis. This quantification scheme is a practical solution to what could otherwise be an unmanageable workload.

The quantification scheme just described works because the majority of samples being screened do not contain any pesticide residues. Experience and knowledge in pesticide residue testing can be valuable in correctly recognizing and interpreting chromatographic results.

Confirmation. Unambiguous determination of pesticide residues is not always necessary, especially for initial screening. However, most pesticide regulatory surveillance and monitoring programs have established standard operating procedures (SOPs) to address the confirmation of initial findings of pesticide residues by addressing regulatory implications. A common approach and the most practical one for confirming a positive chromatographic response has been "the dual column confirmation," a technique that correlates two different retention times of a pesticide under two different chromatographic conditions. This technique is applicable in most situations, especially when differing retention times can be acquired simultaneously using a single chromatographic instrument. This can work well with a dual-column GC. However, it falls short when background matrix interferences become too great and the suspected pesticide residue response cannot be resolved sufficiently from them.

For the unambiguous identification of pesticide residues, MRMs rely on mass spectrometry (MS), another determinative technique that is different from GC. Mass spectrometry is a common choice because it gives direct physical and chemical information about the analyte and is easily coupled to chromatographic techniques. Various criteria for MS confirmation have been proposed for pesticide analysis [41]. As GC/MS and LC/MS become more affordable, MRMs are being developed that are based on the use of MS for both initial and confirmatory determinations.

Recent advances in the MS/MS technique, especially GC coupled with ion trap mass spectrometry, promised an easy one-step technique to detect, quantify, and confirm in a single analysis, but ion trap mass spectrometry was not widely accepted for MRM for several reasons. Although ion trap MS was economical compared to other MS techiques, it was not as user-friendly. The matrix interference is most noticeable when an analyte coelutes with an exceptionally high concentration of matrix interference. Finally, the quantification with ion trap MS was not reproducible and required a separate analysis for quantification.

The confirmation of pesticide residues may involve several steps: reanalysis of the initial extract with MS, re-extraction of the sample using a different method (often a single-residue method), and the use of a different analyst. These steps are taken to ensure against errors from multiple sources. Some laboratories would choose one over the other, whereas others might require all of the steps.

3 SURVEY OF NATIONAL MRMs FROM AROUND THE WORLD

Multiresidue methods, particularly those used for regulatory surveillance and monitoring of fresh fruit and vegetables, represent the pinnacle of pesticide residue methodology. A simple, rapid, and efficient MRM is a wonderful tool, providing analysis of large numbers of common pesticides in many types of samples. The purpose of the following discussion is to examine the variety and assess the performance of different MRMs used in selected countries around the world. The described MRMs are commonly used for the surveillance and monitoring of pesticides in food. This discussion is not meant to be a complete survey of every MRM currently in use throughout the world. We seek rather to demonstrate commonalities and differences among MRMs and to gain insight into the basic principles of pesticide residue analysis in fruits and vegetables through examination of MRMs.

Table 2 summarizes the listed MRMs in terms of (1) sample preparation, (2) removal of the water phase, (3) SPE cleanup, (4) additional cleanup, (5) concentration, and (6) detection. These steps were described in detail in previous sections. Despite the difference in extraction solvents and SPE, the principle of the methods is the same: Water-miscible solvent is used to extract a broad range of pesticides, followed by quick purification using partition chromatography and concentration of solution to increase detection sensitivity.

Lists of pesticides screened and recoveries of individual pesticides vary among methods as well as within an MRM depending on matrices (Table 2). One must use the list as a guide and not as an absolute standard. An MRM is a dynamic method that can change to include or exclude certain pesticides depending on analytical requirements without changing much of the analytical procedure. The fact that a particular pesticide is not listed under an MRM does not mean that the MRM cannot be used to screen for it. It might simply be that the given pesticide was not part of the screening interest and was never evaluated. In order for it to be added to the list, one needs to conduct a brief spike-and-recovery study using the pesticide and matrix of interest. For that reason, the EPA has requested (did not mandate) that all pesticide manufacturers registering the use of a new pesticide test the applicability of known MRMs for the new pesticide.

It is important to realize that no single MRM can be used to screen for all pesticides or quantify them all. What is not addressed in Table 2 is perhaps the most important factor in predicting how a given MRM might perform. This is the "matrix factor," effects that result from varying the matrix and/or conditions. A simple analysis of lettuce can be complicated by varieties (e.g., leaf or Romaine), where and when it was grown, and how it was stored (moisture content). These factors affect the amounts of water and extractable organic matter in the lettuce and contribute to the variations in recoveries of residues from complete (C) to partial (P) to variable (V). Residues known to be recovered by a given MRM but not quantified are denoted by "recovers" (R).

When one MRM does not provide good results for a given matrix, try a different MRM that consists of different steps. For example, acephate and methamidophos are organophosphate pesticides that are commonly found in foods. Most MRMs pose no difficulty in assays for these residues except in matrices with more water than usual. Because of their partition coefficient, recoveries of these residues into miscible organic solvents are poor unless the moisture is eliminated during the initial extraction. Therefore, the Swedish method using ethyl acetate and sodium sulfate yields much better recoveries of these residues than MRMs that use acetonitrile and salts. For example, the Swedish MRM lists com-

TABLE 2 Pesticides Included in International Monitoring Programs

Pesticide	Eur Std CEN-M	Eur Std CEN-N	Eur Std CEN-O	Eur Std CEN-P	Sweden		Netherlands		Canada		PAM-I Meth. 302a (No cleanup)	PAM-I Meth. 302b (Florisil cleanup)	USDA-PDP MS screen	Calif Anal	Korea	Japan (MS)	Australia (Victoria) LOD (fruit)
					Anal	LOQ	Anal	LOD	Anal	LOD							
Acephate		Coll.		R	C	0.02			P	0.31	C			R^{FPD}	R^{NPD}		
Acetochlor											C						
Aclonifen					C	0.02											
Alachlor		R							C	0.02	C			R^{ECD}	R^{ECD}		
Aldicarb					LC[3]		C^{CRB}		CV	0.07				R^{CRB}			0.02
Aldicarb sulfone					LC[3]		P^{CRB}		C	0.04				R^{CRB}			
Aldicarb sulfoxide					LC[3]		S^{CRB}		C	0.03				R^{CRB}	R^{CRB}		
Aldrin	R	R	Coll.		C	0.01					C			R^{ECD}	R^{ECD}		
Allethrin									C	0.02				R^{ECD}			
Allidochlor									P	0.03	C						
Ametryn	R								C	0.02	C						
Amidithion		R															
Aminocarb					C	0.1			CV	0.04	C						
Aminocarb deg. product									CV	0.07							
Amitraz (2,4-DMA)																	
AMPA					LC[6]												
Anilazine		R			C	0.02					C			R^{ECD}			
Anthraquinone		R															
Aramite 1									C	0.03	C						
Aramite 2									C	0.03	C						
Aspon									C	0.02							
Atrazine	R	R							C	0.02	C			R^{ECD}			
Atrazine desethyl									C	0.03							
Azinphos ethyl	R	R		R	C	0.05			C	0.06	C			R^{FPD}			0.06
Azinphos methyl	R	R		R	C	0.05			P	0.09	C			R^{FPD}		R^{NPD}	0.01
Azinphos methyl O-analog											C			R^{FPD}			
Aziprotryne	R																
Azocyclotin																	
Benalaxyl					C	0.05			C	0.02						R^{NPD}	
Bendiocarb					C	0.2			C	0.02	C			R^{CRB}		R^{CRB}	
Bendiocarb deg. prod									C	0.04							
Benefin														R^{ECD}			
Benfluralin		R							P	0.02	C						

Benodanil									C	0.03	C			
Benomyl (as Carbendazim)											C			0.1
Benoxacor											C			
Bensulide											C	R^{FPD}		
Bentazone					C	0.5								
Benzoylprop-ethyl		R							C	0.02				
Beta-cyfluthrin					C	0.1								
BHC-alpha			Coll.						P	0.02	C	R^{ECD}	R^{ECD}	
BHC-beta			Coll.						C	0.02	C	R^{ECD}	R^{ECD}	
BHC-delta											C	R^{ECD}	R^{ECD}	
BHC-epsilon												R^{ECD}	R^{ECD}	
Bifenox		R							P	0.04	C	R^{ECD}	R^{ECD}	
Bifenthrin	R	R			C	0.1			P	0.03	V	R^{ECD}	R^{ECD}	0.01
Binapicryl		R			C	0.3					C			
Biphenyl					C	1					C			
Biteranol		R			C	0.5							R^{NPD}	
Bromacil	R	R							P	0.12	C		R^{ECD}	
Bromide (inorganic)														
Bromophos-ethyl	R	R		R	C	0.04			P	0.02	C			
Bromophos	R	Coll.		R	C	0.5			C	0.03	C	R^{FPD}		
Bromopropylate	R	Coll.			C	0.1			C	0.02	C		R^{ECD}	0.02
Bromoxynyl octanoate		R										R^{ECD}		
Bufencarb							C^{CRB}					R^{CRB}		
Bufencarb deg. product									C	0.04				
Bupirimate	R				C	0.1			C	0.02	C			
Buprofezin					P	0.05								
Butachlor									CV	0.16	C			
Butralin									C	0.02	V			
Butocarboxim														
Butocarboxim sulfoxide							P^{CRB}	0.01						
Butoxycarboxim							S^{CRB}							
Buturon							R^{PHU}	0.01						
Butylate									S	0.03				
Cadusafos					C	0.01					C		R^{NPD}	
Camphechlor (toxaphene)		R												
Captafol deg. product									P	0.22				
Captafol	R	R			C	0.03					C	R^{ECD}	R^{ECD}	
Captan	R	Coll.			C	0.05			PV	0.8	C	R^{ECD}	R^{ECD}	
Captan deg. product									CV	0.39				
Carbanolate							C^{CRB}							
Carbaryl					C	0.1	C^{CRB}	0.01	CV	0.03	C	R^{CRB}	R^{CRB}	0.05

TABLE 2 Continued

Pesticide	Eur Std CEN-M	Eur Std CEN-N	Eur Std CEN-O	Eur Std CEN-P	Sweden Anal	Sweden LOQ	Netherlands Anal	Netherlands LOD	Canada Anal	Canada LOD	PAM-I Meth. 302a (No cleanup)	PAM-I Meth. 302b (Florisil cleanup)	USDA-PDP MS screen	Calif Anal	Korea	Japan (MS)	Australia (Victoria) LOD (fruit)
Carbendazim							R^{BZM}										
Carbetamide									C	0.03							
Carbofuran					C	0.1	C^{CRB}	0.01	C	0.04				R^{CRB}			
Carbofenothion	R	R		R	C	0.05			P	0.03	C			R^{FPD}		R^{NPD}	
Carbofenothion-methyl		R												R^{FPD}			
Carbofenothion O-analog											C			R^{FPD}			
Carbofenothion sulfone											C			R^{FPD}			
Carbosulfan											P						
Carboxin									P	0.02	C					R^{NPD}	
Chinomethionate		R			C	0.05										R^{NPD}	
Chlorbenside	R	R							P	0.04	C						
Chlorbenside sulfone		R															
Chlorbromuron							R^{PHU}	0.05	C	0.05	V						
Chlorbaufam									C	0.02	C						
Chlordane-alpha		R			C	0.02			P	0.02	C			R^{ECD}			
Chlordane-gamma		R			C	0.02			P	0.02	C			R^{ECD}			
Chlordimeform					C	0.1			C	0.02	P						
Chlorfenprop-methyl		R															
Clorfenson	R	R			C	0.05			C	0.03	C			R^{ECD}			
Chlorfenvinphos (e)	R	R		R	P	0.02			C	0.04	C			R^{FPD}		R^{NPD}	
Chlorfenvinphos (z)	R	R		R					C	0.04	C			R^{FPD}		R^{NPD}	
Chlorfluazuron						R^{BPU}	0.02										
Chlorflurenol methyl									C	0.03	C						
Chlorflurenol	R																
Chloridazon		R							C	0.03							
Chlorimuron ethyl ester											P						
Chlormephos		R			C	0.01			P	0.04	C						
Chlormequat																	
Chlornitrofen											C						
Chlorobenzilate	R	R			C	0.5			C	0.02	C			R^{ECD}		R^{ECD}	
Chloroneb		R							P	0.02	C			R^{ECD}			
Chloroprooppylate	R	R			C	0.5			C	0.02	P						

Chlorothalonil		Coll.		C	0.01		C	1.22	S	R^{ECD}		R^{ECD}	
Chlorotoluron		R			R^{PHU}	0.01		C					
Chloroxuron		R							C				
Chlorpropham	R	Coll.		C	0.1		C	0.02	C	R^{ECD}			
Chlorpyrifos	R	Coll.	R	C	0.03		C	0.02	C	R^{FPD}	R^{ECD}		0.005
Chlorpyrifos-methyl		R	R	C	0.03		C	0.02	C	R^{FPD}	R^{NPD}		
Chlorpyrifos O-analog				C	0.05				C				
Chlorsulfuron													
Chlorthal-dimethyl	R	R					C	0.02		R^{ECD}			
Chlorthiamid							S	0.02					
Chlorthiamid deg. product							CV	0.13					
Chlorthion							C	0.04					
Chlorthiophos	R	R	R				C	0.03	C				
Chlorthiophos O-analog									C				
Chlorthiophos sulfone									C				
Chlorthiophos sulfoxide									C				
Chlozinolate				C	0.05		P	0.06					
Chlofentezine									R				
Ciodrin										R^{ECD}			
Clomazone							C	0.02	C				
Clopyralid													
Coumaphos		R	R				C	0.08	C	R^{FPD}			
Coumaphos O-analog									C				
Crotoxyphos		R					C	0.04	C	R^{FPD}			
Crufomate		R					C	0.07	C	R^{FPD}			
Cyanazine	R	R		C	0.1		PV	0.09	C				
Cyanofenphos	R	R	R	C	0.05				C				
Cyanophos	R	R		C	0.02		C	0.02	C				
Cycloate							PV	0.02					
Cyfluthrin 1	R	R		C	0.1		C	0.12	C	R^{FPD}			0.001
Cyfluthrin 4	R	R		C	0.1		C	0.1	C	R^{FPD}			0.001
λ-Cyhalothrin	R	R		C	0.1				C		R^{ECD}		0.001
Cyhexatin													
Cypermethrin (both)	R	R		C	0.2				C	R^{ECD}			0.001
Cypermethrin 1							C	0.1					
Cypermethrin 4							C	0.1					
Cyprazine									C				
Cyproconazole									R				
Cyromazine									S				
2,4-D													
DCPA									C				

TABLE 2 Continued

Pesticide	Eur Std CEN-M	Eur Std CEN-N	Eur Std CEN-O	Eur Std CEN-P	Sweden		Netherlands		Canada		PAM-I Meth. 302a (No cleanup)	PAM-I Meth. 302b (Florisil cleanup)	USDA-PDP MS screen	Calif Anal	Korea	Japan (MS)	Australia (Victoria) LOD (fruit)
					Anal	LOQ	Anal	LOD	Anal	LOD							
Daminozide																	
Danifos					C	0.05											
Dazomet											S						
DDD-0,p	R	R	Coll.	Coll.					C	0.02				R^{ECD}	R^{ECD}		
DDD-p,p		R	Coll.		C	0.01			C	0.02	C			R^{ECD}	R^{ECD}		
DDD-p,p olefin											C						
DDE-o,p	R	Coll.	Coll.								C			R^{ECD}	R^{ECD}		
DDE-p,p	R	Coll.	Coll.		C	0.01			P	0.02	C			R^{ECD}	R^{ECD}		
DDT-o,p	R	Coll.	Coll.						P	0.02	C			R^{ECD}	R^{ECD}		
DDT-p,p	R	Coll.	Coll.						P	0.03	C			R^{ECD}	R^{ECD}		
DEF		R									C			R^{FPD}			
Deltamethrin	R	R			C	0.05			CV	0.33	C				R^{ECD}		0.002
Demeton-S									P	0.04	C			R^{FPD}			
Demeton-O									C	0.04	C						
Demeton-S-methyl		R		R	C	0.02			C	0.02							
Demeton-O-methyl		R		R							C			R^{FPD}			
Demeton-O-methyl sulfone											C						
Demeton-O-sulfone											C						
Demeton-O-sulfoxide											C						
Demeton-S-methyl sulfone		R		R	C	0.03											
Demeton-S-sulfone		R									C						
Demeton-S-sulfoxide		R									C						
Des-N-isopropyl-isofenphos											C						
Desmethyl-norflurazon											V						
Desmethyl-pirimiphos-Me		R															
Desmetryn	R				C	0.1			C	0.02							
Dialifor	R	R		R	C	0.1			C	0.03	C			R^{FPD}			
Diallate 1		R							P	0.02	C						
Diallate 2		R							P	0.02	C						
Diazinon	R	Coll.	Coll.	R	C	0.02			C	0.02	C			R^{FPD}	R^{NPD}		0.05
Diazinon O-analog											C			R^{FPD}			
Dichlobenil	R	R			C	0.05			P	0.04	C			R^{ECD}	R^{ECD}		0.02

Dichlofenthion	R	R		R					C	0.02	C			
Dichlofluanid	R	Coll.			C	0.02			CV	0.21	C		R^{ECD}	
Dichlone											P	R^{ECD}		
Dichlormid									P	0.04				
3,5-Dichloroaniline					P	0.5								
p,p′-Dichlorobenzophenone		R												
Dichloroprop														
Dichlorvos	R	R		R	P	0.01			P	0.03	C	R^{FPD}	R^{NPD}	
Diclobutrazol											C			
Dichlofop methyl		R							C	0.02	C		R^{ECD}	
Dicloran		Coll.			C	0.03			C	0.04	C	R^{ECD}	R^{ECD}	
Dicofol-o,p											C	R^{ECD}		
Dicofol-p,p	R	Coll.			C	0.1			C	0.03	C	R^{ECD}	R^{ECD}	
Dicrotophos		R			C	0.05			C	0.04	C	R^{FPD}		
Dieldrin	R	Coll.	Coll.		C	0.03			C	0.02	C	R^{ECD}	R^{ECD}	
Dethathyl-ethyl											C			
Difenoconazole														0.01
Difenoxuron							R^{PHU}	0.01						
Diflubenzuron							R^{BPU}	0.02						
Dimefox		R												
Dimethipin													R^{ECD}	
Dimethachlor	R	R							C	0.02	C			
Dimethametryn											C			
Dimethoate	R	R		R	C	0.02			C	0.03	C	R^{FPD}	R^{NPD}	0.03
Dimethoate O-analog												R^{FPD}		
Dinitramine		R							C	0.02	C			
Dinobuton		R			C	0.2					C			
Dinocap		R									C			
Dinoseb					C	0.2								
Dinoterb					C	0.1								
Dioxabenzofos											C			
Dioxacarb							R^{CRB}				C			
Dioxathion	R	R		R	C	0.1			C	0.02	V	R^{FPD}		
Diphenamid									C	0.02	V		R^{NPD}	
Diphenylamine					C	0.1			P	0.02	C		R^{NPD}	0.2
Diquat														
Disulfoton	R	R		R					P	0.04	C	R^{FPD}	R^{NPD}	
Disulfoton O-analog												R^{FPD}		
Disulfoton sulfone		R									C	R^{FPD}		
Dislufoton sulfoxide		R									C	R^{FPD}		
Ditalifmos	R	R		R	C	0.05								

TABLE 2 Continued

Pesticide	Eur Std CEN-M	Eur Std CEN-N	Eur Std CEN-O	Eur Std CEN-P	Sweden		Netherlands		Canada		PAM-I Meth. 302a (No cleanup)	PAM-I Meth. 302b (Florisil cleanup)	USDA-PDP MS screen	Calif Anal	Korea	Japan (MS)	Australia (Victoria) LOD (fruit)
					Anal	LOQ	Anal	LOD	Anal	LOD							
Dithianon											NR						
Dithiocarbamates (as CS_2)																	
Diuron											C				R^{ECD}		0.1
DNOC					C	0.2											
Edifenphos		R							C	0.04	C				R^{NPD}		
Endosulfan alpha	R	Coll.			C	0.01			C	0.02	C				R^{ECD}		0.001
Endosulfan beta	R	Coll.			C	0.02			C	0.02	C			R^{ECD}	R^{ECD}		0.001
Endosulfan sulfate	R	Coll.			C	0.02			C	0.08	C			R^{ECD}	R^{ECD}		0.001
Endrin		Coll.	Coll.		C	0.01			C	0.02	C			R^{ECD}	R^{ECD}		
EPN		R			C	0.05			C	0.03	C			R^{FPD}	R^{NPD}		
EPTC									S	0.03							
Erbon									P	0.03							
Esfenvalerate					C	0.1					C			R^{ECD}			
Etaconazole 1									C	0.02	C			R^{ECD}			
Etaconazole 2									C	0.03	C						
Ethalfluralin									P	0.02	C			R^{ECD}	R^{ECD}		
Ethophon											NR						
Ethiofencarb					P	1	C^{CRB}	0.01			C						
Ethiofencarb sulfone							P^{CRB}	0.01							R^{CRB}		
Ethiofencarb sulfoxide							P^{CRB}	0.01									
Ethiolate											C						
Ethion	R	Coll.	Coll.	R	C	0.02			C	0.03	C			R^{FPD}	R^{NPD}		
Ethion O-analog											C						
Ethirimol											P						
Ethofumesate									C	0.02	C						
Ethoprofos	R	R		R					C	0.02					R^{NPD}		
Ethoprop														R^{FPD}			
Ethoxyquin											C						
Ethylan									C	0.02				R^{ECD}			
Etridizaol									P	0.02	C						
Etrimfos	R	R		R	C	0.02			C	0.02	C				R^{NPD}		
Etrimfos O-analog											C						

Ethylenethiourea (ETU)											S			
Famofos		R												
Famphur											C			
Famphur O-analog											C			
Fenamiphos	R	R		R	C	0.02			C	0.03	C	R^{FPD}	R^{NPD}	
Fenamiphos sulfone											C			
Fenamiphos sulfoxide											C			
Fenarimol	R	Coll.			C	0.02			C	0.02	C	R^{ECD}	R^{ECD}	0.02
Fenarimol metab. B											NR			
Fenarimol metab. C											S			
Fenbuconazole											C			
Fenbutatin oxide														
Fenchlorphos	R	R	Coll.	R	C	0.03			C	0.02	C	R^{FPD}		
Fenchlorphos O-analog											C			
Fenfuram											C			
Fenitrothion	R	Coll.		R	C	0.03			C	0.03	C	R^{FPD}	R^{NPD}	0.02
Fenitrothion O-analog											C			
Fenobucarb												R^{CRB}		
Fenoxaprop ethyl ester											S			
Fenoxycarb											C			
Fenpropathrin	R	Coll.			C	0.1			C	0.04			R^{ECD}	
Fenpropimorph											C			
Fenson	R	R			C	0.03			C	0.02				
Fensulfothion	R	R		R	C	0.05			C	0.08	C	R^{FPD}	R^{NPD}	
Fensulfothion O-analog											C			
Fensulfothion sulfone											C			
Fenthion	R	R		R	C	0.05			C	0.03	C	R^{FPD}	R^{NPD}	0.06
Fenthion O-analog											C			
Fenthion sulfone					C	0.05					C			
Fenthion sulfoxide					C	0.1								
Fentin hydroxide														
Fenuron							R^{PHU}	0.01						
Fenvalerate (both)	R	R			C	0.1					C	R^{ECD}		0.001
Fenvalerate 1									C	0.04				
Fenvalerate 2									C	0.03				
Fenvalinate													R^{ECD}	
Flamprop-isopropyl									C	0.02	C			
Flamprop-methyl									C	0.02	C			
Fluazifop butyl ester											C			
Fluazinam					P	0.2								
Flubenzimine		R												

TABLE 2 Continued

Pesticide	Eur Std CEN-M	Eur Std CEN-N	Eur Std CEN-O	Eur Std CEN-P	Sweden		Netherlands		Canada		PAM-I Meth. 302a (No cleanup)	PAM-I Meth. 302b (Florisil cleanup)	USDA-PDP MS screen	Calif Anal	Korea	Japan (MS)	Australia (Victoria) LOD (fruit)
					Anal	LOQ	Anal	LOD	Anal	LOD							
Fluchloralin	R	R							C	0.03	C			R^{ECD}			
Flucythrinate	R	R									C				R^{ECD}		
Flufenoxuron							R^{BPU}	0.02									
Flumethrin																	0.002
Flumetralin									C	0.03							
Fluometuron							R^{PHU}	0.01									
Fluorodifen	R								C	0.03							
Fluotrimazole		R															
Flurochloridone l									C	0.03							
Flurochloridone 2									C	0.02							
Fluroxypyr																	
Flusilazole											C				R^{NPD}		0.03
Fluvalinate	R	R									C				R^{ECD}		0.02
Folpet	R	Coll.			C	0.05			S	0.42	C			R^{ECD}	R^{ECD}		
Fonofos	R	R		R	C	0.02			P	0.02	C			R^{FPD}			
Fonofos-O-analog											V						
Formothion	R	R		R	C	0.1					C				R^{NPD}		
Fuberidazole		R															
Furilazole											C						
Genite		R															
Glyphosate																	
HCH-alpha	R	Coll.			C	0.02	H										
HCH-beta	R	R			C	0.02											
HCH-delta		R			C	0.02			C	0.02							
HCH-gamma (Lindane)	R	Coll.	Coll.		C	0.02			C	0.02	C			R^{ECD}			0.05
HCH-epsilon		R															
Heptachlor	R	R	Coll.		C	0.01			P	0.02	C			R^{ECD}	R^{ECD}		
Heptachlor epoxide	R	Coll.	Coll.		P	0.02			C	0.05	C			R^{ECD}			
Heptachlor epoxide, cis		R															
Heptachlor epoxide, trans		R															
Heptanophos	R	R		R	C	0.05			C	0.02	C						
Hexachlorobenzene		Coll.			C	0.02			S	0.03	C			R^{ECD}			
Hexaconazole											C				R^{NPD}		

Hexaflumuron							R^{BPU}	0.02						
Hexazinone					C	0.1			C	0.05	P			
Hexythiazox														
3-Hydroxycarbofuran							P^{CRB}		C	0.04		R^{CRB}		
Imazalil		R			P	0.1	C^{CZL}	0.05	C	0.08	C	R^{ECD}	R^{ECD}	0.05
Imazamethabenz methyl ester											C			
Iodofenphos	R	R			C	0.05			C	0.03	C			
Ioxynil octanoate		R												
Iprobenfos											C			
Iprodione	R	Coll.			C	0.2			PV	0.09	C	R^{ECD}	R^{ECD}	0.05
Iprodione metabolite isomer											C			
Isazophos									C	0.02	C			
Isobenzan		R												
Isocarbamid		R												
Isodrin		R												
Isofenphos	R			R	C	0.05			C	0.03	C	R^{FPD}	R^{NPD}	
Isofenphos O-analog											C			
Isoprocarb							C^{CRB}					R^{CRB}		
Isopropalin		R							C	0.03	C			
Isoprothiolane											C			
Isoproturon							R^{PHU}	0.01			S			
3-Keto-carbofuran							C^{CRB}					R^{CRB}	R^{CRB}	
d-Keto-endrin		R												
Lenacil		R												
Leptophos		R			C	0.02			P	0.02	C	R^{FPD}		
Leptophos-O-analog											C			
Leptophos photoproduct											C			
Linuron		R					R^{PHU}	0.05	C	0.1	V		R^{NPD}	
Lufenuron							R^{BPU}							
Malathion	R	Coll.	Coll.	R	C	0.03			C	0.04	C	R^{FPD}	R^{NPD}	0.01
Malathion-O-analog	R				C	0.05			C	0.13	C	R^{FPD}		
MCPA-(2-butoxyethyl) ester		R												
Mecarbam	R	Coll.			C	0.05					C		R^{NPD}	
Mecoprop														
Melamine											NR			
Menazon				R										
Mephosfolan		R			C	0.05					C			
Mepiquat														
Merphos		R							P	0.03		R^{FPD}		

TABLE 2 Continued

Pesticide	Eur Std CEN-M	Eur Std CEN-N	Eur Std CEN-O	Eur Std CEN-P	Sweden		Netherlands		Canada		PAM-I Meth. 302a (No cleanup)	PAM-I Meth. 302b (Florisil cleanup)	USDA-PDP MS screen	Calif Anal	Korea	Japan (MS)	Australia (Victoria) LOD (fruit)
					Anal	LOQ	Anal	LOD	Anal	LOD							
Metalaxyl	R	R			C	0.1			C	0.02	C				R^{NPD}		0.01
Metasystox thiol											C						
Metazachlor	R								C	0.02	C						
Methabenzthiazuron		R									C						
Methacrifos																	
Methamidophos		R		R	P	0.02			P	0.1	V			R^{FPD}	R^{NPD}		0.01
Methidathion	R	R		R	C	0.02			C	0.04	C			R^{FPD}	R^{NPD}		0.01
Methiocarb					C	0.1	C^{CRB}	0.01	C	0.06	C			R^{CRB}	R^{CRB}		0.02
Methiocarb sulfone							P^{CRB}	0.01	C	0.06	S				R^{CRB}		
Methiocarb sulfoxide							P^{CRB}	0.01	C	0.04	P			R^{CRB}			
Methomyl							C^{CRB}	0.01	C	0.04				R^{CRB}			0.05
Methoprotryne	R	R							C	0.02	C						
Methoxychlor olefin											C						
Methoxychlor-p,p′	R	R	Coll.		C	0.2			C	0.04	C			R^{ECD}	R^{ECD}		
Methoxuron							R^{PHU}	0.01			V						
Methyl Trithion									C	0.02							
Metobromuron							R^{PHU}	0.05	C	0.02	C				R^{NPD}		
Metolachlor	R	R							C	0.02	C			R^{ECD}	R^{ECD}		
Metolcarb											C						
Metribuzin	R	R			C	0.02			C	0.03	V			R^{ECD}	R^{NPD}		
Metribuzin deaminated metab.											C						
Metsulduron-methyl																	
Mevinphos	R	R		R	C	0.02					C				R^{NPD}		0.01
Mevinphos-cis									C	0.02	C			R^{RPD}	R^{NPD}		
Mevinphos-trans									PV	0.07	C			R^{FPD}	R^{NPD}		
Mexacarbate									C	0.03							
Mexacarbate degrad. prod.									C	0.04							
Mirex		R	Coll.						P	0.02	P						
Monocrotophos		R		R	C	0.03			C	0.02	C			R^{FPD}	R^{NPD}		
Monolinuron		R							C	0.02	C						
Monuron							R^{PHU}	0.01									

Morphothion		R												
Myclobutanil									C	0.02	C			0.05
Myclobutanil alcohol metab.											S			
Myclobutanil diol metab.											NR			
Naled	R	R		R							C			
1-Naphthol												R^{CRB}		
Napropamide											C			
Neburon							R^{PHU}	0.01			C			
Nitralin		R							C	0.03	C			
Nitrapyrin									P	0.02	C		R^{ECD}	
Nitrofen	R	R							C	0.03	C	R^{ECD}		
Nitrothal-isopropyl		R							C	0.02	C			0.08
Nonachlor, cis											C			
Nonachlor, trans											C			
Norea											C			
Norflurazon									C	0.03	V			
Nurarimol									CV	0.07	C			
Octachlor epoxide		R									C			
Octachlorodipropyl ether		R												
Octhilinone											C			
Ofurace											C			
Omethoate		R		R	P	0.02			PV	0.25			R^{NPD}	
O-Phenylphenol					C	0.03					C			
Oxadiazone		R							C	0.02	C	R^{ECD}	R^{ECD}	
Oxadixyl									C	0.02	C		R^{NPD}	
Oxamyl					P	1	P^{CRB}	0.01	C	0.03		R^{CRB}	R^{CRB}	
Oxamyl oxime														
Oxycarboxin									S	0.04	C			
Oxychlordane									C	0.05				
Oxyflurofen									C	0.02	C	R^{ECD}		0.01
Paclobutrazole											C			0.05
Paraoxon	R	R			C	0.03			C	0.05	C	R^{FPD}		
Paraoxon-methyl		R			C	0.05								
Parathion	R	Coll.	Coll.	R	P	0.02			C	0.03	C	R^{FPD}	R^{NPD}	0.01
Parathion methyl	R	R	Coll.	R	C	0.02			C	0.03	C	R^{FPD}	R^{NPD}	
Pebulate									P	0.02	C			
Penconazole					C	0.1			C	0.03	C		R^{ECD}	0.01
Pendimethalin	R	R			C	0.05			C	0.02	C	R^{ECD}	R^{NPD}	
Pentachloroaniline		R			C	0.03					C	R^{ECD}		
Pentachloroanisole		R			C	0.02								

TABLE 2 Continued

Pesticide	Eur Std CEN-M	Eur Std CEN-N	Eur Std CEN-O	Eur Std CEN-P	Sweden		Netherlands		Canada		PAM-I Meth. 302a (No cleanup)	PAM-I Meth. 302b (Florisil cleanup)	USDA-PDP MS screen	Calif Anal	Korea	Japan (MS)	Australia (Victoria) LOD (fruit)
					Anal	LOQ	Anal	LOD	Anal	LOD							
Pentachlorobenzene		R			C	0.03					C			R^{ECD}			
Pentachloronitrobenzene											C			R^{ECD}			
Pentachlorophenyl methylether											C						
Pentachlorophenyl methylsulfide											C						
Permethrin	R	Coll.			C	0.05											0.002
Permethrin-cis									P	0.04	C			R^{ECD}			
Permethrin-trans									P	0.04	C			R^{ECD}			
Perthane	R	R	Coll.								C						
Phenkapton	R	R															
Penmedipham		R															
Phenthoate		R			C	0.05			C	0.02	C				R^{NPD}		
O-Phenylphenol									C	0.02	C						
Phorate	R	R		R					P	0.03	C			R^{FPD}	R^{NPD}		
Photate-O-analog											C						
Phorate-O-analog sulfone											C						
Phorate-O-analog sulfoxide											C						
Phorate sulfone											C			R^{FPD}			
Phorate sulfoxide											C			R^{FPD}			
Phosalone	R	Coll.		R	C	0.05			C	0.05	C			R^{FPD}	R^{NPD}		0.02
Phosalone O-analog											C			R^{FPD}	R^{NPD}		
Phosmet				R	C	0.1			CV	0.07	C			R^{FPD}	R^{NPD}		0.1
Phosmet-O-analog					C	0.1								R^{FPD}			
Phosphamidon		R		R	C	0.05			C	0.05	C			R^{FPD}	R^{NPD}		
Phospholan											C						
Phosphorous acid																	1
Phoxim		R		R							C				R^{NPD}		
Phoxim-O-analog											C						
Piperophos											C						
Piperonyl butoxide		R															
Pirimicarb		R			C	0.04			C	0.03	C				R^{NPD}		

Pirmiphos-ethyl		R	R	C	0.04			C	0.02	C		R^{NPD}	0.01
Pirimiphos-ethyl-O-analog										C			
Pirimiphos-methyl	R	Coll.	R	C	0.05			C	0.02	C	R^{FPD}	R^{NPD}	
Pretilachlor										C			
Probenazole										C			
Prochloraz				C	0.1	R^{CZL}	0.05	P	0.03	C		R^{NPD}	
Procyazine										C			
Procymidone	R	Coll.		C	0.1			C	0.03	C		R^{ECD}	0.03
Prodiamine										C			
Profenofos	R	R		P	0.1			C	0.05	C	R^{FPD}	R^{NPD}	
Profluralin	R	R						C	0.02	V	R^{ECD}		
Promecarb				C	0.1	R^{CRB}				V	R^{CRB}		
Promecarb deg. prod. 1								C	0.06				
Promecarb deg. prod. 2								C	0.02				
Prometon								C	0.02				
Prometryn	R							C	0.02	C		R^{NPD}	
Pronamide								C	0.02	C	R^{ECD}		
Propachlor		R						C	0.02	C			
Propamocarb												R^{NPD}	
Propanil		R						C	0.02	C	R^{ECD}	R^{ECD}	
Propargite				C	0.1			C	0.02	C			0.05
Propanzine	R							C	0.02	C			
Propetamphos								C	0.03	C	R^{FPD}		
Propham	R	Coll.		C	0.05			C	0.02	C	R^{FPD}		
Propiconazole		R		C	0.2					C		R^{NPD}	
Propiconazole 1								C	0.02				
Propiconazole 2								C	0.03				
Propoxur		R		C	0.2	R^{CRB}	0.01	C	0.04	C	R^{CRB}		
Propyzamide	R	R		C	0.1								
Prothiofos	R	R		C	0.05			P	0.05	C			0.01
Prothoate			R							C			
Propylenethiourea													
Praclofos				C	0.05								
Pyrazon										C			
Pyrazophos	R	R	R	C	0.02			C	0.05	C		R^{NPD}	
Pyrethrins	R	R									R^{ECD}	R^{ECD}	
Pyridaphenthion										C			
Quinalophos	R	R		C	0.03			C	0.03	C			
Quintozene	R	Coll.		C	0.02			P	0.02	C		R^{ECD}	
Quizalotop ethyl ester										C			

TABLE 2 Continued

Pesticide	Eur Std CEN-M	Eur Std CEN-N	Eur Std CEN-O	Eur Std CEN-P	Sweden Anal	Sweden LOQ	Netherlands Anal	Netherlands LOD	Canada Anal	Canada LOD	PAM-I Meth. 302a (No cleanup)	PAM-I Meth. 302b (Florisil cleanup)	USDA-PDP MS screen	Calif Anal	Korea	Japan (MS)	Australia (Victoria) LOD (fruit)
Rabenzazole		R															
Resmethrin		R															
Salithion		R															
Schradan									P	0.02	C						
Secbumeton									C	0.02							
Simazine	R	R			C	0.01			C	0.02	C			R^{ECD}			0.01
Simetryn									C	0.02	C						
Strobane T		R															
Sulfallate									P	0.02	C			R^{ECD}			
Sulfotep	R	R		R	C	0.01			P	0.03	C						
Sulfur														R^{ECD}			
Sulphenone											C						
Sulprofos		R							C	0.07	C			R^{FPD}			
Sulprofos-O-Analog sulfone											C						
Sulprofos sulfone											C						
Sulfopros sulfoxide											C						
2,4,5-T																	
2,3,5,6-TCA					C	0.05											
TCMTB									P	0.03	C						
2,3,4,5-TCNB		R			C	0.05											
Tebuconazole											C				R^{NPD}		
Tebufenpyrad																	0.01
Tecnazene	R	R			P	0.01			P	0.03	C				R^{ECD}		
Teflubenzuron							R^{BPU}	0.02									
Temephos				R										R^{FPD}			
TEPP				R	C	0.01					C						
Terbacil	R	R							C	0.05	C			R^{ECD}			
Terbufos	R	R							P	0.02	C			R^{FPD}	R^{NPD}		
Terbufos-O-analog											C						
Terbufos-O-analog sulfone											C						
Terbufos sulfone											C						

Terbumeton								C	0.02	C			
Terbuthylazine				C	0.05			C	0.02	C			
Terbutryn	R	R		P	0.1			C	0.02	C		R^{NPD}	
Tetrachlorovinphos	R	R	R	C	0.05			C	0.07		R^{FPD}		
Tetradifon	R	Coll.		C	0.02			C	0.03	C	R^{ECD}	R^{ECD}	
Tetramethrin	R	R								C	R^{ECD}		
Tetramethrin 1								C	0.02				
Tetramethrin 2								C	0.02				
Tetrasul	R	R		C	0.1			P	0.02	C			
Thiabendazole				C	0.3	R^{BZM}	0.05			C			
Thifensulfuron-methyl													
Thiobencarb								C	0.02	C	R^{ECD}	R^{ECD}	
Thiodicarb												R^{CRB}	
Thiofanox sulfone						R^{CRB}							
Thiofanox sulfoxide						P^{CRB}							
Thiometon			R							C		R^{NPD}	
Thionazin	R	R		C	0.01					C	R^{FPD}		
Thiophanate													
Thiophanate-methyl													
THPI										C			
Tolclofos-methyl	R	Coll.	R	C	0.02							R^{NPD}	
Tolyfluanid	R	R		C	0.05			C	0.16	C		R^{ECD}	
Toxaphene										C			
Tralomethrin										C			
Triadimefon	R	R		C	0.1			C	0.03	C	R^{ECD}	R^{ECD}	0.05
Triadimenol		R		C	0.02			C	0.02	C		R^{NPD}	
Tri-allate								P	0.02	C		R^{ECD}	
Triamiphos		R	R	C	0.04								
Triazophos	R	R	R	C	0.02			C	0.02	C	R^{FPD}	R^{NPD}	
Triazoxide		R											
Tribenuron-methyl													
Tributylphosphate											R^{FPD}		
Trichlorfon		R	R	C	0.05					C		R^{NPD}	
Trichloronat													
Trichloronat-O-analog	R	R	R	C	0.3					C			
2,4,6-Trichlorophenol				C	0.01								
Tricyclazole										C			
Tridiphane										C			

TABLE 2 Continued

Pesticide	Eur Std CEN-M	Eur Std CEN-N	Eur Std CEN-O	Eur Std CEN-P	Sweden		Netherlands		Canada		PAM-I Meth. 302a (No cleanup)	PAM-I Meth. 302b (Florisil cleanup)	USDA-PDP MS screen	Calif Anal	Korea	Japan (MS)	Australia (Victoria) LOD (fruit)
					Anal	LOQ	Anal	LOD	Anal	LOD							
Triflumizole											C						
Triflumuron							R^{BPU}	0.02									
Trifluralin	R	R							P	0.02	C			R^{ECD}	R^{ECD}		
2,3,5-Trimethacarb							R^{CRB}										
3,4,5-Trimethacarb																	
Triflusulfuron methyl ester											V						
Triphenylphosphate											C						
UDMH																	
Vamidothion				R											R^{NPD}		0.05
Vernolate									P	0.02							
Vinclozolin	R	Coll.			C	0.05			C	0.02	C			R^{ECD}	R^{ECD}		0.04
Vinclozolin metabolite B											C						
Vinclozolin metabolite E											C						
Vinclozolin metabolite F											R						
Vinclozolin metabolite S											V						

Codes: BPU, benzoylphenylurea; BZM, benzimidazoles; C, complete (>80% recovery); Coll., collaborative study carried out; CRB, prepared according to "Carbamate" branch of MRM; ECD, prepared according to electron capture detector branch of MRM; ELCD, prepared according to MRM branch ending with electrolytic conductivity detection; FPD, prepared according to MRM branch ending with flame photometric detection; LC, liquid chromatography; NPD, prepared according to MRM ending in nitrogen phosphorus detection; NR, not recovered; P, partial (>50% and <80%); PHU, phenylurea; R, recovered but no quantitative information available; S, small (<50%); V, variable (SD >20%).

TABLE 3 Detection Comparison Between the Combination of GC-FPD and GC-ECD and the Combination of GC-MSD and GC-AED

	Matrix	GC-FPD/GC-ECD, FPD, ECD	Concn. (ppm)	GC-AED, presumptive	Elements	GC-MSD (full scan) hit candidates	Chemists' choice, AED/MSD
1	Strawberries	Carbaryl	0.21	Captafol	N,S	Carbaryl	Carbaryl
		Iprodione	0.4			Iprodione	Iprodione
		Captan	4.7			Captan	Captan
						Myclobutanil	Myclobutanil
2	Potatoes	Chloropropham	3.71	Chloropropham	N,Cl	Chloropropham	Chloropropham
				Chlordimeform		3-Chloroaniline	4-Chloroaniline
				2, 3-Chlorophenoxy		4-Chloroaniline	
				Propionamide		Flutriafol	
						Amitraz	
						2-Methyl phenol	
3	Navel oranges	Ethyl chlorpyrifos	0.02	Ethyl chlorpyrifos	P,S,Cl	N.D.	Ethyl chlorpyrifos
				Quinalphos	P		Quinalphos
				2,3,5-Trichlorophenol	Cl		
4	Blackberries	Dicloran	0.1	Dicloran	N,Cl	Dicloran	Dicloran
		Carbofuran	1			Carbofuran	Carbofuran
		Phosalone	0.5	Phosalone	P,S,Cl	Phosalone	Phosalone
		Esfenvalarate	0.5	Esfenvalerate	N,Cl	Esfenvalerate	Esfenvalerate
				Fenvalerate	N,Cl	Fenvalerate	Fenvalerate
				Malathion	P,S	Malathion	Malathion
5	Kale	*p,p'*-DDE	0.03	*p,p'*-DDE	Cl	*p,p'*-DDE	*p,p'*-DDE
				Ethion	S		Ethion
				Acephate	N,S		Acephate
6	Tangerines	Ethion	0.02	Ethion	P,S	*o*-Phenylphenol	Ethion
				Folpt	N,S		Folpt
							o-Phenylphenol
7	Strawberries	Carbaryl	0.11	Captan	N,S,Cl	Carbaryl	Carbaryl
		Iprodione	0.56			Iprodione	Iprodione
						Myclobutanil	Myclobutanil
		Captan	1.13			Captan	Captan
8	Gala apples	Phosmet	0.05	Phosmet	S,P	Phosmet	Phosmet
		Diazinon	0.01	Diazinon	P,S		Diazinon
				Dicloran	Cl		Dicloran

plete recovery of acephate, whereas the Canadian MRM lists only partial recovery at much higher residue levels (Table 2).

4 CURRENT TRENDS AND THE FUTURE OF TRACE RESIDUE ANALYSIS

The world is becoming smaller and smaller through improved transportation and electronic communication. We share each other's cultures, foods, and even pesticide residues. There are international efforts to harmonize various analytical methods, especially those used for regulatory purposes. The Committee of European Normalization (CEN) [14] has made a great effort to compare and to harmonize MRMs. Their results show that it is not the harmonized method itself but the results that must be comparable in order for harmonized regulation to take place. An appropriate and valid performance evaluation scheme is needed to compare different MRMs and the variations that exist among similar MRMs.

The international regulatory communities demand that laboratories use a validated method recognized internationally [42], but the process of this validation is time-consuming (greater than 5 years) and carries an enormous cost. Except for a few original MRMs, most variations of MRMs are not validated, and it would be difficult to validate them in the future. Several guidelines have been developed to allow a laboratory to validate an analytical method in house and avoid a lengthy and costly collaborative method validation [42,43]. Such guidelines and a performance evaluation scheme would better harmonize pesticide residue methods and allow the comparison of results among laboratories worldwide. Finally, the revolutionary development of semiconductor technology—precise and reproducible control of heating, cooling, and mechanical movement of devices—has become not only possible but also affordable. Newer generations of gas and high performance liquid chromatographs equipped with a powerful desktop computer can make data acquisition, review, comparison, and tabulation faster, easier, and more accurate than ever before. The sorting and interpretation of data that used to take days can now be done with a keystroke in a split second. The comparison of data from different analytical systems can now be normalized and interpreted automatically.

The culmination of the technology is expressed in synchronized uses of gas chromatography coupled with atomic emission detection (GC/AED) and/or GC/MS through the retention time locking (RTL) software presented by Agilent Technologies [35]. By controlling the precise gas flow (via a computerized pressure control device), RTL allows two separate gas chromatographs to synchronize their retention times when they both operate under the same chromatographic conditions. The determination of over 500 pesticides and endocrine disruptors is possible by comparing data to the existing retention time database. The synchronized and simultaneous uses of an atomic emission detector, a specific elemental detector, and the mass spectrometer, a universal detector, could provide nearly unambiguous results.

In the California Department of Food and Agriculture (CDFA) laboratory, both GC/AED and GC/MS were linked using RTL software. The results from two linked instruments were compared with conventional GC MRM (CDFA MRM). Over 3 months, fresh fruit and vegetable samples were selected from the routine CDFA Residue Monitoring Program. Samples were prepared according to the CDFA MRM, and several samples containing incurred pesticide residues were chosen for subsequent assays using RTL-AED and MSD.

Table 3 shows a summary of results from GC/FPD and GC/ECD compared to GC/AED and GC/MS. The GC/AED instrument was operated to detect carbon, nitrogen, phosphorus, sulfur, chlorine, bromine, and fluorine. The GC mass spectrometer was operated in full EI scan mode. The retention time locking (RTL) software listed possible detections of pesticides based on comparison to the internal retention time library of 560 pesticides (relative to the reference compound, methyl chlorpyrifos). The presumptive findings were evaluated by comparing two detection systems and further confirmed by reviewing the supporting mass spectral evaluation.

The combined GC/AED and GC/MS screen showed the presence of more pesticides than the detections by conventional GC detectors using the same extracts. This does not necessarily mean that the combination of GC/AED and GC/MS is superior to the GC/FPD and GC/ECD combination. In fact, it is more than likely that an automated screen such as RTL allows objective analyses from two universal detection systems and removes any possible bias introduced by analysts who are using specific detectors. For example, small chromatographic responses or peaks with no recognized retention times are frequently ignored by analysts, whereas the automated RTL screen will flag any and all possible fits to the known library of information. A detailed study on RTL coupling of AED and mass spectrometer is published elsewhere [44].

What lies ahead for pesticide residue analysis includes developments in the miniaturization of mechanical devices through MicroElectroMechanical System (MEMS) [45]. Miniaturization of analytical devices allows for portable laboratories and near real-time analyses of samples. The continuing challenge for the future is not instrumentation but the development of representative sampling and sample preparation techniques.

REFERENCES

1. Codex Alimentarius. Joint Publication by the Food and Agriculture Organization of the United Nations (FAO) and the World Health Organization (WHO), Roma, Italia.
2. Code of Federal Regulations, Title 40, Parts 150–189. Office of the Federal Register. Washington, DC: Natl Arch Records Admin, 1997.

3. Farm Chemicals Handbook 2001, Willoughby, OH: Meister, 2001.
4. U.S. EPA. The Federal Insecticide, Fungicide and Rodenticide Act as Amended. US Environ Protect Agency. Revised 1988.
5. FDA. The Federal Food, Drug and Cosmetic Act, as Amended and related laws. US Food and Drug Admin. Revised 1989.
6. FDA. Pesticide Analytical Manual, Vol II, Index. Washington, DC: US Food and Drug Administration, 2000. Website: http://www.cfsan.fda.gov/~frf/pam2.html
7. Pesticide Use Reporting. Sacramento, CA: Calif Dept Pesticide Regulations, 1989–2000. Website: http://www.cdpr.ca.gov/docs/pur/purmain.htm
8. USDA. Pesticide Data Programs: Annual Reports. Washington, DC: USDA Agric Marketing Services, Sci Technol Div 1992–1999. Website: http://www.ams.usda.gov/science/pdp/index.html
9. AOAC. AOAC Official Method 985.22, Organochlorine and Organophosphorus Pesticide Residues—Gas Chromatographic Method (10.1.02). In: Official Methods of Analysis of AOAC International, 17th ed. Gaithersburg, MD: AOAC, 2000.
10. R-C Hsu, I Biggs, NK Saini. Solid-phase extraction cleanup of halogenated organic pesticides. J Agric Food Chem 39(9):1658–1666, 1991.
11. RG Luchtefeld. Multiresidue method for determining substituted urea herbicides in foods by liquid chromatography. J Assoc Off Anal Chem 70(4):740–745, 1987.
12. A de Kok, M Hiemstra, CP Vreeker. Improved cleanup method for the multiresidue analysis of N-methylcarbamate in grains, fruits and vegetables by means of HPLC with postcolumn reaction and fluorescence detection. Chromatographia 24:469–476, 1987.
13. Pesticide Analytical Manual, Vol I. Washington, DC: US Food and Drug Admin, October 1999. Website: http://www.cfsan.fda.gov/~frf/pamil.html
14. European Standard. Non-fatty foods—Multiresidue methods for the gas chromatographic determination of pesticide residues. Part 2: Methods for extraction and clean-up. Document prEN 12393-2. Brussels, Belgium: CEN (European Committee for Standardization), January 1998.
15. European Standard. Non-fatty foods—Multiresidue methods for the gas chromatographic determination of pesticide residues. Part 3: Determination and Confirmatory Tests. Document prEN 12393-3. Brussels: CEN (European Committee for Standardization), January 1998.
16. Rapport 17/98 Pesticide Analytical Methods in Sweden, Part 1. Uppsala, Sweden: Livsmedelsverket (Natl Food Admin), 1998.
17. General Inspectorate for Health Protection's Working Group for the Development and Improvement of Residue Analytical Methods. Analytical Methods for Pesticide Residues in Foodstuffs. 6th ed. Ministry of Public Health and Sport, The Netherlands, 1996.
18. Standard Operating Procedures of the Central Science Laboratory. Sand Hutton, York, UK: Dept for Environment, Food & Rural Affairs, 1998. Individual procedures as follows: a. M Duff. PGB/016: Ethyl Acetate Extraction of Pesticides from Fruit and Vegetables for the Multi-Residue Method. b. K Jackson. PGB-/086: Acetone Extraction of Pesticides from Fruit and Vegetables for the Multi-Residue Method. c. P Harrington. PGB-045: Determination of Polar Organophosphorus Pesticides in Fruit and Vegetable Samples by Gas Chromatography. d. T Griffiths. PGB-

048: Summary of the Procedures Used in the Multi-Residue Analysis of Pesticides in Fruit and Vegetables. e. R Fussell. PGB-068: Clean-Up Procedure Using High Performance Gel Permeation Chromatography (HPGPC). f. M Duff. PGB-035: Neutral Alumina Clean-Up Procedure for Gas Chromatography Mass Spectroscopy Using Ion Trap Detection. g. K. Jackson. PGB-042: Oxidation of Sulphides and Sulphoxides to Sulphones and Subsequent Analysis by Gas Chromatography.

19. Ottawa Pesticide Laboratory. Multiresidue Method for the Determination of Pesticides in Fruits and Vegetables by GC/MSD and HPLC Fluorescence Detection. Document P-RE-023-96(7.1)-FV. Ottawa: Lab Services Division of Agriculture and Agri-Food Canada, 1996.
20. M Luke. Pesticide multiresidue method using solid phase extraction with only two solvents, acetone and petroleum ether. Presented at 11th Annu Calif Pesticide Residue Workshop, Sacramento, CA, Mar 20–25, 1999.
21. Multiresidue Screening (CDFA Method). CDFA-CAC-MRM-1. Calif Dept Food and Agriculture, Center for Analytical Chemistry, Sacramento, CA, October 1999.
22. K Sasaki, Y Nakamura, T Ninomiya, T Tanaka, M Toyoda. Application of the bulletin method for rapid analysis of pesticide residues on the analysis of 10 pesticides notified in 1997. J Food Hyg Soc Jpn 39:448–452, 1998. (In Japanese.)
23. Y Akiyama, M Yano, T Mitsuhashi, N Takeda, M Tsuji. Simultaneous determination of pesticides in agricultural products by solid-phase extraction and gas chromatography-mass spectrometry. J Food Hyg Soc Jpn 37:351–371, 1996. (In Japanese.)
24. C Cook. Victorian Produce Residue Monitoring, Results of Residue Testing, 1997. ISSN 1039-3846. East Melbourne, Australia: Dept Nat Resources and Environment, 1998.
25. H-G Kang. Monitoring pesticide residues in agricultural products in Korea. Presented at 12th Annu Calif Pesticide Residue Workshop, Yosemite, CA, Mar 25–31, 2000.
26. Food and Nutrition Information Center, Natl Agric Library/USDA, Beltsville, MD.
27. SM Lee, AS Fredrickson, GR Hunter, TF Joe. Paper presented at 196th ACS Meeting, Los Angeles, CA, Agrochemicals Division, 1988, Abstr 0077.
28. WB Wheeler, NP Thompson, P Andrade, RT Krause. Extraction efficiencies for pesticides in crops I. [^{14}C]Carbaryl extraction from mustard greens and radishes. J Agric Food Chem 26:1333–1337, 1978.
29. SJ Lehotay. Pesticides in Nonfatty Foods Using SFE and GC/MS. A multilaboratory collaborative method for AOAC Official Methods. Gaithersburg, MD: AOAC Intl. In press.
30. SM Lee, ML Papathakis, H-M Feng, GF Hunter, JE Carr. Multipesticide residue method for fruits and vegetables: California Department of Food and Agriculture. Fresenius J Anal Chem 339:376–383, 1991.
31. J Fillion, R Hindle, M Lacroix, J Selwyn. Multiresidue determination of pesticides in fruits and vegetables by gas chromatography mass-selective detection and liquid-chromatography with fluorescence detection. J Assoc Off Anal Chem 78(5):1252–1266, 1995.
32. MA Luke, JE Froberg, HT Masumoto. Extraction and cleanup of organochlorine, organophosphate, organonitrogen, and hydrocarbon pesticides in produce for determination by gas-liquid chromatography. J Assoc Off Anal Chem 58(5):1020–1026, 1975.

33. JN Seiber. Determination methods. In: WG Fong, HA Moye, JN Seiber, JP Toth, eds. Pesticide Residues in Foods. New York: Wiley, 1999, pp 63–102.
34. G Zweig, J Sherman. Analytical Methods for Pesticides and Plant Growth Regulations, Vol VI, Gas Chromatographic Analysis. New York: Academic Press, 1972.
35. V Giarrocco, B Quimby, M Klee. Retention Time Locking: Concepts and Applications. Appl Note 228–392. Little Falls, VA: Hewlett-Packard, December 1997.
36. HA Moye. High performance liquid chromatographic analysis of pesticide residues. In: HA Moye, ed. Analysis of Pesticide Residues. New York: Wiley, 1981, pp 157–197.
37. RT Krause. Liquid chromatographic determination of N-methylcarbamate insecticides and metabolites on crops. I. Collaborative study. J AOAC Int 68(4):726–733, 1985.
38. MA Brown, ed. Liquid Chromatography/Mass Spectrometry: Applications in Agricultural, Pharmaceutical, and Environmental Chemistry. Washington, DC: Am Chem Soc, 1990.
39. W Jennings. Analytical Gas Chromatography. Orlando, FL: Academic Press, 1987.
40. KF Ivie. High performance liquid chromatography (HPLC) in pesticide residue analysis. In: G Zweig, J Sherma, eds. Analytical Methods for Pesticides and Plant Growth Regulators, Vol IX, Updated General Techniques and Additional Pesticides. New York: Academic Press, 1980, pp 55–78.
41. T Cairns, EG Sigmund, JJ Stamp. Mass Spectrom Rev 8:93–117, 1989.
42. AOAC® Official Methods℠ Program Manual: A Policies and Procedures Guide for the Official Methods Program OMA. Gaithersburg, MD, AOAC Intl, Website: http://216.55.34.180/vmeth/omamanual/omamanual.htm
43. Report of the AOAC/FAO/IAEA/IUPAC Expert Consultation on Single-Laboratory Validation of Analytical Methods for Trace-Level Concentrations of Organic Chemicals, 8–11 Nov 1999, Miskolc, Hungary. Vienna, Austria: FAO/IAEA, 1999.
44. SM Lee, CK Meng. The uses of retention time locked (RTL) of GC-MSD and GC-AED for the analysis of trace level pesticides in foods. Paper 12: Agrochemicals Division Symposium on Innovative Analytical Techniques for Residues in ppb Concentrations. 223rd ACS National Meeting, Orlando, FL April 7–11, 2002.
45. W Higdon. Silicon micro-machining technology. Presented at the 11th Annu Calif Pesticide Residue Workshop, Sacramento, CA, Mar 20–25, 1999.

7

Pest Management Issues on Minor Crops

Richard T. Guest*
IR-4 Project
Technology Centre of New Jersey
Rutgers University
New Brunswick, New Jersey, U.S.A.

Paul H. Schwartz
Agricultural Research Service
U.S. Department of Agriculture
Beltsville, Maryland, U.S.A.

1 BACKGROUND

The variety of agricultural commodities commonly referred to as "minor crops" is nearly limitless. They include the vegetables, fruits, nuts, herbs, and an ever-increasing variety of ethnic produce that are commonly found in the fresh foods section of the local supermarket. But minor crops also include commercially grown ornamental plants such as trees, shrubs, flowers, and turf grass that are products of the rapidly growing "green" industry. Minor crops suffer from the same pest depredations as the large-acreage major crops and often require specialized pest management practices, including pesticides, to produce a healthy, attractive, and nutritious product for the consumer.

Historically, producers of agricultural commodities have depended upon the agricultural chemical industry to provide them with safe and effective chemi-

* Retired.

cal pesticides that supplement their pest control practices in order to maintain crop yields and protect the health of animals. As the cost of meeting regulatory requirements has increased, pesticide registrants have concentrated their registration efforts in areas where they could obtain sufficient economic returns to justify their research and development costs. This resulted in greater registrations of pesticides for the large-acreage crops such as corn (72.6 million acres), cotton (10.7 million), soybeans (70.8 million), and wheat (59.0 million) [1]. Producers of fruits, nuts, vegetables, and specialty crops such as cranberries, flax, hops, mint, sunflowers, and ornamentals found that they had fewer and fewer pesticides available to them compared to growers of the major crops. When minor crops are considered individually, the acreage of most of them is relatively small. However, the combined acreage of these crops in the United States exceeds 11 million acres, which represents an annual value of more than $39 billion and accounts for 40% of all U.S. crop sales [2]. Twenty-seven states have minor crop sales exceeding 50% of their total crop sales (see Appendix 1). Among these are California with greater than $14.3 billion, Florida with $4.7 billion, Washington with $2.3 billion, Oregon and Pennsylvania with $1.4 billion, and Georgia with $1.0 billion.

There has been general agreement over the years that a minor use of a pest control product is any use for which the volume is insufficient to justify the cost to a commercial registrant to obtain a registration. This may relate to the general or frequent use of a product on a low volume crop, or it may apply to the infrequent or localized use of a product on a high volume crop. In either case the problem of obtaining clearances for the minor crop/minor use market is primarily one of economics. Traditionally, all crops except corn, cotton, soybeans, and wheat have been considered minor crops in the United States. However, recent legislation enacted by Congress clearly defines minor use in terms of crop acreage.

The recently amended Federal Insecticide, Fungicide and Rodenticide Act (FIFRA) defines the term "minor use" as any use of a pesticide on a commercial agricultural crop where the total U.S. acreage for the crop is less than 300,000 acres or the Administrator of the U.S. Environmental Protection Agency (USEPA) determines that the use does not provide sufficient economic incentive to support the initial or continuing registration of that pesticide [3]. The definition further states that the Administrator may determine that a minor use exists if there are insufficient alternatives available for use on the crop, that the alternatives pose greater risk to the environment or human health, or that the minor use pesticide plays a significant part in the management of pest resistance or in integrated pest management systems. A list of crops grown in the United States on less than 300,000 acres is shown in Appendix 2.

Limiting the acreage of a minor crop to less than 300,000 acres initially excludes certain crops that were formerly considered to be minor crops. These include, for example, sunflowers for seed, dry edible beans, white potatoes, sorghum, tomatoes, apples, grapes, almonds, and pecans (see Appendix 3).

Although federal legislation now contains provisions for expediting the registration and reregistration of pest control products for minor uses, the economics of obtaining initial registrations and retaining registrations through the reregistration process has been and will continue to be a serious threat to the production of an abundant and diverse supply of high quality commodities in the United States. The significant time and expense required to develop data to support the registration of new chemicals and to defend existing uses leave pest control producers fewer resources for minor use registrations.

This situation was exacerbated by the enactment of the 1988 amendments to FIFRA, which required that all pesticides and their uses registered prior to November 1984 be reregistered by the end of 1997. At that time, experts estimated that 25% of existing tolerances for pesticides registered for use on food crops would not be supported by their registrants. This was forecast to have particularly serious implications for growers of minor crops and for minor uses of pesticides on major agronomic crops. This scenario came to be known as the "minor use dilemma" and focused attention on the need to accelerate the development of pest management alternatives on minor crops.

Subsequently, the passage of the Food Quality Protection Act (FQPA), in August 1996, which amended FIFRA, contained provisions that will ultimately further limit the availability of pesticides for minor uses. The new law established a single health-based safety standard and required the USEPA to use up to an extra tenfold safety factor to protect infants and children. The act further required the USEPA to reassess all existing tolerances and exemptions for both active and inert ingredients within 10 years.

It is clear that conventional pesticides will continue to play a primary role in agricultural crop protection for both major and minor crops. It is equally clear that new chemicals being developed will need to address current environmental concerns. The agricultural chemicals industry is making significant strides in developing effective pesticides that exhibit greater safety for the environment and nontarget species and are generally used at very low rates of application. In many instances the industry is initiating research to include these products for use on a variety of minor crops in addition to the more lucrative major crop markets. In other cases, industry, crop producers, and the public sector are forming partnerships to extend registrations to minor crop markets where distinct environmental benefits exist compared to currently registered pest control products.

Although still a very small segment of the commercial pest management industry, biological pest control agents, including microbial and naturally occurring biochemicals such as pheromones, are increasingly attractive alternatives to conventional pesticides. Collectively known as biopesticides, they generally exhibit a high degree of safety, low environmental impact, and excellent compatibility with integrated pest management (IPM) programs. However, they tend to be very selective in their spectrum of pests controlled. This often results in a low volume of use, which is unattractive for commercial development, particularly for minor crops, despite their typically lower registration costs.

It has long been recognized that public sector research is needed to complement the private sector in providing for safe and effective pest management. This has been especially true for minor crops because of the economic considerations of registering pesticides for low volume uses. Consequently, in 1962 the State Agricultural Experiment Station (SAES) directors responded to grower needs for assistance in the area of minor crop pesticide registrations and asked the U.S. Department of Agriculture's (USDA) Cooperative State Research Service (CSRS), now known as the Cooperative State Research, Education and Extension Service (CSREES), to initiate an interregional research project to coordinate research activities within the agricultural community to obtain registrations for minor use needs. This project, which has become known as the IR-4 Project, was established in 1963 and encompasses the following objectives:

To obtain minor and specialty use pesticide clearances and assist in the maintenance of current registrations

To further the development and registration of microbial and specific biochemical materials for use in pest management systems

Administered by CSREES and funded by both CSREES and the Agricultural Research Service (ARS) and with the cooperation of state land grant institutions, the agricultural chemical industry, the USEPA, and commodity organizations, IR-4 is the only public research program in the United States created to assist with the registration of pest control agents for minor uses. Figure 1 shows the relationships among the elements of IR-4.

The role of the IR-4 Project is that of expanding existing pesticide product labels to include minor crop uses. In order to do this, IR-4 gathers information on pest management needs for minor crops, including fruits, vegetables, and ornamentals; develops priorities to address the most important uses first; coordinates and funds both field and laboratory research among state and federal scientists; and prepares and submits appropriate tolerance and registrant documents to the USEPA. All research conducted by IR-4 on food crops is compliant with good laboratory practice (GLP). IR-4 works cooperatively with pesticide registrants in order to access, by letter of authorization, the basic registration information used to support major crop registration. To accomplish this task, IR-4 interacts with the crop producers to ensure that research and registration programs are relevant to current needs, with the USEPA to ensure that there are no major impediments that could unduly delay registrations, and with the agricultural chemicals industry to ensure that the intended uses will be commercially registered and offered for sale.

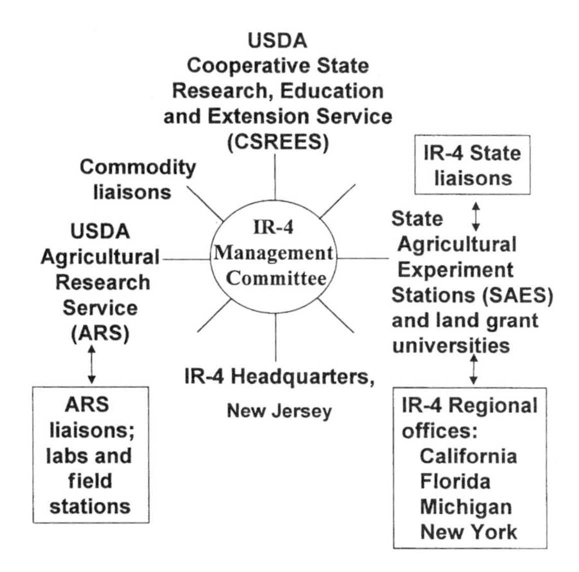

FIGURE 1 Relationships of the various entities in the IR-4 program.

The IR-4 is justifiably proud of its accomplishments in assisting minor crop producers with registration needs. As shown in Table 1, IR-4 has contributed significantly to food and ornamental pesticide clearances and to the advancement of registrations for biological pest control materials. IR-4 also has met the challenges of FIFRA-88 by supporting minor use registrations that would have other-

TABLE 1 Progress of the IR-4 Project

Project accomplishment—pesticide clearances 1963–1998	
4745	food crop clearances
5142	ornamental clearances
107	biopesticide clearances
FIFRA 88 responses	
Reregistered 700 minor uses on food crops	
Reregistered over 2000 ornamental uses	
Obtained 10 biopesticide tolerance exemptions on 56 crops	
FQPA responses	
1997:	45 reduced risk studies out of 150 total studies
1998:	78 reduced risk studies out of 163 total studies
1999:	82 reduced risk studies out of 139 total studies

Source: Ref. 22.

wise been lost in the reregistration process and the FQPA by focusing its research efforts and the registration of pesticides classified by the USEPA as reduced risk products. In fact, nearly 45% of all recent pesticide research projects sponsored by the IR-4 averaged over a 3 year period involve reduced risk pesticides.

Because research for the purpose of establishing the registration of a pest control product is generally beyond the purview of state and federal agricultural scientists as well as most commodities-based organizations, there is abundant opportunity for these segments of the agricultural community to interact with the IR-4 Project to identify needs and participate in the research process with supplemental funds available from the IR-4. More often, research carried out by state agricultural experiment station and federal research scientists forms the basis of new and innovative pest management systems and techniques that benefit minor crop producers. Information developed by these scientists often applies directly to nonchemical control methods that may supplement or in some instances replace traditional pesticide-based control measures or that may identify integrated pest management strategies that require the new registration of a pesticide or biologically based product to achieve implementation. In such instances, public and private sector scientists work closely together with the IR-4 to respond to these research needs.

The issue of pest management on minor crops is not limited to the United States. The Federal Republic of Germany convened a symposium in 1993 to study the issue of expanded pesticide labeling to include off-label crops and to explore ways to harmonize the use of pesticides among the European Community nations. A discussion of the need for expanded pesticide labeling for pesticides on minor crops was included in a pesticide residue workshop in Tokyo in 1996. In addition, the Canadian government established a minor use program that works closely with the USDA's minor use program. These countries, together with Mexico, are concerned with the need for properly labeled safe and effective pesticides for use on low volume crops or for the occasional use on major crops where pest outbreaks are sporadic or geographically limited.

Through the USDA's IR-4 Minor Use Program, the United States has joined efforts with Germany's Federal Biological Research Centre for Agriculture and Forestry and with the Canadian minor use program to sponsor research programs and share research information on projects of mutual interest. Testing protocols and good laboratory practice compliance procedures have been implemented to enable the exchange and use of data by the respective regulatory agencies. There are presently about 20 cooperative projects under way with the Canadian government and several additional research projects involving Germany and Mexico. Data resulting from these trials will be combined and used by the respective countries to support new pesticide labels.

Pest management on minor crops is clearly a global problem. Although different countries are approaching the issue in varying ways, the growing trend toward international cooperation will hasten the registration of newer and safer pest control products for a variety of minor crops while benefiting growers by reducing the associated research and development costs. This approach will likely expand to include partnerships with the agricultural chemicals industry as the availability of reduced risk products increases.

2 IMPORTANCE OF ECONOMIC LOSSES

Crops and livestock are attacked by about 50,000 species of fungi that cause more than 1500 diseases. About 15,000 species of nematodes attack crop plants, and more than 1500 of these cause serious damage. More than 10,000 species of insect pests cause losses to crops, livestock, forests, structures, and stored products. About 600 species of insects cause heavy losses to crops each year. About 30,000 species of weeds compete with crops, and about 1800 of these cause serious losses each year. Losses caused by insects, plant pathogens, nematodes, and weeds continue to reduce the maximum potential yield of crops grown throughout the world. In the United States, preharvest losses to pests have been estimated at about 37%, with insects accounting for 13% of the losses and plant pathogens and weeds each accounting for 12%. Postharvest losses to pests are estimated to be about 9%. These losses occur despite the fact that good agricultural practices with pest control technologies are followed [4].

Research on pests is a very important component of the budgets of the universities, SAES, and the USDA. In 1997, federal funds from the USDA and federal plus non-federal funds were about $174 million and $407 million, respectively, to support research on pests [5]. Table 2 indicates the distribution of these funds.

Justification for expenditures of research dollars in the public and private sectors on pests and their control is based in part on losses caused by pests, acreage of the crop grown, and the extent of pesticide use. To some extent, the magnitudes of pest losses also influence what studies an investigator will undertake and the ability to obtain increases in research budgets.

TABLE 2 Allocation of Federal Funds in Fiscal Year 1997 to Pest Control Research

	Funds ($million)	
Program	Federal	Total
Control of fruit and vegetable pests	80.5	185.8
Control of field crop pests	93.3	220.8

Estimates of crop losses can also assist in determining the constraints in crop production that may be overcome by the application of more expensive technologies in integrated pest management (IPM) programs. Some of these technologies may require considerable research and development expenditures before they are ready for commercialization. Loss estimates are useful in estimating the effects of pest density on yield. These data are most often used to construct equations or mathematical models to predict losses for various pest densities. This information can then be used in the decision-making process of when or when not to apply measures to reduce the pest population density to avoid an economic loss and is useful in the development of pest management programs. An economic loss is defined here as the production value of the estimated loss of the commodity as a result of a pest infestation. These data help provide some insight as to the capacity of the pest to cause a loss and the conditions under which that loss occurs. They also can serve as a gauge to measure the effectiveness of different pest control measures.

Crop losses are not the major driving force behind the decisions of pesticide manufacturers to develop and label new pesticides or expand the labels of existing pesticides. These decisions are based primarily on the market size, which is governed by the acreage of the crop grown, the number of pesticide applications to control the pest, and crop liability in the case of product failure or crop destruction. Pesticide manufacturers generally target their products for the major crops such as corn, cotton, soybeans, and wheat. These four crops were planted to 213 million acres in 1998, which accounted for about 68% of the total cropland harvested for food and feed crops in the United States, whereas commercial acreage of fruits and vegetables was about 7 million acres. Development of data to register minor uses is generally left up to publicly funded programs such as the IR-4 program in the United States and similar programs in Canada and Germany.

The basic philosophy of managing pests to prevent or reduce losses is different for each of the major categories of pests. With few exceptions, the way weeds cause major losses in crops is by interference [6]. This includes weed competition with the crop for environmental factors contributing to plant growth such as light, moisture, and nutrients. Allelopathy plays a role in some species of weeds. Therefore, the major strategy to prevent losses from weeds has been to eliminate the weeds from the crop environment by either mechanical or chemical methods or a combination of the two. More recently, genetic engineering has come to play a significant role in weed control for the major crops through the introduction of herbicide-resistant genes. Sethoxydim, glyphosate, and glufosinate-ammonium are some of the herbicides used with transgenic crops such as cotton, corn, and soybeans. Approximately one-third of U.S. soybean acreage was planted to the Roundup Ready variety in 1998. The future trend will be to have more acreage planted to transgenic plants in the major crops for weed control. However, public resistance to this new technology may delay its application on a large scale for the minor use food crops.

Diseases, insects, and nematodes, on the other hand, are dependent on the host plant at some stage in their life cycles. These pests cause losses that can be attributed to a parasitic relationship. For the most part, these organisms are held in check by biotic factors. It is only when pest outbreaks occur or are likely to occur that pest control measures are applied. There are times, particularly in the case of insects, when pest control measures cause outbreaks of other pests. The insect pathogen *Bacillus thuringiensis* was introduced into corn, cotton, and potatoes in 1995 and 1996 and offers an effective way of controlling lepidopterous pests with minimum disruption to beneficial insects. The predominant method to prevent disease and nematode losses has been and continues to be the use of resistant cultivars and the treatment of seed with fungicides, with the use of fungicides and nematicides as preventive or curative measures.

The per-acre value of the crop is an important consideration when methods to control pests to reduce losses are considered. Vegetables, fruits, and nuts are worth about 3.5–16-fold more in value per acre than cotton, corn, soybeans and wheat. (See Table 3.)

It is worthy of note that losses to minor crops represent a much greater value than do losses to major crops at the same percentage reduction in yield. This most likely influences the degree of acceptance of losses and the extent to which control measures are applied to prevent or reduce losses. It also influences to some extent the crops that pesticide registrants will add to their labels because of the liability incurred if crop damage or product failure occurs.

The USDA National Agricultural Statistics Service (NASS) periodically conducts surveys to determine pesticide usage on various commodities. Their surveys conducted in 1996 and 1997 for vegetables and fruits, respectively, show that a high percentage of acres is treated for most of these crops (Appendixes 4

TABLE 3 Crop Value per Acre, 1996–1998

Crops	Production value per acre ($)	Acres (millions)
Vegetables, fresh	4535	1.9
Vegetables, processed	949	1.4
Fruits	3326	3.2
Nuts	1458	1.2
Miscellaneous minor crops	537	1.2
Major crops	272	247.6

Source: Refs. 1, 2.

and 5). On average, insecticide acreages are the highest, herbicides next, followed by fungicides. The percentage of acres treated ranges from 52% to 72% for vegetables and from 75% to 77% for fruits. Multiple applications of insecticides and fungicides are generally used to maintain the quality that the U.S. consumer is used to and expects in the marketplace and to meet the marketing standards of the Agricultural Marketing Service (AMS) and USDA and Food and Drug Administration standards for pest parts in processed foods.

In the United States, there has been no attempt to develop comprehensive national data on losses for all pest classes since the publication of the USDA handbooks on losses in 1951 [7] and 1965 [8]. The Weed Science Society conducted a survey in 1979 on the percentage average annual losses due to weeds for 64 crops [9]. Their data suggest that loss of potential production can range as high as 20% for fruits and vegetables. Sometimes losses can be so severe that growers are compensated by the Farm Service Agency of the USDA. For instance, in 1996 and 1997 growers were paid about $9 million each year for losses due to Kamal bunt fungus [10].

In 1988, an extensive survey was conducted in North Carolina to estimate losses caused by plant diseases and nematodes [11]. Losses for the vegetables and fruits and nuts categories were estimated at 24.7% and 22.8%, respectively, while losses for field crops were estimated at 14.9%. Overall, the losses to crops in North Carolina attributed to diseases and nematodes in 1988 were $500.1 million for crops valued at $3.3 billion. These losses accounted for 15% of the economic value of the crops. For vegetables and fruits and nuts, the economic losses were much higher, representing 32.7% and 29.5%, respectively.

The following provides specific information on pest losses as compiled and reported by the Pesticide Impact Assessment Program (PIAP) and reported in their Crop Profiles page on the Web.

According to the crop profile for walnuts [12], California produces 99% of the walnuts grown in the United States on approximately 177,000 acres. Production averages about 235,000 tons/year and was valued at $314 million in 1995. Approximately 60% of the acreage is susceptible to damage by the codling moth and requires from one to three treatments per year. If uncontrolled, damage can exceed 40%. Other pests of walnuts grown in California and the potential for damage are listed in Table 4.

Mushrooms are Pennsylvania's largest cash crop, with a farm gate value of $272 million in 1996, and account for 45% of the nation's total production [13]. Scarid fly larvae can limit the yield of mushrooms by as much as 70%, whereas Phorid flies cause crop losses as vectors of certain mushroom pathogens. Losses from viral epidemics can range from 10% to 100%. Bacterial blotch, which causes a discoloration of the mushroom cap, reduces the crop value by 30–80%. Fungal diseases also take their toll of mushroom yields. Trichoderma green mold currently causes losses of 5–10%, but when it was at its worst it

Table 4 Insect and Disease Pests of Walnuts[a]

Pest	Percent yield loss
Insect	
Navel orangeworm	20
Walnut husk fly	50
Walnut aphid	25
Mites (two-spotted, European red, Pacific)	25
Armored scales, walnut, San Jose	10
Soft scales, frosted, European fruit Lecanium	10
Fall webworm	20
Redhumped caterpillar	5
Diseases	
Amillaria root rot	25
Crown gall	50
Phytophthora root and crown rot	50
Walnut blight	50–70
Nematodes	
Lesion, ring, and root knot	50–nearly complete

[a] Weeds cause serious problems in walnut production, but no percent loss was provided.

caused losses of 20–80%. Verticilium spot and dry bubble routinely cause crop losses between 15% and 60%, at times reaching 100%.

3 PEST MANAGEMENT STRATEGIES FOR MINOR CROPS

A great deal of concern has been expressed over the past decade about the effects of federal pesticide legislation on the continued ability of minor crop farmers to produce quality products. Initially, the 1988 amendment to FIFRA focused attention on the vulnerability of the minor crop industry to an inadequate supply of safe and effective pest control products. National surveys and workshops conducted by the IR-4 Minor Use Program in 1989 suggested that about 1000 minor use registrations important to the agricultural community would be lost because of the economics associated with the cost of reregistration for minor crops [14].

Losses of pest control agents of this magnitude in the minor use market, together with associated losses in the major crop market, were forecast as having significant effects on U.S. agricultural production. The most profound of these would be a substantial increase in the cost of foods, which would result in the greatest hardship being borne by the lower income population. Along with increased costs there would be decreased quality of produce. When coupled, these factors would lead to reduced consumption of U.S. produce with a concomitant increase of imported foods. Fortunately, IR-4 researchers, with additional funding provided by Congress and with the help and cooperation of state and federal researchers, private industry, and commodity producers, were able to present data to defend about 700 of the most important minor crop reregistrations.

The 1996 Food Quality Protection Act (FQPA) presented additional and more complex challenges to the minor crop industry. In addition to establishing new standards to protect the health of the public in general and that of infants and children in particular, the act also contains provisions that will encourage private sector registrations for minor crops. These provisions will likely focus increased commercial interest on minor use clearance needs. Nevertheless, the reassessment of upward of 10,000 tolerances by 2006 will result either in certain uses being voluntarily canceled or in the USEPA mandating additional exposure data. This will increase the cost to registrants to maintain these registrations, which will be passed on to the consumer in the form of higher costs for fresh fruits and vegetables. Moreover, because minor crops utilize a disproportionately greater percentage of the risk cup,* the economic disincentive inherent to minor uses will be further exacerbated. It is clear that many, if not all, of the nearly 400 minor crops grown in the United States will be affected and will require either additional data to support registration of pesticides for minor crops or the registration of new lower risk pest management alternatives.

It is likely that the "green" industry will be similarly affected by FQPA. Any decisions resulting in the decreased availability of pest control products for food crops will affect the availability of products that are registered for application on nursery, floral, and forestry crops and turf grass. Conversely, new safer pest control products developed for use for the major crop market will, in all likelihood, be available for use on nursery, floral, and turf crops.

Clearly, some of the older pesticide products will be lost to the minor crop market. Fortunately, many new products have been or will be introduced that will provide effective and environmentally safe pest management. One of the provisions of the FQPA is that of mandating the USEPA to expedite the review of "safer" or reduced risk pesticides. The law requires that the USEPA develop criteria and procedures for expediting the consideration of applications for safer pesticide products that will enhance public health and environmental protection, thus helping them to reach the marketplace more rapidly as replacements for older and potentially riskier chemicals.

---

* The USEPA establishes the total level of acceptable risk from the lifetime exposure for each pesticide, which is represented by the pesticide's population-adjusted dose. This is commonly known as the "risk cup."

Expedited review is clearly a powerful incentive that has encouraged the agricultural chemicals industry to explore new and safer pesticide chemistries and will continue to do so. Although these efforts will focus primarily on the major crops, minor crop producers will also benefit.

In 1997 the IR-4 Project adopted a risk reduction strategy to accelerate the registration of new pest control products for both minor food and ornamental commodities. This strategy promised to

1. Promote reduced risk pest management for minor crops
2. Develop risk mitigation measures for existing pesticide registrations
3. Assist with the registration of biologically based pest control products for minor crops
4. Register and maintain pesticides essential to IPM systems

Utilizing established partnerships with agricultural chemical companies and crop producer organizations, IR-4 is moving forward rapidly to target products that are eligible for the USEPA's Reduced Risk classification. In 1999, over 60% of IR-4 food use research was focused on reduced risk pesticide chemistry, with the goal of $2^1/_2$ years from project initiation to the submission of the tolerance petition to the USEPA. To further expedite the search for safe and efficacious products, it is likely that IR-4 will need to expand its research program beyond its traditional objective of GLP-compliant residue testing to non-GLP performance evaluations to assess the value of new products for specific minor (and, particularly, ultraminor) crop pest management needs.

Working closely with product managers at the agricultural chemical companies, IR-4 has developed and continually updates a list of new products and recent introductions that may be useful to minor crop producers. A discussion of these products is presented to state and federal research scientists, extension personnel, and crop producers at annual minor use workshops in order to elicit broad-spectrum input into the usefulness of the products.

Some newer products that may be beneficial for minor crop pest control are listed in Table 5. It is important that all producers of minor crops become aware of the registration status of pesticide products used in their pest management programs by contacting state agricultural extension service and pesticide manufacturer representatives. Although registration objectives are generally directed toward the more lucrative large volume crops, pesticide manufacturers are more aware of the minor crop market than in the past and are more likely to direct developmental research efforts in this direction. The IR-4 Minor Use Program can be of significant value in assisting with the registration of new products. Because it generally requires 3–5 years from initiation of research to labeling, it is important that minor crop producers initiate early clearance requests through their state IR-4 representative.

TABLE 5 Pest Management Products with Potential for Use on Minor Crops

Common name	Trade name	U.S. manufacturer	Action[a]	Use	Comments
Abamectin	Avid, Clinch, Zephyr	Novartis	I, M	Colorado potato beetle, mites, pinworms, leafminers	IPM pesticide
Ampelomyces quisqualis	AQ-10	Ecogen	F	Powdery mildew	Biopesticide
Azadiractin	Azatin, Bullwhip	Thermo Trilogy	I	Insect IGR	Biopesticide
Azoxystrobin	Abound, Quadris	Zeneca	F	Broad-spectrum fungicide	Reduced Risk candidate
Bacillus sphaericus	Vectolex	Abbott	I	Effective against *Culex* larvae	Biopesticide
Bacillus subtilus	Kodiak, Quantrum	Gustafson	F	*Rhizoctonia, Fusarium*	Biopesticide
Bacillus thuringiensis, various strains	Various	Many	I	Lepidopterous insects	Biopesticide
Beauveria bassiana strain TBI	Naturalis	Troy Biosciences	F	Turf and ornamental diseases	Biopesticide
Beauveria bassiana strain GHA	BotaniGard, Mycotrol	Mycotech	F	Ornamental turf, nursery disease	Biopesticide
Bifenzate	Flomite	Uniroyal	M	Spider mites on ornamentals	Reduced Risk registration
Buprofezin	Applaud	AgrEvo	I	Chitin inhibitor	IGR
Carfentrazone-ethyl	Aim, Shark	FMC	H	Postemergence broadleaf weed control	Reduced Risk registration
Chlorfenapyr	Alert, Pirate	Am Cy	I, M	Broad-spectrum contact and stomach poison	Low use rates
Cyprodinil	Chorus, Unix, Vangard	Novartis	F, ST	Broad-spectrum fungicide	Reduced Risk candidate
Diflufenzopyr	Distinct	BASF	H	Postemergence annual and perennial weed control	Reduced Risk candidate

Emanamectin benzoate	Proclaim, Strategy	Novartis	I	Controls larval lepidoptera	IPM potential
Fenbuconazole	Enable, Govern, Indar	Rohm & Haas	F	Broad-spectrum systemic fungicide	Low use rates
Fenhexamid	Elevate	Toman	F	Protectant fungicide	Reduced Risk candidate
Fipronil	Regent	Rhône-Poulenc	I	Broad-spectrum systemic insecticide	Low use rates
Fluazinam	Shirlan	Zeneca	F	Broad-spectrum fungicide	Acaricidal activity
Fludioxonil	Celect, Maxim	Novartis	F	Seed treatment with foliar disease control potential	Low toxicity
Flumiclorac-phenyl ester	Resource	Valent	H	Postemergence herbicide	Reduced Risk candidate
Fluthiacet-methyl	Action	Novartis	H	Postemergence broadleaf herbicide	Low use rates
Flutolanil	Folistar, Prostar	AgrEvo	F	Systemic fungicide	Low use rates Low toxicity
Gliocladium virens	SoilGard	Thermo Trilogy	F	Microbial soil fungicide	Biopesticide
Halosulfuron-methyl	Manage, Permit	Monsanto	H	Pre- and postemergence control of broadleaf weeds, nutsedge	Biopesticides
Harpin protein	Messinger	Eden Bioscience	F	Systemic control of certain bacterial and fungal disease	Biopesticide
Imazamox	Raptor	American Cyanamid	H	Annual grasses, broadleaf weeds	Reduced Risk candidate
Imidacloprid	Admire, Provado	Bayer	I	Seed, soil, and foliar use against many insects	IPM potential
Indoxacarb	Avaunt, Steward	DuPont	I	Controls lepidopterous insects	IPM potential; low use rates
Isoxaflutole	Balance	Rhône-Poulenc	H	Preemergence control of annual grasses, broadleaf weeds	Low use rates

Table 5 Continued

Common name	Trade name	U.S. manufacturer	Action[a]	Use	Comments
Kaolin	Suprex, Nuflo	Engelhard	I/M	Controls insects and mites	Biopesticide
Methoxyfenozil	Intrepid	Rohm & Haas	I	Controls lepidopterous larvae	Reduced Risk candidate; IPM potential
Methyl iodide		MIF Partners	F/I	Contact fumigant	Possible methyl bromide replacement
Milsana		KHH Bio Sciences	F	Unique mode of action	Biopesticide; IPM potential
Mycobutanil	Nova, Rally	Rohm & Haas	F	Many fungi	Low use rates
Myrothecium verncaria	DiTera	Abbott	N	Biological nematicide	Biopesticide
Oxasulfuron	Dynam, Expert	Novartis	H	Postemergence broadleaf weed control	Low use rates
Primicarb	Pirimor, Rapid	Zeneca	I	Aphicide, labeled on many crops	IPM potential
Paecilomyces fumosoroseus	PRF 97	Thermo Trilogy	I	Various insects on ornamentals	Biopesticides
Potassium bicarbonate	Armicarb, Kaligreen	Church & Dwight	F	Powdery mildew	Biopesticide
Prohexadione-calcium	Viviful, Apogee	BASF	PGR	Unique mode of action	Reduced Risk candidate
Propiconazole	Tilt, Orbit, Banner	Novartis	F	Systemic and eradicative properties	Low use rates
Prosulfuron	Peak	Novartis	H	Postemergence broadleaf weed control	Low use rates
Pseudomonas cepacia, Wisconsin strain	Intercept	Soil Technologies	F/N	Nematicides, soilborne diseases	Biopesticide
Pymetozine	Fulfill	Novartis	I	Controls sucking insects	Reduced Risk candidate
Pyridaben	Pyramite	BASF	I/M	Residual control of various insect mites	Reduced Risk candidate

Pyriproxyfen	Knack, Nemesis	Valent	I	IGR effective against wide range of insects	IPM potential
Rimsulfuron	Matrix, Shadeout	DuPont	H	Controls annual grasses, broadleaf weeds	Reduced Risk candidate
Sodium tetrathiocarbonate	Enzone	Entex	F/I	Contact fumigant	Possible methyl bromide replacement
Spinosyn	Success	Dow	I	Fermentation product	Reduced Risk candidate
ASpinosynD	Naturalyte, Tracer	Dow	I	Fermentation product	Reduced Risk candidate
Spodoptera exigua	Spod-X	Thermo Trilogy	I	Controls beet armyworm	Biopesticide
Sulfentrazone	Authority	FMC	H	Controls broadleaf weeds, grasses	Low use rates
Tebuconazole	Elite, Folicur	Bayer	F	Systemic broad-spectrum fungicide	Low use rates
Tebufenozode	Elite, Folicur	Bayer	F	Systemic broad-spectrum fungicide	Low use rates
Thiamethoxam	Cruiser, Actara	Novartis	I	Systemic broad-spectrum insecticide	Reduced Risk candidate
Tralkoxydim	Grasp	Zeneca	H	Postemergence control of grasses	Reduced Risk candidate
Triazamate	Aphistar	Rohm & Haas	I	Foliar and subterranean control of aphids	IPM potential low use rates

[a] I = insecticide; IGR = insect growth regulator; M = miticide; F = fungicide; H = herbicide; N = nematicide; PGR = plant growth regulators; ST = seed treatment.

4 IMPEDIMENTS TO EFFECTIVE PEST MANAGEMENT

To reduce or eliminate pest damage at minimal cost to the grower, there must be an arsenal of weapons to fight the destruction caused by insects, nematodes, plant pathogens, weeds, and other pests. Research and the communication of research results to the grower are the keys to developing and improving methods to fight pests. Minor crop growers generally do not fare well in obtaining research for minor crops. Upon examination of the federal dollars allocated to federal agencies and the state agricultural experiment stations to conduct research on the various commodity groups, it is apparent that minor crops receive about 40% as many dollars as are allocated to major crops. The research dollars for corn alone approximate the dollars allocated to research on any one of several commodity groups such as fruits and nuts, citrus/subtropical fruits, or vegetables. These commodity groups may contain upward of 25 or more crops per group. Table 6 summarizes the research dollars allocated by the public sector for various commodity groups in fiscal year 1997 [15]. A similar picture exists for total funds, which include federal and non-federal funds allocated by the public sector.

Approximately equal amounts of public sector dollars were used in fiscal year 1997 for research to control pests of minor crops ($186 million) and major crops ($221 million). This represents about one-half the research funds allocated to minor crops. The bulk of these funds are used in the public sector to support research on fundamental pest biology; nonpesticide control methods; toxicology, metabolism, and fate of pesticides; and other chemicals and economics. For example, USDA appropriations were approximately $298 million for pest research and control programs in FY 1997. Of this, approximately one-third was directed toward pesticide chemical research to improve pesticide use patterns, toxicology, pathology, metabolism and fate of pesticides, and economics [16]. About 2.8% of the funds ($8.3 million) was used for field studies and residue analysis of pesticides to support the registration of conventional chemical pesticides on minor crops through the IR-4 program.

Pests as defined in Table 7 by the Current Research Information System (CRIS) include insects, mites, snails, and slugs. Diseases include nematodes, weeds, and other hazards such as climatic extremes, birds, rodents, and other mammals. Field crops include corn, cotton, rice, soybeans, wheat, pasture, forage, and other major crops.

With respect to pest control dollars, the minor crops fared about as well as the major crops. Research on specific pests may be of benefit to more than one crop or group of crops. This is particularly true for pesticide chemicals, where the label specification for pests may apply to a number of different commodities for which there are tolerances for the pesticide product. The greater impediment to developing control methods is in research on nonchemical methods, where there is generally a significantly greater expenditure of dollars and resources to

Table 6 USDA Funding for Research on Crops in Fiscal Year 1997

Commodity	Research dollar appropriations ($million) Federal	Total
Citrus/subtropical	26.0	52.9
Corn	51.8	100.5
Cotton	40.0	73.1
Cotton seed	3.1	3.5
Forage crops	24.7	59.0
Fruits and nuts	48.5	117.2
Grains and sorghum	4.3	15.8
Miscellaneous new crops	13.4	30.0
Ornamentals and turf	15.5	60.2
Other small grains	11.0	30.6
Other oilseed crops	9.4	18.4
Pasture	6.4	11.3
Peanuts	8.0	18.1
Potatoes	17.3	39.2
Rice	8.3	23.4
Soybeans	34.0	84.4
Sugar crops	11.9	19.4
Vegetables	57.5	148.5
Wheat	36.4	83.4

Table 7 USDA Funding for Research on Pests in Fiscal Year 1997

	Research dollar appropriations ($million)			
Problem area	Federal		Total	
Control of pests of fruits and vegetables	38.7		76.2	
Control of diseases of fruits and vegetables	37.2		96.4	
Control of weeds of fruits and vegetables	4.6		13.2	
Subtotal fruits and vegetables		80.5		185.8
Control of pests of field crops	40.2		84.0	
Control of diseases of field crops	31.2		84.4	
Control of weeds of field crops	21.9		52.4	
Subtotal field crops		93.3		220.8
Total		173.8		406.6
Percent pest control on minor crops	53.9%		51.9%	

control specific pests than for the broader spectrum of action for chemical pesticides.

It is generally recognized that chemical pesticides are the main line of defense for pest control on minor crops. The data in Appendixes 4 and 5 illustrate the use of pesticides on fruits and vegetables. There are approximately 3 million acres each of fruits and vegetables. Of these, approximately three-fourths are treated one or more times with herbicides and insecticides and about one-half to three-fourths are treated with fungicides. This is a small portion of the total crop acreage compared to the major crops. Likewise, it represents a smaller amount of pesticide product sales to the registrant. These crops are also two to more than ten times as high in value per acre as the major crops (Table 3).

High value per acre can increase the liability for products and, coupled with low sales, can be a disincentive for registration of new pesticides. American Crop Protection Association member sales of pesticides for use on cropland for 1997–1998 in the United States was $7303.3 million. Corn, cotton, and soybeans accounted for 62.6% of these sales. If the other major crops are included, they account for a total of 82% of pesticide sales [17]. There are about 300 species of food and feed crops grown in the United States. Of these, 27 are major crops as defined as being grown on 300,000 or more acres [18].

Recent federal legislation is exerting a significant impact on minor crop pest management. The FIFRA was amended with the passage of the FQPA on August 3, 1996. A number of provisions of the act can potentially decrease the availability of pesticides for minor uses. The FQPA establishes a single health-based safety standard for pesticide tolerances and requires the USEPA to use up to an extra tenfold safety factor to take into account potential pre-and postnatal developmental toxicology and completeness of the data with respect to exposure to infants and children. The USEPA establishes the total level of acceptable risk from the lifetime exposure for each pesticide, which is represented by the pesticide's population-adjusted dose. This is commonly known as the "risk cup." Each use of the pesticide contributes a specific amount of exposure that adds a finite amount of risk to the cup. When the risk cup is full, new uses involving the establishment of tolerances are not permissible. Generally, minor crops use a disproportionately greater share of the risk cup than the major crops, which is a disincentive to pesticide manufacturers to seek registration for these uses. This situation is further aggravated by FQPA's requirement to consider aggregate exposures from all nonoccupational sources such as dietary intake, water, air, and residential and other uses for the pesticide in question. The effects of cumulative exposure to the pesticide and other substances with common mechanisms of toxicity as well as effects of in utero exposure and the potential for endocrine-disrupting effects must be considered where information is available. The act further provides for tolerance reassessment. Under the new law, the USEPA is required to reassess all existing tolerances and exemptions from tolerances for both active and inert ingredients within 10 years. Currently there are over 9000 tolerances in effect. One way to reduce the risk of pesticide use is to reduce or eliminate the number of tolerances on agricultural crops [3]. Because pesticides used on minor crops are typically uneconomical in terms of return on investment for the chemical companies and contribute disproportionately to the risk cup, they are likely to sustain the greatest impact from tolerance reassessment.

It is estimated that the implementation of the new act will result in the loss of a large number of minor uses. Of immediate concern is the possibility of losing the use of some of the older pesticides such as the carbamates and the organophosphates and critical uses of pesticides identified by the USEPA as B2 carcinogens. Some of the consequences of losing minor uses due to cancelation, lack of registration, or to the inability to register new uses can be briefly summarized as follows [14]:

1. Growers who need the use may be at an unfair advantage with growers of major crops for which viable alternatives are available. This applies to competition among growers within the United States and among U.S. growers and foreign competitors.
2. For some crops, growing areas or regions will change to avoid pests that lack adequate control methods. This could result in more foreign production of minor crops.
3. Financial hardships may occur or major adjustments may be required of individual growers who are unable to successfully compete in the marketplace owing to a lack of adequate pest control methods.
4. Production costs will increase owing to changes in horticultural practices to accommodate changes in pest control practices.
5. Pest-induced losses will increase, thereby decreasing the supply of the commodity or requiring additional acreage to obtain the same yield if adequate pest control measures were available.
6. Pest resistance will likely increase as a result of fewer choices of pesticides.
7. Options for use in IPM programs will be reduced.
8. Supplies of minor crops will diminish, become more costly, and move to foreign production. A result will be a less diversified diet for all U.S. citizens and a less nutritious diet for children and the economically disadvantaged.

The production of minor crops relies heavily on the use of chemical pesticides (Appendixes 4 and 5). Data requirements and the costs to obtain clearances for these uses has risen considerably since the enactment in 1989 of the requirement for compliance with GLP regulations for magnitude of the residue studies. As shown in Figure 2, the cost to obtain clearances increased over the 11 year period 1989–1999. The program budget rose about fourfold, while the number of trials conducted rose about twofold during the same period. Approximately 90–94% of the funds are used to obtain food use clearances. Two factors played

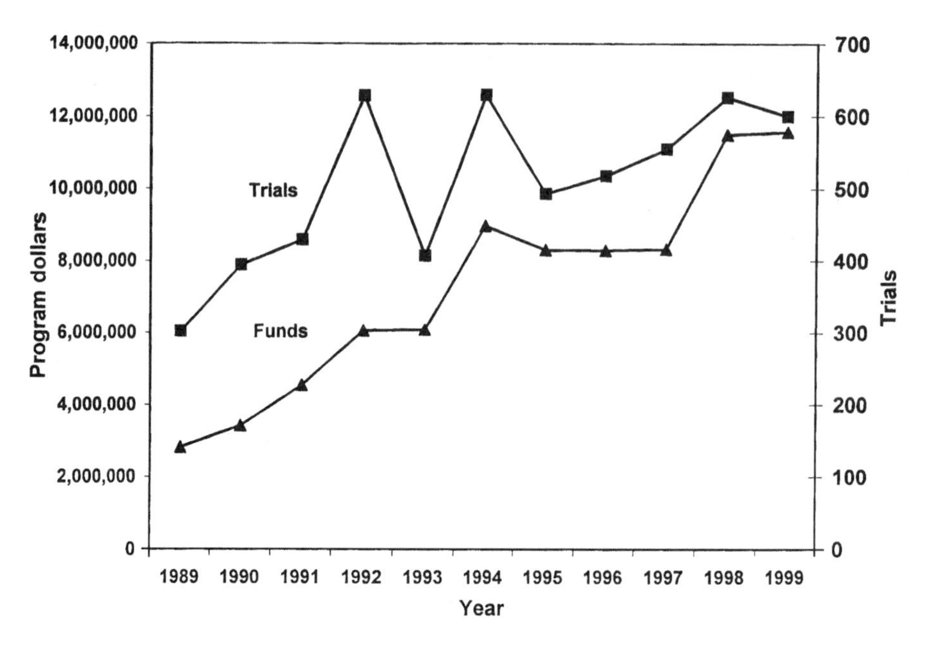

FIGURE 2 Trends of the IR-4 budget versus the number of field trials conducted.

a major role in the cost increases. The first was the implementation of the GLP regulations. Two surveys conducted by the IR-4 showed that approximately 36–39% of the resources for the field and laboratory were required to comply with these regulations. There was an increase in paperwork as well as an increase in personnel required to conduct quality assurance. The second factor was the USEPA's establishment of a system to define the numbers and geographic distribution of field trials to obtain a crop tolerance. Although this system took the guesswork out of deciding where to place trials and the number of trials to be conducted, it disproportionately increased the number of field trials required for minor crops compared with the acreage of the crops. Under this scheme, wheat grown on 59 million acres is required to have 20 trials to establish a tolerance, whereas broccoli grown on 133,000 acres must have eight trials. Part of this scheme is based on per capita consumption. The consumption of wheat flour in the United States in 1997 was 150 lb per person, and that of all fresh vegetables was 186 lb per person [1]. This comparison generally holds true for the other major and minor crops. The number of trials for minor crops, which constitute a substantial portion of most diets, is necessary to ensure that the food supply is free of harmful pesticide residues. Any residues remaining in the food are considered safe when they are within established tolerances.

5 FUTURE DIRECTIONS OF PEST MANAGEMENT ON MINOR CROPS

The 1988 Amendment to FIFRA focused attention on minor crops and the minor crop pest management dilemma. Prior to that time, Congress had little awareness of the plight of farmers and nurserymen who produced the wide assortment of agricultural commodities that are aggregated under the term "minor crops," producers of these commodities were not organized or well represented, and the IR-4 Minor Use Program was struggling on a meager research budget. All that changed beginning in the late 1980s. Congress recognized the needs of minor crop producers, public monies for research and development increased dramatically, and a dialogue with agricultural interests began. More recently, the FQPA has provided incentives to industry to explore newer, safer, more environmentally compatible chemistry. As a result there has been more activity in the research and development of pesticides during the past several years than at any other time since the 1940s and early 1950s. The extremely high throughput of new chemical screening programs has evolved a large number of products with new and unique modes of action.

Although progress has been made in providing safe and effective pesticides for minor crops, many needs still exist. The problem of insufficient pest management options for minor crops will be compounded as a result of FQPA risk considerations. The number of minor crop registrations that may be lost through tolerance reassessment cannot be realistically estimated until the implications of a common mode of toxicity are better understood. However, it is clear that with more than 9000 pesticide tolerances to be reassessed by the year 2006, risk reduction actions will result in the loss of a significant number of minor crop registrations.

With FQPA emphasis on reduced risk pesticides and the need for IPM-compatible pest control products, there is a need to emphasize efforts to assist with the registration of microbial and biochemical pesticides. Frequently overlooked is the lack of research to support the labeling of pesticide products in an industry where effective pest management is lacking. Focusing on reduced risk pesticides and biologicals that adapt to the unique cultural practices of the minor crop industry and that address worker safety concerns should be a high priority for public and private sector research programs. The need for reduced risk pesticides and products compatible with IPM programs will continue to increase as the availability of older pesticides diminishes because of the reregistration mandates of the FQPA.

An increasingly attractive alternative to conventional pesticides is provided by biopesticides, which include both microbial and naturally occurring biochemical agents such as pheromones. These products generally combine a high degree of human safety with low environmental impact and excellent compatibility with IPM programs. However, they tend to be very selective in the spectrum of pests they control. This leads to a low volume of use that is unattractive for commercial development. It is noteworthy that the USEPA had registered about 187 biochemical pesticides as of the end of 1999 (R. Torla, EPA/OPP/BPPD, personal communication, 2000).

Much of the research and development of biochemicals and microbials is being done in the public sector by state and federal scientists. With few exceptions, microbial and biochemical pesticides are species-specific, and biochemicals, in particular, are used in very small quantities on a per-acre basis. Although these features make biopesticides environmentally attractive, they also generally make them unattractive for commercial development. State and federal agricultural scientists have little professional interest in going beyond the research stage of determining efficacy and use in the development of biopesticides. Moreover, the vast array of studies and regulatory requirements for product registration are well beyond the expertise and resources of the public sector scientists who performed the initial efficacy and use studies.

To encourage the development of biopesticides for minor uses, it may be necessary to provide financial impetus or other incentives to researchers to carry the research process through to registration. One of the objectives of the IR-4 Project is to facilitate the registration of biopesticides. To fulfill this objective, the IR-4 has a modest research program that funds public sector scientists to conduct research on promising biopesticides and to provide both guidance and hands-on assistance in preparing data packages to submit to the USEPA in support of registrations. Although this program is vastly underfunded considering the magnitude of the problem, the IR-4 has been responsible for supporting research on more than 50 biopesticide products, which has resulted in over 175 crop use registrations from 1982 to 1998 [22].

Recent major achievements in crop pest management by industry have been in the field of genetic engineering. Although the incorporation of the *Bacillus thuringiensis* endotoxin into corn, which renders the plant toxic to pests such as the European corn borer, and the introduction of herbicide resistance genes into crops such as soybeans, cotton, and corn have been controversial, they signaled the beginning of a new era in agricultural pest management. Although this technology is adaptable to minor crops, the pace with which genetic engineering to achieve the management of pests in these crops advances will undoubtedly be much slower, because of reduced economic incentives.

Major strides have been made both in understanding the significance of the need for pest management for the large number of agricultural commodities collectively known as minor crops and in increasing public and private sector interest in finding solutions to these needs. Clearly, the unprecedented attitude of cooperation between federal and state agencies and research scientists, the agricultural chemicals industry, and the crop producer community has been important to the success of effective pest management on minor crops and, indeed, all crops. What was regarded as a minor crop "dilemma" a decade ago is now viewed as a work in progress with many satisfactory solutions at hand.

REFERENCES

1. USDA, NASS. Agricultural Statistics 1999. Washington, DC: US Govt Printing Office, 1999, Table 9–21, p IX 16.
2. USDA, NASS. 1997 Census of Agriculture, United States Summary AC 97-A-51. United States Summary and State Data, Table 2, pp 193–210.
3. Congress of the United States. Food Quality Protection Act. Public Law 104-170, 1996.
4. D Pimentel. Introduction. In: D Pimentel, ed. Handbook of Pest Management in Agriculture, Vol. 1. 2nd ed. Boca Raton, FL: CRC Press, 1991, pp 3–11.
5. USDA, CSREES. Inventory of Agricultural Research Fiscal Year 1997. CSREES, Sci Educ Resources Dev. Washington, DC: USDA, 1998, Table II-D.
6. RF Norris. Why control weeds. Fremontia 13(2):10–12, 1985.
7. USDA. Losses in Agriculture—A Preliminary Appraisal for Review. AARS 1–20. Beltsville, MD: Agric Res Service, USDA, 1954.
8. USDA. Losses in Agriculture. Agric Handbook 291. 1965.
9. JM Chandler, AS Hamill, AG Thomas. Crop Losses Due to Weeds in Canada and the United States. Champaign, IL: Weed Sci Soc Am, 1984.
10. USDA, NASS. Agricultural Statistics 1999. Washington, DC: US Govt Printing Office, 1999, p XI 10.
11. CE Main, SK Gurtz, eds. Estimates of Crop Losses in North Carolina Due to Plant Diseases and Nematodes. Spec Pub 8. Raleigh, NC: Dept of Plant Pathology, NC State Univ, 1989.
12. USDA, OPMP, PIAP. Crop Profile for Walnuts in California. 1999. pestdata.ncsu.CropProfiles/Detail.CFM?FactSheets.
13. USDA, OPMP, PIAP. Crop Profile for Mushrooms in Pennsylvania. 1999. pestdata.ncsu.CropProfiles/Detail.CFM?FactSheets.
14. CAST. Pesticides: Minor Uses/Major Issues. Ames, IA: Council for Agric Sci Technol, 1992.
15. USDA, CSREES. Inventory of Agricultural Research Fiscal Year 1997. Washington, DC: CSREES, Sci Educ Resources Dev, USDA, 1998.
16. RM Faust. Keynote Address: The ARS Program in Pest Management. Proc ARS Reducing Pesticide Risk Natl Workshop, Riverdale, MD, 1997, pp 8–15.
17. Anonymous. US ACPA members' sales up 1.1% in 1998. ASGROW, No. 330. PJB Pub, 1999.
18. GM Markle, JJ Barron, BA Schneider. Food and Feed Crops of the United States. 2nd ed. Willoughby, OH: Meister, 1998, p xi.
19. USDA, NASS, ERS. Agricultural Chemical Usage. 1997 Fruits Summary. Washington, DC: Natl Agric Stat Service and Econ Res Service, 1998.
20. USDA, NASS, ERS. Agricultural Chemical Usage. 1996 Vegetables Summary. Washington, DC: Nat Agric Stat Service and Econ Res Service, 1997.
21. USDA, NASS. 1997 Census of Agriculture—United States Data. Washington, DC: Nat Agric Stat Service, 1997, Tables 42–44.
22. IR-4. Annual Report 1998. New Brunswick, NJ: IR-4 Project 1999 (6 pp mimeo).
23. RT Meister. Farm Chemicals Handbook, 1999, Vol. 85. Willoughby, OH: Meister, 1999, pp. c5–c429.

Appendix 1 1997 Sales of All Crops and Selected Minor Crops, United States

State	All crop sales ($1000)	% U.S. sales	Minor crop sales ($1000)[a]					
			Vegetables, sweet corn, melons	Fruits, nuts, berries	Nursery, greenhouse crops	Other crops[b]	Total value minor crops	% Value all crop sales
Alabama	632,978	0.7	21,352	7,812	178,216	112,577	319,957	51
Alaska	15,968	0.2	719	21	10,017	1,822	12,579	79
Arizona	1,222,891	1.3	398,469	118,542	131,519	17,212	665,742	54
Arkansas	2,188,026	2.2	18,879	9,659	27,167	1,614	57,319	3
California	17,033,417	17.4	4,019,298	7,822,769	2,210,574	297,312	14,349,953	84
Colorado	1,326,944	1.4	110,992	10,029	211,743	149,127	481,891	36
Connecticut	263,799	0.3	17,291	11,920	172,371	54,308	255,890	97
Delaware	174,845	0.2	38,591	1,993	16,806	9,193	66,583	38
Florida	4,817,261	4.9	1,083,921	1,493,470	1,449,951	692,566	4,719,908	98
Georgia	1,920,598	2.0	273,090	108,875	219,370	491,512	1,092,847	57
Hawaii	401,411	0.4	33,702	174,573	83,159	109,180	400,614	100
Idaho	1,773,699	1.8	50,636	24,408	57,189	813,238	945,471	53
Illinois	6,567,164	6.7	63,709	14,133	299,936	—[c]	377,778	6
Indiana	3,246,617	3.3	43,622	11,885	110,877	50,751	217,135	7
Iowa	6,187,269	6.3	8,568	3,627	73,208	1,725	87,128	1
Kansas	3,221,766	3.3	3,454	1,621	49,302	2,187	56,564	2
Kentucky	1,578,861	1.6	7,984	2,450	56,018	828,038	894,490	57
Louisiana	1,411,472	1.4	8,351	5,513	72,586	361,328	447,778	32
Maine	212,229	0.2	16,974	47,118	29,852	101,745	195,689	92
Maryland	458,719	0.5	41,679	12,153	120,007	23,210	197,049	43
Massachusetts	357,377	0.4	37,438	148,247	128,192	31,133	345,010	97
Michigan	2,199,721	2.2	183,645	231,595	478,448	189,603	1,083,291	49
Minnesota	4,200,970	4.3	97,155	8,990	153,313	414,408	673,866	16
Mississippi	1,291,365	1.3	6,209	4,034	35,366	15,066	60,675	5
Missouri	2,307,009	2.4	18,718	13,339	89,056	19,810	140,923	6
Montana	903,822	0.9	1,546	1,172	20,173	78,076	100,967	11

Nebraska	3,798,462	3.9	3,163	556	21,791	84,245	109,755	3
Nevada	151,717	0.2	22,222	440	15,629	13,014	51,305	34
New Hampshire	73,728	0.1	8,614	10,760	44,957	2,246	66,577	90
New Jersey	592,713	0.6	150,508	89,768	277,957	6,386	524,619	89
New Mexico	462,178	0.5	88,776	43,560	48,409	37,830	218,575	47
New York	1,000,417	1.0	206,866	185,078	290,722	—[c]	682,666	68
North Carolina	2,595,213	2.7	73,707	30,137	318,203	202,089	624,135	24
North Dakota	2,193,672	2.2	1,291	86	8,673	292,811	302,861	14
Ohio	2,827,924	2.9	97,189	20,634	402,118	14,510	534,451	19
Oklahoma	907,865	0.9	13,671	9,622	109,004	46,854	179,151	20
Oregon	2,114,196	2.2	213,101	307,917	676,429	231,918	1,429,365	68
Pennsylvania	1,282,526	1.3	64,658	93,252	639,778	48,301	845,989	66
Rhode Island	39,423	0.0	3,873	1,889	30,962	1,504	38,228	97
South Carolina	791,104	0.8	59,313	33,037	114,313	215,382	422,045	53
South Dakota	1,654,044	1.7	1,395	195	21,621	7,678	30,889	2
Tennessee	1,143,674	1.2	49,478	5,792	213,365	191,616	460,251	40
Texas	4,293,474	4.4	251,967	87,630	486,918	242,984	1,069,499	25
Utah	247,443	0.3	12,068	10,859	70,160	5,286	98,373	40
Vermont	59,592	0.1	6,549	10,287	18,588	9,108	44,532	75
Virginia	780,099	0.8	45,704	34,606	166,411	352,340	599,061	77
Washington	3,251,291	3.3	270,260	1,240,242	271,580	543,090	2,325,172	
West Virginia	64,907	0.1	1,727	13,806	19,332	5,220	40,085	62
Wisconsin	1,640,283	1.7	149,443	140,140	157,348	206,811	653,742	40
Wyoming	173,216	0.2	158	20	4,132	48,235	52,545	30
U.S. total	98,055,656	100%	8,401,697	12,660,262	10,942,816	8,787,806	39,650,968	40%

[a] Excludes grains, corn for grain, wheat, soybeans, sorghum for grain, barley, oats, other grains, cotton and cotton seed, hay, silage, and field seeds.

[b] Includes tobacco.

[c] Data withheld in census.

Source: Ref. 21.

APPENDIX 2 Minor Crops (Less Than 300,000 Acres)

Acerola
Allspice
Aloe vera (Barbadensis)
Amarath, Chinese
Anise
Anon
Apricot
Arracacha
Arrowroot
Arrugula
Artichoke
Artichoke, Jerusalem
Asparagus
Atemoya
Avocado
Balsam pear
Banana
Barbados cherry
Basil
Bay
Bean, dry lima
Bean, fava
Bean, guar
Bean, lima
Bean, long
Bean, mung
Bean, snap
Beechnut
Beet
Blackberry
Blueberry
Bok choy
Boysenberry
Brazil nut
Breadfruit
Broccoli
Broccoli raab
Broccoli, Chinese
Brussels sprout
Buckwheat
Burdock
Burnet
Bushnut
Butternut
Cabbage
Cabbage sini
Cabbage, Chinese
Cacao
Cactus fruit
Cactus pad
Calabaza
Calamondin
Canarygrass
Caneberry
Canistel
Canola
Cantaloupe
Carambola
Caraway
Cardoon
Carob
Carrot
Casaba
Cashew
Cassava
Cassia
Cauliflower
Celeriac
Celery
Ceriman
Chayote
Cherimoya
Cherry
Cherry, sour
Cherry, sweet
Chervil
Chestnut
Chicory
Chinese broccoli
Chinese mustard
Chinquapin
Chives
Choy sum
Chufa
Cidra
Cinnamon
Citron
Clove
Cocoa bean
Coconut
Coffee
Collard
Corazon
Coriander
Corn, sweet
Crabapple
Crambe
Cranberry
Crenshaw
Cress, upland
Cucumber
Cumin
Currant
Curry leaf
Daikon
Dandelion
Dasheen
Date
Date, Chinese
Dewberry
Dill
Eggplant
Elderberry
Endive
Escarole
Evening primrose
Evergreenberry
Feijoa
Fennel
Fenugreek
Fig
Filbert
Garbanzo
Garlic
Genip
Ginger
Ginseng
Gooseberry
Gourds, edible
Grapefruit
Greens, rape
Greens, turnip
Guanabana
Guar
Guava
Hazelnut
Herbs and spices
Hickory nut
Honey + beeswax
Honeydew
Hops
Horseradish
Huckleberry
Jaboticaba
Jicamba
Jujube
Juneberry
Kai choy
Kale
Kiwifruit
Kohlrabi
Kumquat
Langsat
Lavender
Leek
Lemon
Lemon grass
Lentil
Lettuce, head
Lettuce, leaf
Lime
Loganberry
Longan
Loquat
Lotus root
Lychee
Macadamia
Mace (nutmeg)
Mache
Malanga
Mamey sapote
Mandarin
Mango
Maple sap
Marigold
Marionberry
Marjoram
Mayhaw
Melon
Melon, Chinese
Millet
Millet, Proso
Mint
Mizuna
Mulberry
Mushroom
Mushroom, shiitake
Mustard (oilseed)
Mustard greens

Appendix 2 (continued)

Mustard, Chinese
Naranjilla
Nasturtium
Nectarine
Nutmeg (*see* Mace)
Okra
Olallieberry
Olive
Onion
Onion (dry bulb)
Onion (green)
Orchard grass (seed)
Oregano
Oyster plant
Pak choy (mustard cabbage)
Pak toy
Pakchoi
Papaya
Paprika
Parsley
Parsley root
Parsley, Chinese
Parsnip
Passion fruit
Paw paw
Pe tsai
Pea
Pea, Austrian
Pea (edible podded)
Pea, green
Pea, pigeon
Pea, southern
Pea, succulent
Peach
Pear
Pepinos
Pepper
Pepper, bell
Pepper, Bohemian
Pepper, cherry
Pepper, chili
Pepper, cubanelle
Pepper, hot banana
Pepper, jalapeno
Pepper, non-bell
Peppermint
Perennial peanuts
Persimmon
Pimento
Pine nut
Pineapple
Piñon
Pistachio
Pitanga cherry
Plantain
Plum
Poke greens
Pomegranate
Poppy
Prickly pear cactus
Prune
Pumpkin
Quince
Quinoa
Radicchio
Radish
Radish, Japanese
Raisin
Rapeseed
Rapini
Raspberry
Rhubarb
Rosehip
Rosemary
Rutabaga
Safflower
Sage
Sainfoin
Salsify
Sapodilla
Sapote
Savory, summer/winter
Sesame
Shallot
Sorrel
Soursop
Spearmint
Spinach
Squash, summer
Squash, winter
Stone fruits
Strawberry
Sugar apple
Sweet sop
Sweetpotato
Swiss chard
Tamarind
Tangelo
Tangerine
Tanier
Taro
Tarragon
Thyme
Tomatillo
Tomato
Towelgourd
Turmeric
Turnip (roots)
Turnip (tops)
Ung choi
Walnut
Water chestnut
Watercress
Watermelon
Wild rice
Yam
Yam bean (tuber)
Yautia
Youngberry
Yucca
Yuquilla

APPENDIX 3 Major Minor Crops (300,000 to 1 Million Acres)

Almond
Apple
Bean (adzuki)
Bean (dry)
Bean (field)
Beet (sugar)
Citrus
Corn, sweet (processing)
Corn (popcorn)
Grape
Orange
Pea (dry)
Peanut
Pecan
Potato
Sorghum
Sugarcane
Sunflower
Tobacco
Tomato (processing)

APPENDIX 4 Fruit Acreage Treated One or More Times with Pesticides

Crop	Bearing acres	% Acres treated[a]			Acres treated[a]		
		H	I	F	H	I	F
Apples	350,800	60	96	90	210,480	336,768	315,720
Apricots	20,300	30	62	52	6,090	12,586	10,556
Avocados	62,500	44	33	12	27,500	20,625	7,500
Blackberries	5,510	94	79	87	5,179	4,353	4,794
Blueberries	34,200	67	83	88	22,914	28,386	30,096
Grapefruit	159,000	91	91	71	144,690	144,690	112,890
Grapes	893,600	75	60	87	670,200	536,160	777,432
Kiwifruit	6,150	41	20	15	2,522	1,230	923
Lemons	49,000	78	73	66	38,220	35,770	32,340
Limes	2,100	98	100	100	2,058	2,100	2,100
Nectarines	38,000	73	82	79	27,740	31,160	30,020
Olives	37,400	53	16	30	19,822	5,984	11,220
Oranges	832,900	91	88	65	757,939	732,952	541,385
Peaches	135,900	54	82	84	73,386	111,438	114,156
Pears	67,900	57	90	85	38,703	61,110	57,715
Plums	44,000	74	85	69	32,560	37,400	30,360
Prunes	100,500	48	71	58	48,240	71,355	58,290
Raspberries	13,200	90	90	95	11,880	11,880	12,540
Sweet cherries	48,000	61	84	80	29,280	40,320	38,400
Tangelos	13,300	96	97	91	12,768	12,901	12,103
Tangerines	38,700	80	79	56	30,960	30,573	21,672
Tart cherries	32,400	78	98	99	25,272	31,752	32,076
Temples	6,700	24	98	94	1,608	6,566	6,298
Total	2,992,060				2,240,011	2,308,059	2,260,585
Percent treated		75	77	76			

[a] H = herbicides; I = insecticides; F = fungicides.
Source: Ref. 19.

APPENDIX 5 Vegetable Acreage Treated One or More Times with Pesticides

Crop[a]	Bearing acres	% Acres treated[b]			Acres treated[b]		
		H	I	F	H	I	F
Asparagus	72,000	88	56	33	63,360	40,320	23,760
Beans, lima, Fr.	5,500	93	47	34	5,115	2,585	1,870
Beans, lima, Pr.	30,500	49	60	18	14,945	18,300	5,490
Beans, snap, Fr.	66,500	49	75	73	32,585	49,875	48,545
Beans, snap, Pr.	134,200	90	72	49	120,780	96,624	65,758
Broccoli	106,000	64	96	37	67,840	101,760	39,220
Cabbage, Fr.	63,800	62	94	57	39,556	59,972	36,366
Cabbage, Pr.	6,200	95	89	11	5,890	5,518	682
Carrots	108,000	89	40	78	96,120	43,200	84,240
Cauliflower	43,800	31	97	18	13,578	42,486	7,884
Celery	26,100	68	97	86	17,748	25,317	22,446
Corn, sweet, Fr.	145,500	79	89	42	114,945	129,495	61,110
Corn, sweet, Pr.	416,600	90	74	11	374,940	308,284	45,826
Cucumbers, Fr.	49,200	60	68	77	29,520	33,456	37,884
Cucumbers, Pr.	71,500	76	36	34	54,340	25,740	24,310
Eggplant	2,500	33	89	84	825	2,225	2,100
Lettuce, head	194,900	52	98	76	101,348	191,002	148,124
Lettuce, other	73,700	52	86	73	38,324	63,382	53,801
Melon, water	163,800	43	41	65	70,434	67,158	106,470
Melons, other	113,000	36	85	47	40,680	96,050	53,110
Onions, bulb	127,400	88	83	89	112,112	105,742	113,386
Peas, green, Pr.	221,700	89	35	2	197,313	77,595	4,434
Peppers, bell	64,800	67	88	75	43,416	57,024	48,600
Spinach, Fr.	12,300	56	72	45	6,888	8,856	5,535
Spinach, Pr.	7,800	61	63	50	4,758	4,914	3,900
Strawberries	44,500	37	85	86	16,465	37,825	38,270
Tomatoes, Fr.	88,700	54	93	90	47,898	82,491	79,830
Tomatoes, Pr.	318,000	78	71	90	248,040	225,780	286,200
Total	2,778,500				1,979,763	2,002,976	1,449,151
Percent treated		71	72	52			

[a] Fr. = fresh; Pr. = processed.
[b] H = herbicides, I = insecticides, and F = fungicides.
Source: Ref. 20.

8

Arthropod Resistance to Pesticides: Status and Overview

David Mota-Sanchez, Patrick S. Bills, and Mark E. Whalon
Center for Integrated Plant Systems
Michigan State University
East Lansing, Michigan, U.S.A.

1 INTRODUCTION

In the early part of the twentieth century, the first pesticide-resistant arthropod species, the San Jose scale, *Quadraspidiotus perniciosus* (Comstock), was discovered to be resistant to lime sulfur in deciduous fruits in the state of Washington [1]. By the year 2000, there were 533 arthropod species reported to be resistant to one or more pesticides. Our work updates that of Georghiou and Lagunes-Tejeda [2], whose widely reported tabulation of 504 species exhibited an increase in pesticide resistance of just over 6% in 10 years. This count is based upon an examination of over 2600 peer-reviewed journal articles, which supplements the 1263 references cited in previous reviews of Georghiou and others (Table 1). Our information currently resides in an electronic database at the Michigan State University Center for Integrated Plant Systems that is available via the Internet at http://www.cips.msu.edu/resistance.

This review is a summary of the contents of that database, and it includes our initial analysis of the pesticide resistance problem. Because it deals with

TABLE 1 Documented Cases of Arthropod Resistance

	Georghiou and Lagunes-Tejeda, 1989	MSU updated database, 1999	Percent change
Species: Arthropod species that are resistant to one or more pesticides	504	533	5.8%
Compounds: A unique pesticide active ingredient to which one or more arthropod species is resistant	231	305	32.0%
Cases: A case of a unique species resistant to a unique compound, e.g., unique (species, compound) pairs	1640	2574	57.0%
National cases: Case of resistance unique to any one country, e.g., unique (species, compound) country	4458	4682	5.0%
Regional cases: Species–compound–region combination. May include multiple identical cases from the same country (e.g., different states or provinces)	Not reported	5630	—
Referenced documents: Reports of new regional cases (e.g., new species, compounds, or regions of occurrence)	1263	1468	16.2%
Total documents reviewed (peer-reviewed journal articles)	Not reported	2589	—

arthropods, this chapter focuses mainly on insecticides and acaricides, but resistance to fungicides, herbicides, and other pesticides exhibits many of the same features and as such is equally as important in the scope of pest management. We begin with a brief summary of the issues surrounding pesticide resistance in arthropods, specifically for the species resistant to the largest number of compounds. This work is not intended to be a complete literature review, nor could it be for such an expansive topic. However, our database and its analysis should provide a measure of the importance of pesticide resistance for pest managers in agriculture, human health protection, and elsewhere.

2 DEFINITIONS OF RESISTANCE

Resistance is the microevolutionary process of genetic adaptation through the selection of biocides [3]. One consequence of resistance is the failure of a plant protection tool, tactic, or strategy to control a pest where such failure is due to a genetic adaptation in the pest. This definition has traditionally been applied to insect populations that escape the effects of a chemical insecticide. However, nearly all classes of organisms provide an example of resistance to pest management measures, chemical or otherwise.

Just as resistance evolves over time, the definition of resistance has been developed and refined. A panel of World Health Organization (WHO) experts defined resistance as "the development of an ability in a strain of insects to tolerate doses of toxicants which would prove lethal to the majority of individuals in a normal population of the same species" [4]. This definition was the operational definition for years. After more than 60 years of synthetic insecticide applications, insect populations all over the world have been exposed to, and selected by, one or more pesticides, making it very difficult to find a normal population. In addition, the WHO definition is for populations rather than individuals, a distinction with more significance today because new biochemical and physiological techniques facilitate the detection of resistance in single individuals. Pest populations in crop systems deploying plant pesticides, such as *Bacillus thuringiensis* (Bt) toxin producing crops, are screened to detect resistant alleles present in very low frequencies. If detected, this would not fit the WHO definition.

In 1960, J. F. Crow presented a more inclusive definition of resistance that considers single individuals as well as populations. He proposed that "resistance marks a genetic change in response to selection" [5]. This definition is not restricted to high resistance levels or dependent upon the failure of an insecticide in the field. Incipient resistance is included in this definition as well. However, perhaps the most significant consequence of pesticide resistance is missing: field failure. In 1987, R. M. Sawicki improved upon Crow's definition by adding the significance of field failure to the definition as follows: "Resistance marks a genetic change in response to selection by toxicants that may impair control in the field" [6]. Note that Sawicki was careful to consider the possibility that resistance may or may not impair control of the organism in real-world applications. By this definition, strains of organisms that are selected for pesticide resistance in the laboratory are considered resistant.

The agrochemical industry has not been idle in the effort to understand, define, monitor, and manage pesticide resistance. The exponential increase in the worldwide cases of resistance during the first three-quarters of the twentieth century, combined with scientific and public pressure, led the pesticide industry to form various "resistance action committees" including ones for insecticides (IRAC), fungicides (FRAC), and herbicides (HRAC). These action committees worked in various aspects of resistance management, specifically monitoring programs. The criteria developed by IRAC for defining resistance include the following circumstances [7]:

An insect should be viewed as resistant only when

The product for which resistance is being claimed carries a use recommendation against the particular pest mentioned and has a history of successful performance.

Product failure is not a consequence of incorrect storage, dilution, or application and is not due to unusual climatic or environmental conditions.

The recommended dosages fail to suppress the pest populations below the level of economic threshold.

Failure to control is due to a heritable change in the susceptibility of the pest populations of the product.

Based on the above criteria, IRAC pointed out that the term "resistance" should be used only when field failure occurs and this situation is confirmed. Although the IRAC criteria were sufficient to ensure that a pest population had truly developed resistance, the definition is still problematic for the early detection of resistance, setting the stage for anecdotal reporting and crisis rather than prevention and management. Detection of low frequencies of resistant alleles in a population does not warrant a claim of resistance.

Why is detection important? Because of the transition from anecdotal reporting to resistance management, monitoring efforts can now include the detection of resistant alleles sufficiently early to change management as well as to avert and ameliorate resistance development. Consider a case in which resistant individuals are present in small numbers and the recommended dose suppresses the pest population below the economic threshold. In this instance, there is no detected "field failure" and by definition there is no resistance. Potentially, the frequency of resistant individuals in future generations will increase, leading to failure to control the pest. On the other hand, it could be argued that even with this increase in resistance, a correct insecticide application could guarantee reduction of pest populations below an economic threshold.

Even so, there are additional factors aside from pesticide application that may affect reduction of pest population levels. These factors could include the impacts of predators and parasites, pest spatial distribution, crop phenology, weather, life stage of the pest (e.g., larval instar), and frequency of resistant individuals [8]. Therefore, special care has to be taken in the interpretation of the resistance definition. By the time it is determined that field applications have failed to control a pest population, it is likely too late to implement strategies for the management of resistance to this pesticide (and other pesticides the insect may be cross-resistant to) owing to the high frequency of resistant individuals. Clearly, early detection of resistance is an important aspect missing from this definition.

Most documented studies of resistance fall in the area of physiological resistance. However, behavior plays an important role in resistance. The term "behavioral (or "behavioristic") resistance" describes the development of the ability of individuals within a population to avoid a dose of pesticide that would otherwise prove lethal [4]. There are, however, limited examples of behavioral resistance. In at least one case, behavioral resistance was confounded with an unidentified and undifferentiated sibling species. Initially, resistance workers believed that a species of *Anopheles* mosquito in Africa avoided residues inside houses by remaining outdoors [9]. Later, this "behaviorally resistant" population was demonstrated to be a complex of sibling species [10]. One example of true behavioral resistance can be seen in the sheep bowfly, *Lucillia cuprina* (Wiedemann), in which the oviposition of the fly was selected for behavioral resistance to cycloprothrin [11]. Genetic studies of this insect have shown that this resistance is partially dominant and that the origin is polygenic. To demonstrate behavioral resistance it is necessary to show genetic differences as they occur in physiological resistance, rather than present only observations of insects avoiding pesticides [12].

More recently exposed putative behavioral resistance to pest management strategies have been observed in the corn root worm, *Diabrotica vigifera vigifera* (LeConte) [13], which overwinters as a larva, emerges, and then feeds on corn rootstock. In Illinois, by laying eggs in soybean fields, this insect appears to have overcome crop rotation, the dominant strategy of keeping population levels low. In the following season, the fields with *D. vigifera* larvae are sown with corn. If this oviposition behavior is a result of a genetic change in the population, selected for by the pest management strategy, then perhaps this case meets Whalon and McGaughey's definition. However, there is some debate about the cause of this newly observed behavior, and the possibility exists that it is not a change in the organism itself but that the agroecological landscape has changed. Perhaps the overwhelming majority of acreage devoted to corn–soybean rotation has given *D. vigifera* no other choices for ovipositional sites.

Because of the few cases of behavioral resistance, the myriad of factors affecting insect behavior, the lack of accepted tests, and other issues making proof extremely difficult, this chapter focuses only on cases of physiological resistance. However, future developments of bioassays to detect behavioral resistance together with genetic studies certainly would be an important area for the detection of resistance.

3 THE IMPACT OF PESTICIDE RESISTANCE

The global economic impact of pesticide resistance has been estimated to exceed $4 billion annually [14]. Other estimates have been lower, but most scientists, agrochemical technical personnel, and agricultural workers agree that resistance is a very important driver of change in modern agriculture. There are many examples of production systems that have been incredibly vulnerable to the development and devastating effects of pesticide resistance.

In potato agroecosystems, the Colorado potato beetle, *Leptinotarsa decemlineata* (Say), has developed resistance to more than 38 insecticides (see Table 2 in Sec. 6). This insect is a strong candidate for the archetype of multiply resistant species. Because of the evolution of resistance to nearly all chemical classes of insecticides in Maine, Pennsylvania, Michigan, Wisconsin, and New York (Long Island), farmers in these states have even employed alternative tactics, including the radical use of propane flamers and plastic-lined ditches to stop the devastation of their crops by this pest.

Animal agriculture is another production system that has been affected by resistance. Famous instances include the dairies of Denmark, farms of California, and other regions of the world where populations of housefly, *Musca domestica* (Linneus), had developed dramatic levels of resistance to many insecticides [15]. Cattle ticks, *Boophilius annulatus* (Can.), and the sheep bowfly, *Lucilia cuprina* [16,17], are other significant examples of resistance development that have resulted in long-term economic problems. Both the transmission of diseases and the direct damage to livestock by cattle ticks have necessitated frequent pesticide treatments for many producers [16]. Indeed, resistance is one of the most significant challenges facing production agriculture, human and animal health protection, and structural and industrial pest management.

We usually think first of large-scale crops, such as cotton or staple foods, with resistance. Specialty crops, or those crops with less than 300,000 acres in production (162,000 hectares), which are defined by U.S. legislation to be a "minor use" for pesticides, are not immune to the impacts of resistance. In crucifer production systems (e.g., cabbage, broccoli, and other crops in the family Brassicae), the diamondback moth, *Plutella xylostella* (L.), has developed resistance throughout its cosmopolitan range [18]. Lack of control has resulted in the presence of immature stages in the heads of crucifers at the end of the season with the consequent rejection of the harvest due to the regulation of insect parts in food.

Economic failure and crop displacement are not the only effects of insecticide resistance. Misguided efforts to control resistant pests include the overuse of pesticides, which contributes to externalities such as environmental pollution, residues in food, and human exposure. For instance, high levels of insecticide resistance in tandem with high temperature, frequent rain, and high pest incidence in cotton led to applications of more than 29 liters (36.6 quarts) of active ingredient per hectare in Tapachula, Chiapas, southern Mexico [19].

Indian cotton production was severely curtailed initially due to resistance to chlorinated hydrocarbons (e.g., DDT), then resistance to organophosphates, and finally resistance to synthetic pyrethroids [20]. The cotton resistance situation became so severe in Andhra Pradesh in 1989 that it was widely reported that cotton producers in several villages committed suicide when their crops failed due to insecticide-resistant pest damage. Such acute human suffering resulting from pesticide resistance is unusual, but, regrettably, regional crop devastation is not as rare.

The onset of pesticide resistance has certainly contributed to the increase in severe human suffering from the mosquito *Anopheles*, the malaria vector, which is resistant to many different insecticides. Therefore, induced pesticide resistance can challenge not only agriculture but also national and international health institutions.

4 RESISTANCE MANAGEMENT, MONITORING, AND DETECTION

Resistance management attempts to ameliorate the development of resistance through strategies, tactics, and tools that reduce selection pressure. Management steps are deployed to reduce resistance evolution by

1. Diversifying mortality sources with strategies of managing resistance such as sequencing, rotating, or alternating pesticides with differing modes of action and the use of other strategies of integrated pest management including biological control, resistant varieties, cultural control, and pheromone disruption, among others
2. Monitoring to detect low frequency resistant alleles
3. Modeling to predict resistance development

and/or

4. Facilitating the survival or immigration of susceptible individuals that will dilute the frequency of homozygous resistant individuals in pest populations

Resistance exhibits many of the characteristics described by Garret Hardin in his article "Tragedy of the commons" [21]. His concept relates to a public animal grazing area known as a "commons." Many families could benefit from this single resource by careful management and equal sharing. However, overgrazing by even a single user could upset the balance of regrowth and destroy it for all. Hardin's argument, oversimplified, is that individuals are compelled to do this. Much like the grass in those fields, the proportion of individual pests in a population that is susceptible to a pesticide is a precious commodity held in common. Such a statement may sound surprising, but the susceptible genes can be "overgrazed" by a single individual who continues to apply an insecticide that only serves to establish a resistant population. The now abundant resistant individuals will disperse and establish in other fields. In short order this pesticide would no longer be effective in that region. Very little incentive exists for an individual producer to manage resistance on his or her farm if a neighbor ignores resistance management principles and thus selects a resistant strain, especially if in practice this results in increased crop losses [22]. Perhaps some of the 5630 documented regional cases of arthropod resistance are a result of this lack of incentive.

To complicate the resistance management issue, very little resistance reporting has *not* been anecdotal. Early on, many resistance episodes were attributed to poor spray coverage, ineffective timing, and rain wash-off. Therefore resistance evolution from the early 1950s to the 1980s was often described as a pesticide applicator problem. Various stakeholders, including industry, government and state agencies, and university representatives, sought other explanations for insecticide failure. Because resistance monitoring was difficult, expensive, and of questionable value, widespread and effective monitoring programs have not generally been supported by the private and/or public sectors. Ironically, monitoring had been suggested by scientists and government agencies and welcomed as a resistance management strategy. This contrast reflects the uncertain nature of deploying a monitoring strategy with adequate efficiency to allow the implementation of alternative resistance management tactics. As a result, resistant pest populations have become established before pest managers have even suspected a problem; thus their reporting has been anecdotal. Some might say that for implementation of resistance management in the field, it is better to assume that resistance must be present rather than to waste time and money in monitoring because it can be economically impractical. Rather than taking action only after monitoring procedures declare that the pest population is resistant, it is not unreasonable to recommend the prevention of resistance by implementing a resistance management strategy whenever pesticides are used.

5 COUNTING RESISTANT ARTHROPODS

As early as 1957, J. R. Busvine published a tally of resistant arthropods in the *Bulletin* of the WHO [23]. Following Busvine's initiative, W. A. Brown published tables of resistance cases for the WHO and other agencies in the 1950s until the early 1970s. These early reviews focused on human and animal disease vectors, which were the initial targets of worldwide pesticide application [9]. In the 1980s, Brian Croft and Karen Theiling began to collect documentation of resistance of arthropod biocontrol agents such as insect predators and parasites [24]. Their novel approach involved using pesticide resistance as an advantage by determining compatible natural enemies and pesticides to manage pests within an agroecosystem [25]. Croft's database was subsequently updated, and portions are available from Oregon State University [26].

The United Nations and national governments have long been interested in ascertaining the resistance situation. A 1984 study initiated by the U.S. Board on Agriculture of the National Research Council made 16 recommendations, one of which stated that "federal agencies should support and participate in the establishment and maintenance of a permanent repository of clearly documented cases of resistance" [27]. This recommendation was made law by the Food, Agriculture, and Trade Act in 1990, which called for a "national pesticide resistance monitoring program." The U.S. Food Quality Protection Act of 1996 (FQPA) invoked resistance as one of four conditions defining a pesticide as a "minor use." Specifically, a pesticide registration may be declared a "minor use" when the U.S. Environmental Protection Agency (USEPA), the U.S. Department of Agriculture (USDA), and the pesticide registrant determine that the pesticide use "does not provide significant economic incentive to support the initial registration or continuing registration" and that the use "plays or will play a significant part in managing pest resistance" (FQPA, 1996). A "minor use" pesticide is given special provisions that reduces the pesticide registration burden, for otherwise the registrant has little to gain economically despite the fact that the pesticide may be important for the continued production of specific crops.

The penultimate publication delineating the scope of the resistance problem was authored by Dr. George Georghiou and was initiated at the request of the United Nations Food and Agriculture Organization (FAO). His thorough review of resistant arthropod research with Angel Lagunes-Tejeda culminated in a database, published in book form in 1991 [2]. Their text included 504 species that are resistant to one or more compounds in one or more regions (states, provinces, and countries), covering over 200 pesticide compounds (Table 1) and based on 1263 cited references.

We used these references as our starting point for the construction of our electronic database and added records based upon the review of over 2500 refereed journal articles. Like previous efforts, the database discussed herein is the result of a review of published accounts of resistance. As has been stated previously, a report from the field that an insecticide has failed is not a good indication of the presence of resistant individuals. Many factors contribute to the effectiveness of a pesticide in the field. As a result, scientists and resistance workers that require empirical proof may view an undocumented claim of resistance by a farmer with skepticism, even when such a claim is true. Therefore, for the Michigan State University (MSU) database we referred only to peer-reviewed journals.

However, there may be as many ways as there are authors to observe and document a pesticide-resistant population of insects. Standardized methods for resistance detection do exist. In fact, FAO has been publishing standardized tests for species affecting human health since 1969. Nevertheless, lab techniques are constantly improving, and authors often interpret and report results of standardized tests differently. Even within these established standards there are many factors that might cause misunderstanding, and it is difficult for any reviewer to determine the veracity of such diverse data. Our strategy was to rely upon the expertise of the reviewers of manuscripts and the editorial boards of publications as well as upon our own review of the values of the median lethal dose (LD_{50}), median lethal concentration (LC_{50}), median lethal time (LT_{50}), median knockdown (KD_{50}), and discriminating doses.

The primary objective involved examining the statistical differences between resistant populations and a susceptible reference colony for previously unreported species, compounds, and/or regions. A very commonly reported measure of resistance is the resistance ratio (RR), which is the ratio of dose-mortality of the tested strain (defined by the statistic used, e.g., LD_{50}, LC_{50}, KD_{50}, or TL_{50}) to that of a known susceptible strain. We used reports of RR of 10 or greater as a general threshold for declaring a "case" of resistance. However, in some cases we also included reports with RR smaller than 10 when the authors were clear that this was high enough to cause significant resistance. This allowed consistency with previous efforts, specifically Georghiou's. We also considered cases of resistance developed in the laboratory, as they are important demonstrations of the potential for the development of resistance in the field. This is consistent with our working definition of resistance that may or may not lead to field failure. Factors used in deciphering a resistance report included the Whalon and McGaughey definition of resistance [3], several intrinsic and extrinsic factors of the test itself [28], and the type of statistic used to report the resistance level.

Confounding the categorization of the literature was variability among definitions of a pest "population." The catalog of resistance would not be complete without a spatial definition of pesticide-resistant populations. Researchers often collected individuals from multiple reproductively isolated locations but, unfortunately, reported aggregate bioassay results. Populations were described with vague spatial definitions or overlapping boundaries. This is not surprising, because the sampling and bioassay requirements for mapping the boundaries of a population are expensive. We used a coarse geographic resolution to circumvent these problems and thus limited distinction of regional cases to the national, state, or provincial level.

We made every effort to include all reported cases of resistance, but we are hesitant to say that we have uncovered all cases in our review given the scope of this worldwide phenomenon. We reviewed journals published principally in English and some in Spanish, French, and Italian. However, very probably there are other documented cases of resistance published in languages other than those that are most common in the western hemisphere. We view the enumeration of resistant arthropods as a dynamic process, not only as new populations develop resistance, but also as past reports from around the world are counted. As cases are brought to our attention, we incorporate them into our database.

6 TOP TWENTY RESISTANT ARTHROPODS

Using the database, we ranked arthropods based upon the number of unique compounds to which documented resistance occurred somewhere in the world at least once (we define a "case" of resistance this way: an organism resistant to a compound reported in at least one population). Table 2 reports the 20 most resistant arthropods according to this ranking. The list reads like the billing for the 20 worst arthropod pests on the globe. With new resistance reported steadily from 1943 through to the present, all of these species still present very significant economic and/or health challenges. We should stress, however, that exclusion from this ranking does not indicate that the status of an arthropod's resistance is not important. Many others of the 533 pesticide-resistant species share some of the genetic, biological, and operational factors for the resistance developments of these "top 20." Indeed, every case of resistance is important and should be observed in the context of the system production, human health protection, geographic area, and other factors.

The two-spotted spider mite, *Tetranychus urticae* (Koch), and the diamondback moth, *Plutella xylostella*, are tied for the greatest number of reported cases at 69. These species are closely followed by the green peach aphid, *Myzus persicae* (Sulzer), with 68 cases reported.

Genetic, biological, and operational factors significantly influence the development of resistance [29]. Most of the species listed have similar biological and ecological characteristics, including high generation turnover, great mobility and migration, and large numbers of offspring per generation, as well as operational factors such as high selection pressure and sequential application of related groups of pesticides. In the Homoptera order, there are four species that have developed resistance to many conventional and novel compounds: *Myzus persicae*, *Aphis gossypii*, *Phorodon humuli*, and *Bemisia tabaci*. Besides the common biological and ecological characteristics distinctive to this order, low economic thresholds due to virus transmission, especially in *M. persicae* and *B. tabaci*, have led to repeated insecticide treatments. In addition, frequent treatments in multiple hosts often cause a great deal of selection of individuals for resistance. Conversely, the damson-hop aphid *Phorodon humuli* is different in that it remains during the summer only in hops and wild hops, stays close to the crop, is monophagous and highly fecund, and is the most important pest in hops [30]. These conditions are pointed out by Denholm et al. [30] as "the worst case scenario" for the development of resistance. In the case of the diamondback moth, consumer demands for perfect cosmetic standards and a stricter restriction of "insect parts" in food force producers to lower economic thresholds. This insect, therefore, causes qualitative damage in addition to quantitative costs. The use of Bt has reduced the proliferation of conventional insecticides in crucifers. How-

TABLE 2 Top 20 Resistant Arthropods, Ranked by Number of Unique Compounds

Rank	Species	Family	Order	Number of compounds with reported resistance	Number of references in the MSU database	Year of first reported case	Example hosts	Arthropod common name
1	*Tetranychus urticae*	Tetranychidae	Acari	69	232	1943	Cotton, flowers, fruits, vegetables	Two-spotted spider mite
2	*Plutella xylostella*	Plutellidae	Lepidoptera	69	168	1953	Crucifers, nasturtium	Diamondback moth
3	*Myzus persicae*	Aphididae	Homoptera	68	247	1955	Fruits, vegetables, trees, grains, tobacco	Peach-potato aphid
4	*Boophilus microplus*	Ixodidae	Acari	40	87	1947	Cattle	Cattle tick
5	*Blattella germanica*	Blattellidae	Orthoptera	40	162	1956	Humans (urban pests)	German cockroach
6	*Heliothis virescens*	Noctuidae	Lepidoptera	39	94	1961	Chickpea, corn, cotton, tobacco	Tobacco budworm
7	*Leptinotarsa decemlineata*	Chrysomelidae	Coleoptera	38	124	1955	Eggplant, pepper, potato, tomato	Colorado potato beetle
8	*Panonychus ulmi*	Tetranychidae	Acari	38	173	1951	Fruit trees	European red mite
9	*Culex pipiens pipiens*	Culicidae	Diptera	33	117	1961	Humans (disease vector)	Mosquito

10	*Bemisia tabaci*	Aleyrodidae	Homoptera	32	85	1981	Greenhouse, cotton	Whitefly
11	*Spodoptera littoralis*	Noctuidae	Lepidoptera	32	50	1962	Alfalfa, cotton, potato, vegetables	Egyptian cotton leafworm
12	*Phorodon humuli*	Aphididae	Homoptera	32	64	1965	Hop, plum	Dawson aphid
13	*Culex quinquefasciatus*	Culicidae	Diptera	28	173	1952	Humans (disease vector)	Mosquito
14	*Aphis gossypii*	Aphididae	Homoptera	27	37	1965	Cotton, vegetables	Cotton/melon aphid
15	*Musca domestica*	Muscidae	Diptera	26	58	1947	Humans (urban and veterinary)	House fly
16	*Helicoverpa armigera*	Noctuidae	Lepidoptera	25	74	1969	Cotton, corn, tomato	Bollworm, earworm
17	*Tribolium castaneum*	Tenebrionidae	Coleoptera	25	100	1962	Stored grain, peanuts, sorghum	Red flour beetle
18	*Lucilia cuprina*	Calliphoridae	Diptera	24	31	1958	Cattle, sheep	Sheep blowfly
19	*Rhizoglyphus robini*	Acaridae	Acari	22	22	1986	Ornamental plants, stored onions	Bulb mite
20	*Anopheles albimanus*	Culicidae	Diptera	21	72	1964	Humans (disease vector)	Malaria mosquito (Central America)

ever, intense use of this compound has led to the development of field resistance to Bt [18,31].

Some species with high resistance found in the Lepidoptera order, including *Heliothis virescens*, *Spodoptera littoralis*, and *Helicoverpa armigera*, have been heavily treated with insecticides in cotton. However, treatments in other hosts have increased the selection pressure. In the past, industrial cotton had been the recipient of more than 40% of the applied insecticides produced in the world, making it a significant source of pesticide-resistant species.

Mites of agricultural importance, such as *Tetranychus urticae*, *Panonychus ulmi*, and *Rhizoglyphus robin*, maintain distinctive aspects that lead to pesticide resistance, including high reproductive rate, many generations per year, many alternative hosts, and high selection pressure. Conversely, the Colorado potato beetle, *Leptinotarsa decemlineata*, fails to follow the biological characteristics of having many generations per year, a trait that occurs with the majority of the top 20 species. Instead, this insect usually has from one to three generations per year. However, this insect has a tremendous capacity to colonize a wide range of hosts. Adaptation to defensive secondary metabolites produced by species of the Solanacea family may have allowed the Colorado potato beetle to increase its range of hosts from the original wild hosts to those of the cultivated potato. Adaptation through thousands of years has given this insect formidable ability to break down xenobiotics, a trait that may have extended to insecticides. Another important factor in the development of resistance is reduced migration, leading to local selection [32]. In local selection, individuals stay in the same area, elevating the frequency of individuals with resistant alleles.

Another species, the cattle tick, *Boophilus microplus*, is ranked number 4 in the list of top 20 arthropods, its high ranking related to the particular method of application. Total coverage of cattle by immersion in insecticide solutions increases the resistant selection, and individuals with resistant alleles are rapidly screened.

Insecticide resistance is also a problem in urban areas. For example, the house fly is a significant pest in veterinarian circles. In most farms, high selection pressures for resistance resulting from insecticide treatments occur in areas where the treatments are concentrated, the residuality of the insecticides is long, and the populations are relatively isolated [15]. In addition, the common practice of screening windows and doors to avoid immigration also has led to rapid selection and an increase in resistant individuals [29]. Protection of human health has led to an intense use of insecticides. As a result, there are three principal species of mosquito—*Culex pipens* (ranked 9th), *Culex quinquefasciatus* (ranked 13th), and *Anopheles albimanus* (ranked 20th)—that have developed resistance to many insecticides and have become vectors of diseases. Billions of people in the world's tropics are at risk of contracting malaria from such vectors [33]. In fact, malaria has caused the infection of 300–500 million cases per year, and every year about 2 million individuals die from the disease, half of them under the age of 5 [34]. *Anopheles albimanus* is one vector of this disease that has developed resistance to insecticides used to curb the spread of malaria. Other species in the genus *Anopheles* have developed resistance to insecticides as well, yet *A. albimanus* has maintained the greatest resistance in comparison with these other malaria vectors. One reason for this higher resistance of *A. albimanus* to multiple compounds is the intense insecticide selection pressure exerted over the complex of insect pests in cotton [35], which also indirectly selected immature stages in breeding sites and adult stages in resting sites.

Tribolium castaneum, red flour beetle, is a principal pest of stored grains where complete coverage by insecticide treatment is a common practice to control insects. High selection pressure and low migration are two of the causes that have led this insect to become resistant to many insecticides.

7 DATABASE ANALYSIS

The overwhelming majority of reported cases of resistance are arthropods resistant to organophosphate (44%) and organochlorine (32%) insecticides (Table 3). This is not surprising, because these classes of compounds include the most popular pesticides to date, and many have been in use for over half a century. Pyrethroids and carbamates together constitute only about 16% of resistance cases. Bacterial pesticides, primarily those produced from species of *Bacillus thuringiensis* (Berliner) (Bt), represent a mere 2% of cases, and all other remaining chemical classes combined have led to the development of less than 2% of resistance cases, as reported in the literature.

A unique addition to this field, in our database and analysis, is the tracking of U.S. pesticide registrations by use site and resistance development. We use USEPA data to compare the historical growth of U.S. pesticide registrations with pesticide resistance cases in this country (Fig. 1). The total number of unique insecticide and miticide use sites registered by the USEPA is further broken down by chemical class for those pesticides with resistance in Figure 2. Note that actual registrations started in 1947 with the passage of the Pesticide Labeling Act (Fig. 1). There is an obvious and positive correlation between resistance cases reported and the number of pesticides registered at any one time. Our research confirms that a strong relationship has existed between the cumulative number of active ingredients registered by the USEPA over time and the number of reported resistance cases in the United States for that time (Pearson's correlation coefficient = 0.97) [36].

However, this resistance is probably also correlated with quantity and method of pesticide use as well as scientific interest, demonstrated by the number of scientists reporting resistance. The USEPA defines use site as a unique active ingredient registration on a particular application site. For instance, a given syn-

TABLE 3 Summary of Documented Cases of Arthropods Resistant to Pesticides

Compound mode of action or chemical class	No. of compounds with resistance	Category of resistant arthropods					Total cases by chemical class
		Agricultural, forest, and ornamental plant pests	Medical, veterinary, and urban pests	Predators/ parasites	Other/misc. arthropods	Pollinators	
Organophosphates	112	715	358	52	10		1135 (44.1%)
Organochlorines	26	484	329	10	15	2	840 (32.6%)
Pyrethroids	33	133	74	11	1		219 (8.5%)
Carbamates	35	132	57	14	1		204 (7.9%)
Bacterials	38	42	4				46 (1.8%)
Miscellaneous	30	37	8	1			46 (1.8%)
Fumigants	6	21					21 (0.8%)
Insect growth regulators	10	16	2		3		21 (0.8%)
Organotins	3	8					8 (0.3%)
Formamidines	2	4	2				6 (0.2%)
Arsenicals	2	2	11				13 (0.5%)
Avermectins	2	2	3	1			6 (0.2%)
Chloronicotinoids	1	2	1				3 (0.1%)
Rotenone	1	2					2 (0.1%)
Dinitrofenols	1	1					1 (0.0%)
Sulfur compounds	2	1		1			2 (0.1%)
Phenylpyrazoles	1		1				1 (0.04%)
Total cases by arthropod category		1602 (62.2%)	850 (33.0%)	90 (3.5%)	30 (1.2%)	2 (0.1%)	2574

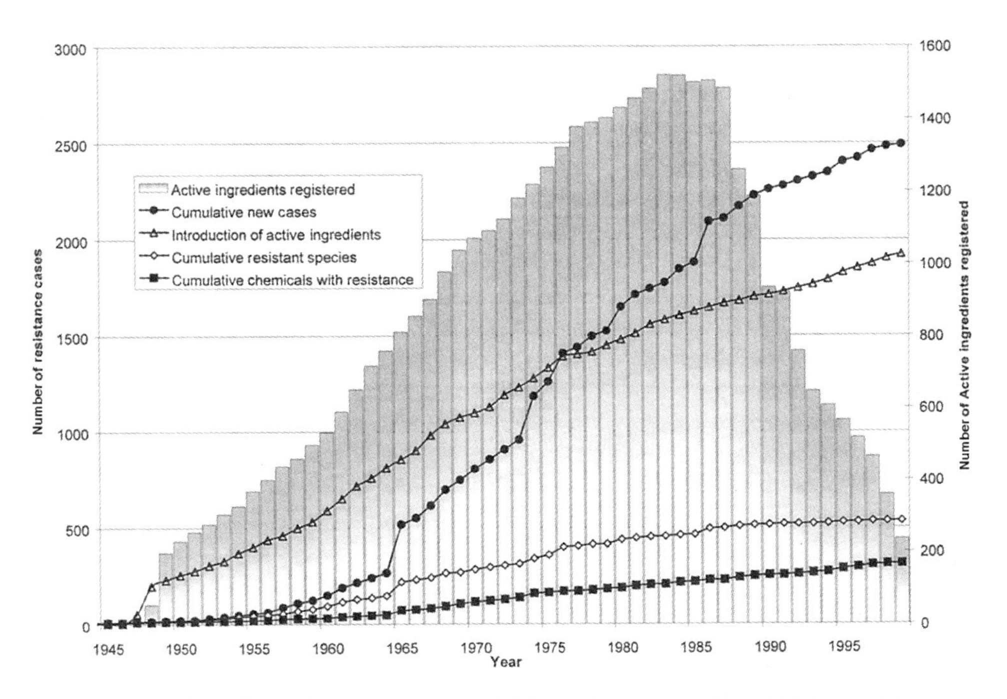

FIGURE 1 Timeline of arthropod pesticide resistance and pesticide registrations in the United States.

thetic pyrethroid X may have a registration on sweet corn, seed corn, and field corn. Thus there are three "use sites" for pyrethroid X. A timeline of registration of use sites for each pesticide class is given in Figure 2 for comparison. The critical question here is whether or not, and how, the number of available modes of action and the number of use sites relate to the rate of resistance development. Registration information is not enough to predict the onset of resistance, and pesticide usage patterns were not well documented before 1990.

8 CAUSES OF RESISTANCE

Of the estimated 10,000 arthropod pests, 533 are reported to have resistance to insecticides (Table 1). Our competition with these species for food and their transmission of disease are the principal reasons why we control them. In addition, control is exacerbated due to markets of higher cosmetic standards. These high qualitative standards have caused farmers to lower economic thresholds and increase the number of pesticide applications. The introduction of integrated pest management (IPM) in the 1970s probably slowed insecticide selection pressure

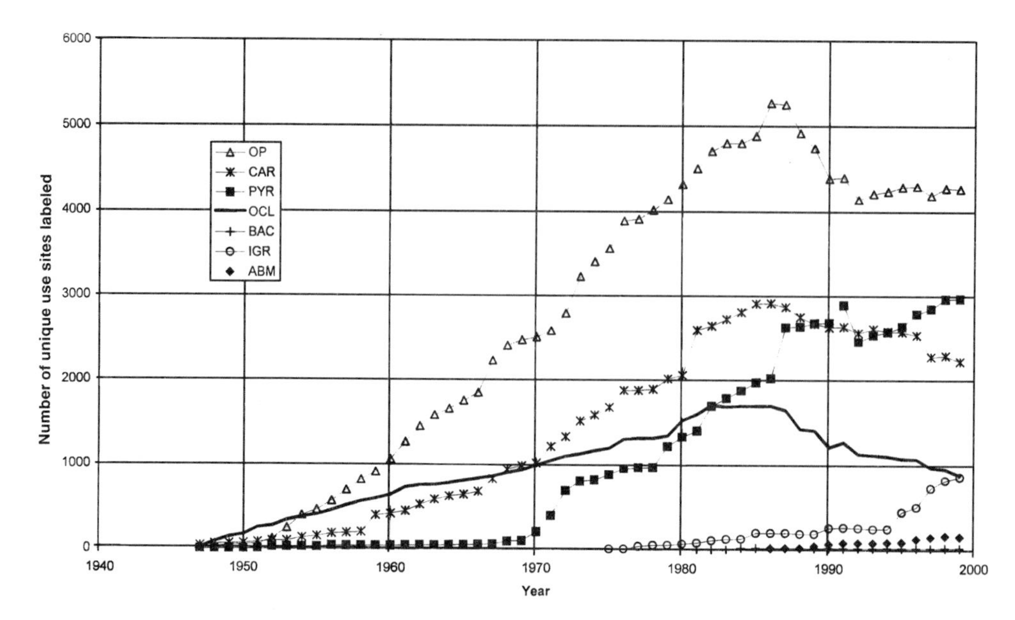

FIGURE 2 Timeline of U.S. pesticide registration for resistant pesticides: use sites per active ingredient chemical class. OP, organophosphates; CAR, carbamates; PYR, pyrethroids; OCL, organochlorines; BAC, bacterials; IGR, insect growth regulators; ABM, avermectins.

and reduced the trends in the development of resistance that we have seen in our results. In addition we must remember that most of the insecticides to which these pests have developed resistance have been taken off the market because of their environmental and human health effects. However, one of the collateral effects of resistance is the presence of a diversity of resistance mechanisms employed by the species with reported resistance that could be cross-resistant to existent and new compounds. More insecticides will probably be canceled due to stricter regulations imposed by legislation such as the FQPA of 1996. Therefore, we will likely see a reduced arsenal deployed against insect pests that already have a high frequency of alleles resistant to many pesticides. Although it is difficult to segregate the reported cases of resistance into application categories, the MSU database demonstrates that more than 62% of the cases occurred in agricultural, forest, or ornamental plant pest management (Table 3). Another 33% occurred in medical, veterinary, and urban pest management. Only 3.5% of the cases reported described the development of resistance in natural enemies such as predators and parasites.

Table 1 compares the efforts at MSU to estimate the number of resistant species with the results garnered from the immense efforts of George Georghiou at the University of California at Riverside from the 1960s to the late 1980s [2]. The MSU database builds on the Georghiou literature and provides a summary of the pesticide resistance literature through 1999. Since Georghiou's last report published in 1991 (which included data up to 1989), the number of resistant species has increased by almost 6% whereas the literature has grown by just over 16%. Georghiou identified 231 compounds against which resistance had developed, and our analysis shows an expansion of almost 32%, or 305 compounds (Table 1). These 305 compounds were documented in over 2574 references.

When one considers the number of species in comparison to the number of chemical compounds and the number of countries or regions reported, there are greater than 4682 cases from the literature. A decrease in the tendency of new resistant species (just a 5.8% increase) from 1990 to 2000 (Table 1) may be due to the fact that pesticide applications in modern agricultural systems have nearly exhausted the total number of arthropod pest species. Nonetheless, there is a trend to increased resistance to both new and old compounds by species already reported. It may be possible to find new resistant species in one or more of the following cases:

1. A shift in pest classification from secondary to primary
2. A change in taxonomy that separates one species into two or more sibling species
3. The dedication of more resources to taxonomic classification of pest species in developing countries
4. An increase in resource allocation for the detection of resistance in developing countries
5. A widening of the host range of wild herbivores to include cultivated species
6. An increase in the importance of minor cultivated species that results in a greater market pressure to improve the quality of harvested products

These possibilities are perhaps demonstrated in the large increase in the number of "cases" of resistance—new species–compound combinations—a greater than 50% increase. However, the current tendency is again the development of resistance to additional compounds as well as an increase in the geographical distribution of species previously reported as pesticide-resistant.

9 THE STATE OF RESISTANCE FOR CHEMICAL CLASSES

Most of the cases of resistance that have occurred with so-called conventional insecticides are classified into either the organochlorine (OC), organophosphate (OP), carbamate (CB), or pyrethroid (PY) chemical groupings (Table 3). Conventional pesticides are those that have controlled a broad spectrum of species, have worked as contact nerve toxins, were easy to use, and have been in use for many

TABLE 4 Rank of Pesticide Classes by Number of Cases of Arthropod Resistance (Species X Compound) per Time Period

Rank by no. of cases of resistance	Prior to 1980		Prior to 1990		Prior to 2000	
1	Organochlorines	757	Organophosphates	1050	Organophosphates	1136
2	Organophosphates	669	Organochlorines	838	Organochlorines	844
3	Carbamates	89	Carbamates	169	Pyrethroids	224
4	Miscellaneous	31	Pyrethroids	166	Carbamates	202
5	Pyrethroids	26	Miscellaneous	39	Bacterials	46
6	Fumigants	18	Fumigants	20	Miscellaneous	46
7	Arsenicals	13	Arsenicals	13	Fumigants	21
8	Nicotinoids	3	Insect growth regulators	7	Insect growth regulators	21
9	Bacterials	2	Formamidines	6	Arsenicals	13
10	Formamidines	2	Organotins	5	Organotins	8
11	Dinitrofenols	1	Bacterials	4	Avermectins	6
12	Rotenone	1	Nicotinoids	3	Formamidines	6
13	Sulfur compounds	1	Sulfur compounds	2	Nicotinoids	6
14			Avermectins	1	Rotenone	2
15			Dinitrofenols	1	Sulfur compounds	2
16			Rotenone	1	Dinitrofenols	1
17					Phenylpyrazoles	1

decades. Few cases of resistance have been detected outside of conventional pesticides such as agonists and antagonists of GABA receptors, insect growth regulators, *Bacillus thuringiensis* (Bt) protoxins, and neonicotinoid compounds. Yet the appearance of insecticide resistance has followed a loose chronological pattern following the deployment of most insecticides. Generally, the first cases of resistance have been reported within 3–5 years after the compound was extensively used. From the time of the first case of DDT resistance in 1947 up until the 1980s the majority of resistance cases resulted from the use of organochlorines, followed by organophosphates, carbamates, and pyrethroids. Adverse human health effects and negative environmental impacts from organochlorine compounds led to the cancelation of almost all of the insecticides in this class.

Reduced registration and use reflects the decline in reported resistance cases in the organochlorine compounds; only 0.7% of the total known cases were reported between 1990 and 2000 (Table 4). To date, the only organochlorine compounds remaining in use are DDT, endosulfan, lindane, and dicofol, and their uses are severely curtailed. In the future, we will see even fewer cases of resistance to organochlorines reported, and even then, perhaps only for endosulfan and cases with mosquitoes, for which DDT is still used as a control agent in many parts of the world. When organochlorines were replaced by organophosphates, carbamates, and pyrethroids, more cases of resistance to these replacement compounds ensued. Although some uses of organophosphates and carbamates have been canceled because of adverse human health effects, these groups are still widely used [37–39].

10 ARTHROPOD RESISTANCE TO COMPOUNDS WITH NOVEL MODES OF ACTION

Insecticides with novel modes of action are relatively new, and many have a comparably narrow spectrum in that they often target small taxonomic groups. Others such as insect growth regulators have their principal effect on immature stages. Most are also more expensive than conventional insecticides. One would expect that pesticides with these characteristics would not have the same selection pressure as broad-spectrum pesticides, and, in fact, we see fewer reports of resistance for such compounds. All of these conditions convey a relaxation in the selection of individuals that carry resistant alleles for novel compounds. However, this relaxed development of resistance is changing with time, and our data suggest a tendency toward a greater rate of increase in the number of cases of resistance. This trend is particularly noticeable if we compare cases of resistance prior to 1989 with those of the 1990s for bacterials, IGRs, avermectins, and nicotinoids (Table 4). In the following text we discuss specific resistance cases for each “novel mode of action” chemical group.

Fipronil is an antagonist of GABA receptors, with its mode of action similar to that of the cyclodienes. However, cross-resistance has not been found between fipronil and cyclodienes, perhaps because they act at different target sites on the GABA receptor. Populations of German cockroaches, *Blatella germanica*, and the house fly, *Musca domestica*, exposed to conventional insecticides, expressed low levels of cross-resistance to fipronil [40]. However, this fact may not limit the use of fipronil against these pests.

GABA receptor agonists such as the avermectins and ivermectins have an important use in controlling populations of the diamondback moth, *Plutella xylostella*; the Colorado potato beetle, *Leptinotarsa decemlineata*; and other insects, mites, and ticks that are highly resistant to other pesticides [41]. There are reports of resistance in both laboratory [42–44] and field [45–47] conditions. Resistance to avermectins has been selected for in laboratory populations of the western predatory mite, *Metaseiulus occidentalis* [42], *L. decemlineata* [43], and *M. domestica* [44]. In the latter case selection leads to a more than 60,000-fold level of resistance. Ivermectin is a semisynthetic version of avermectin B1 for veterinary and medical use [48]. Insects of veterinarian importance such as the hornfly, *Haematobia irritans* [49], and the Australian sheep blowfly, *Lucilia cuprina* [50], have also been selected in the laboratory for resistance to these compounds. Low levels of resistance to abamectin also occur in field populations of *B. germanica* collected in Florida [44] as well as in *P. xylostella* from Malaysia [47] and *Tetranychus urticae*, the two-spotted spider mite from California [46].

Bacillus thuringiensis (Bt) and related bacterial insecticides are grouped with novel compounds, although Bt was discovered early in the twentieth century. Up to now, seven insect species have developed resistance to species of *Bacillus* in the field (Table 5). The first report of field resistance to *Bacillus popilliae* occurred in *Popillia japonica*, the Japanese beetle, and *Anomala orientalis*, the oriental beetle [51]. The second report was of the Indianmeal moth, *Plodia interpunctella* [52]. Grain treated with Bt caused the development of low levels of resistance of *P. interpunctella* [52]. In Hawaiian vegetable production, heavy treatments of Bt used to control *Plutella xylostella* selected populations with high levels of resistance [31]. Resistance of this insect to Bt was also detected a few years later in Florida, the Philippines, Thailand, Malaysia, and Central America [18,47,53]. Resistance to another *Bacillus* species, *Bacillus sphaericus*, has been reported in *Culex pipiens*, the house mosquito, in France [54], *Culex quinquefasciatus*, which developed low levels of resistance in Brazil after 37 treatments in 2 years [55], and in India [56]. High values of the LC_{50} of Cry IA(c), Cry IA(b), and *Bacillus thuringiensis* subsp. *kurstaki* to *Spodoptera exigua*, the beet armyworm, were found in populations collected in the United States [57]. Laboratory selection with commercial products or isolated individual toxins from *Bacillus thuringiensis* or *B. sphaericus* led to resistant strains of 13 species (Table 5). Bt toxins are incredibly diverse in form. For some of these insects, as many as 40 different Bt protoxin crystals were tested for resistance, with varying degrees of mortality. Therefore, in our database we have considered any unique preparation or combination of crystals as a distinct "case" of resistance.

Resistance to microbials and chemical insecticides has been common in recent decades. However, resistance to viral insecticides appears to be rare. We have found only one case: In Brazil the *Anticarsia gemmatalis* nuclear polyhedrosis virus (AgMNVP) occurs naturally and is used extensively as a microbial pesticide [58]. Populations from Brazil selected in the laboratory have developed resistance ratios of more than 1000-fold. Conversely, populations from the United States have developed ratios of resistance (lethal dose for resistant population/ lethal dose for susceptible population) of only approximately fivefold [58].

We found reports of field resistance to neonicotinoid compounds, such as imidacloprid, in three insect species: *Bemisia tabaci* [59], *Bemisia argentifolii* [60], and *Leptinotarsa decemlineata* [61,62]. Not only has *Leptinotarsa decemlineata* developed resistance to imidacloprid, but field populations from Long Island, New York, have also demonstrated low levels of cross-resistance to second-generation neonicotinoid compounds such as thiamethoxam [62]. Also, the aphid species *Myzus nicotianae* has expressed low levels of resistance to imidacloprid in comparison with its sibling species *Myzus persicae* [63]. In addition, strains of German cockroach, *Blatella germanica*, and house fly, *Musca domestica*, which are multiresistant to other pesticides, express low levels of cross-resistance to imidacloprid [64].

Insect growth regulators (IGRs) are a diverse group based upon a general physiological effect rather than chemical family or target site. IGRs include hormonal disrupters such as juvenile hormone analogs and ecdysone agonists; chitin synthesis inhibitors such as benzoylureas and buprofexin; and cyromazine, which also inhibits chitin synthesis (mode of action still unknown). Insect resistance has been reported in all of these chemical groups. Georghiou and Lagunes-Tejeda reported as early as 1991 that fruit flies, *Drosophila melanogaster*, were resistant to the juvenile hormone analog methoprene and the chitin synthesis inhibitor cyromazine. They also reported IGR resistance in *Plutella xylostella* to the benzoylphenylureas (chlorfluazuron, diflubenzuron, teflubenzuron, and triflumuron) and in *Boophilus microplus* to chloromethiuron. Since the work of Georghiou and Lagunes-Tejeda in 1991, there has been a continuous increase in the number of species resistant to IGRs. Resistance to methoprene has recently been reported for *Aedes taeniorhynchus* (mosquito) [65]. Also, the whitefly, *Bemisia tabaci*, has been reported to have high levels of resistance to the new juvenile analog pyriproxyfen in rose greenhouses in Israel [66]. However, in spite of the similarity of pyriproxyfen's chemical structure to that of fenoxycarb, another juvenile hormone analog, there is no clear evidence of cross-resistance between these compounds [67]. The larvae of the house fly, *Musca domestica*, have also developed resistance to pyriproxyfen [68]. Buprofexin is a new IGR that inhibits chitin biosynthesis (through an unknown mechanism) yet is structurally unrelated to the benzoylphenylureas. Buprofexin-resistant *B. tabaci* have been detected in

TABLE 5 Species of Insects That Have Developed Resistance to Microbial Compounds in the Laboratory and/or the Field

Arthropod species	Arthropod common name	*Bacillus* spp., various toxins (Bt = *Bacillus thuringiensis*)	Field detection/ laboratory selection	Ref.
Heliothis virescens	Tobacco budworm	Several/various toxins	Lab	75–77
Heliothis zea	Corn earworm	Bt var *kurstaki* (cry IAc)	Lab	57
Leptinotarsa decemlineata	Colorado potato beetle	Bt subsp. *Tenebrionis*	Lab	78
Popillia japonica	Japanese beetle	*Bacillus popilliae*	Field	51
Anomala orientalis	Oriental beetle	*Bacillus popilliae*	Field	51
Ostrinia nubilalis	European corn borer	Bt subsp. *kurstaki*	Lab	79
Culex pipiens	House mosquito	*Bacillus sphaericus*	Field	54
			Lab	80
Culex quinquefasciatus	Mosquito	*Bacillus sphaericus*	Lab	81, 82
			Field	55, 56
Aedes aegypti	Yellow fever mosquito	Bt subsp. *israelensis*	Lab	83

Trichoplusia ni	Cabbage looper	Cry IA(b)	Lab	84
Spodoptera littoralis	Cotton leafworm	Cry IC and low levels to Bt subsp. *aizawaii*	Lab	85
Spodoptera exigua	Beet armyworm	Cry IA(c)	Lab	86
		Cry IA(c) Bt subsp. *kurstaki*	Field	57
Cadra cautella	Almond moth	Bt subsp. *kurstaki*	Lab	52
Plodia interpunctella	Indian mealmoth	Bt subsp. *kurstaki*	Low levels in the field	52, 87
		Bt subspp. *kurstaki*, *aizawaii*, *entomocidus*, various toxins of Bt	Lab	88
Chrysomela scripta	Cottonwood leaf beetle	Bt *tenebrionis*		89, 90
Plutella xylostella	Diamondback moth	Bt subsp. *kurstaki*, Bt subsp. *aizawaii*, and different isolated proteins	Field Lab	18, 31, 47, 91
Anticarsia gemmatalis	Velvet caterpillar	Nuclear polyhedrosis virus (AgMNPV)	Lab	58

glasshouses in the Netherlands [59], where 22 sprays in just 10 months led to 47-fold levels of resistance. In another instance, intense use of buprofexin led to the development of more than 300-fold resistance in *Trialeurodes vaporariorum*, the greenhouse white fly, in tomato greenhouses in Belgium [69].

Resistance to the benzoylphenylureas has also occurred in lepidopteran insects. In the Italian southern Tyrol, the codling moth, *Cydia pomonella*, developed resistance to diflubenzuron [70,71]. A similar situation occurred in southern France, where *C. pomonella* developed more than 370-fold resistance to diflubenzuron [72]. Cross-resistance in this species was reported in three other growth regulators: teflubenzuron (7-fold), triflumuron (102-fold), and tebufenozide (26-fold) [72]. However, flufenoxuron and fenoxycarb remain highly effective [73]. Resistance to tebufenozide also has been detected in the greenhead leafroller, *Planotortrix octo* (Dugdale), in New Zealand [74].

To date, the appearance of species resistant to compounds with novel modes of action has not been as dramatic as in so-called conventional pesticides (chlorinated hydrocarbons, organophosphates, carbamates, pyrethroids, etc.). We cannot say that this will be true in the future. This trend may be due to the limited use of these sometimes costly novel compounds, especially in developing countries where profits are low and regulation of cheaper alternatives is less stringent than in Europe and the United States. An increasing emphasis on integrated pest management as a key component of sustainable agroecosystems will also increase the use of the compounds that fit with this strategy: those with characteristics of rapid degradation, a narrow spectrum of pest species, and minimal toxic effects (acute or chronic) to humans and other nontarget organisms. Government restrictions of conventional pesticides, such as through risk assessment mandated by the FQPA, will very likely increase the use of these compounds as alternatives. Increased use of, and hence intensified selection pressure exerted by, these novel compounds on pest populations will most certainly result in additional cases of resistant arthropods. The latest deployment of compounds to combat pest resistance looks promising, but we should be extremely cautious. Herbivorous arthropods have a long history of evolving mechanisms to defeat toxins such as the defensive secondary plant chemistry of their hosts—perhaps as long as 350 million years. It is not surprising that in the last 100 years 533 species have evolved resistance to pesticides by human selection. The evolutionary endpoint is the same: resistance. This should be a warning to any pest manager, and the introduction of novel alternative pesticides should be in the context of a truly integrated pest management program.

ACKNOWLEDGMENTS

We thank those who made this chapter possible: Mr. Marek Ulicny for assistance with the literature search; Ms. Robbie Wong for coordinating the literature review, performing the laborious data entry, and organizing the bibliography of this chapter; and Ms. Erin Vidmar for exceptional editing assistance. We thank Willis Wheeler for his editing and continued support. We also thank the USDA CSREES, specifically Michael Fitzner and Rick Meyer, for support and funding for the development of the resistance database.

REFERENCES

1. AL Melander. Can insects become resistant to sprays? J Econ Entomol 7:167–173, 1914.
2. GP Georghiou, A Lagunes-Tejeda. The Occurrence of Resistance to Pesticides in Arthropods: An Index of Cases Reported Through 1989. Rome: Food and Agriculture Organization of the United Nations, 1991.
3. ME Whalon, WH McGaughey. *Bacillus thuringiensis*: Use and resistance management. In: I Ishaaya, D Degheele, eds. Insecticides with Novel Modes of Action: Mechanism and Application. Delhi, India: Springer-Verlag, 1998, pp 106–137.
4. World Health Organization. Expert Committee on Malaria. 7th Report. WHO Tech Rep Ser 125, 1957.
5. JF Crow. Genetics of insecticide resistance: general considerations. Misc Pub Entomol Soc Am 2:69–74, 1960.
6. R Sawicki. Definition, detection and documentation of insecticide resistance. In: MG Ford, DW Holloman, BPS Khambay, RM Sawicki, eds. Combating Resistance to Xenobiotics: Biological and Chemical Approaches. London: Ellis Horwood, 1987, pp 105–117.
7. CDS Tomlin. The Pesticide Manual. Farnham, UK: Br Crop Protection Council, 1997.
8. RT Roush, JA McKenzie. Ecological genetics of insecticide and acaricide resistance. Annu Rev Entomol 32:361–380, 1991.
9. AWA Brown. Insecticide Resistance. Geneva: World Health Org, 1958.
10. M Coluzzi, A Sagatini, V Petrarca, MA DiDeco. Behavioural divergence between mosquitoes with different inversion karyotypes in polymorphic populations of the *Anopheles gambiae* complex. Nature 226:832–833, 1977.
11. HA Mariath, CJ Orton, CJ Shivas. Resistance to oviposition suppressants in *Lucilia cuprina*. In: JA McKenzie, PJ Martin, JH Arunde, eds. Resistance Management in Parasites of Sheep. Melbourne: Australian Wool Corp, 1990, p 52.
12. RT Roush, JC Daly. The role of population genetics in resistance research and management. In: RT Roush, BE Tabashnik, eds. Pesticide Resistance in Arthropods. New York: Chapman & Hall, 1990, pp 97–152.
13. ME O'Neal, ME Gray, CA Smyth. Population characteristics of a western corn rootworm (Coleoptera: Chyrsomelidae) strain in east-central Illinois corn and soybean fields. J Econ Entomol 92(6):1301–1310, 1999.
14. D Pimentel, L McLaughlin, A Zepp, B Lakitan, T Kraus, P Kleinman, F Vancini, WJ Roach, E Graap, WS Keeton, G Selig. Environmental and economic impacts of reducing U.S. agricultural pesticide use. In: D Pimentel, ed. Handbook of Pest Management in Agriculture. Boca Raton, FL: CRC Press, 1991, pp 679–718.

15. J Keiding. Resistance in the housefly in Denmark and elsewhere. In: DL Watson, AWA Brown, eds. Pesticide Management and Insecticide Resistance. New York: Academic Press, 1977, pp 261–302.
16. DO Drummond. Resistance in ticks and insects of veterinary importance. In: DL Watson, AWA Brown, eds. Pesticide Management and Insecticide Resistance. New York: Academic Press, 1977, pp 303–319.
17. PB Hughes, JA McKenzie. Insecticide resistance in the Australian sheep blowfly, *Lucilia cuprina*: Speculation, science and strategies. In: MG Ford, DW Holloman, BPS Khambay, RM Sawicki, eds. Combating Resistance to Xenobiotics: Biological and Chemical Approaches. London: Ellis Horwood, 1987, pp 162–177.
18. AM Shelton, JL Robertson, JD Tang, C Perez, SD Eigenbrode, HK Preisler, WT Wilsey, RJ Cooley. Resistance of diamondback moth (Lepidoptera: Plutellidae) to *Bacillus thuringiensis* subspecies in the field. J Econ Entomol 86(3):697–705, 1993.
19. GP Georghiou, R Mellon. Pesticide resistance in time and space. In: GP Georghiou, T Saito, eds. Pesticide Resistance to Pesticides. New York: Plenum Press, 1983, pp. 1–46.
20. AWA Brown, R Pal. Insect Resistance in Arthropods. Geneva: World Health Org, 1971.
21. G Hardin. The tragedy of the commons. Science 162:1243–1248, 1968.
22. GG Kennedy, ME Whalon. Managing pest resistance to *Bacillus thuringiensis* endotoxin: Constraints and incentives to implementation. J Econ Entomol 88(3):454–460, 1995.
23. JR Busvine. Resistance of insects to insecticides: The occurrence and status of insecticide resistant strains. Chem Ind (Rev) 42:1190–1194, 1956.
24. KM Theiling, BA Croft. Pesticide side-effects on arthropod natural enemies: a database summary. Agric Ecosys Environ 21:191–218, 1988.
25. B Croft. Arthropod Biological Control Agents and Pesticides. New York: Wiley, 1990.
26. P Jepsen, P Hennigan. SELECTV database of natural enemies resistant to pesticides. Internet Site, Oregon State Univ Dept of Entomology, 2000: http://www.ent3.orst.edu/Phosure/database/selctv/selctv.htm
27. MJ Dover, BA Croft. Integration of policy for resistance management. In: EH Glass, PL Adkinsson, GA Carlson, BA Croft, DE Davis, JW Eckert, GP Georghiou, WB Jackson, HM LeBaron, BR Levin, FW Plapp, RT Roush, HD Sisler, eds. Pesticide Resistance: Strategies and Tactics for Management. Washington, DC: Natl Acad Press, 1986, pp 422–435.
28. JR Busvine. Resistance to organophosphorus insecticides in insects. Int Symp Phytopharm Phytiat Pap, 1968, pp 605–620.
29. GP Georghiou, CE Taylor. Factors influencing the evolution of resistance. In: EH Glass et al., eds. Pesticide Resistance: Strategies and Tactics for Management. Washington, DC: Natl Acad Press, 1986, pp 157–312.
30. I Denholm, AR Horowitz, M Cahill, I Ishaaya. Management of resistance to novel insecticides. In: I Ishaaya, D Degheele, eds. Insecticides with Novel Modes of Action: Mechanism and Application. Berlin: Springer-Verlag, 1998, pp 260–281.
31. BE Tabashnik, NL Cushing, N Finson, MW Johnson. Field development of resis-

tance to *Bacillus thuringiensis* in diamondback moth (Lepidoptera: Plutellidae). J Econ Entomol 83(5):1671–1676, 1990.

32. E Grafius, E Morrow, A May. Biology and Control Strategies for Insect Pests of Potatoes. Report to the Potato Industry Commission, 1984.
33. E Marshal. A renewed assault on an old and deadly foe. Science 290(5491):428–430, 2000.
34. P Martens, L Hall. Malaria on the move: human population movement and malaria transmission. Emerging Infect Dis 6(2):103–109, 2000.
35. GP Georghiou. The magnitude of the resistance problem. In: EH Glass, PL Adkinsson, GA Carlson, BA Croft, DE Davis, JW Eckert, GP Georghiou, WB Jackson, HM LeBaron, BR Levin, FW Plapp, RT Roush, HD Sisler, eds. Pesticide Resistance: Strategies and Tactics for Management. Washington, DC: Natl Acad Press, 1986, pp 14–43.
36. Anonymous. Pesticide Product Information System (PPIS). Online. US Environ Protect Agency, Office of Pesticide Programs, Infor Resources and Services Division. June 5, 2000. www.epa.gov/opppmsd1/PPISdata/index.html
37. U.S. Department of Agriculture. Agriculture Chemical Usage: 1997 Fruits Summary. USDA NASS, ERS, 1998.
38. U.S. Department of Agriculture. Agricultural Chemical Usage: 1998 Vegetable Summary. USDA NASS, ERS, 1999.
39. U.S. Department of Agriculture. Agricultural Chemical Usage: 1999 Field Crops Summary. USDA, NASS, 2000.
40. Z Wen, JG Scott. Toxicity of fipronil to susceptible and resistant strains of German cockroaches (Dictyoptera: Blattellidae) and house flies (Diptera: Muscidae). J Econ Entomol 90(5):1152–1156, 1997.
41. RK Jansson, RA Dybas. Avermectins: Biochemical mode of action, biological activity and agricultural importance. In: I Ishaya, D Degheele, eds. Insecticides with Novel Modes of Action: Mechanisms and Application. Indianapolis: Springer-Verlag, 1998, pp 152–170.
42. MA Hoy, YL Ouyang. Selection of the western predatory mite, *Metaseiulus occidentalis* (Acari: Phytoseiidae), for resistance to abamectin. J Econ Entomol 82(1):35–40, 1989.
43. JA Argentine, JM Clark. Genetics and biochemistry of abamectin resistance in Colorado potato beetle. 7th Intl Congr Pesticide Chem 04E-14. 1990, p 418.
44. Y Konno, JG Scott. Biochemistry and genetics of abamectin resistance in the house fly. Pestic Biochem Physiol 41:21–28, 1991.
45. DG Cochran. Abamectin resistance potential in the German cockroach (Dictyoptera: Blattellidae). J Econ Entomol 87(4):899–903, 1994.
46. F Campos, RA Dybas, DA Krupa. Susceptibility of *Tetranychus urticae* (Acari: Tetranychidae) populations in California to abamectin. J Econ Entomol 88(2):225–231, 1995.
47. M Iqbal, RHJ Verkerk, MJ Furlong, PC Ong, SA Rahman, DJ Wright. Evidence for resistance to Bt subsp. *kurstaki* HD-1, Bt subsp. *aizawai* and abamectin in field populations of *Plutella xylostella* from Malaysia. Pestic Sci 48:89–97, 1995.
48. AS Perry, I Yamamoto, I Ishaaya, RY Perry. Insecticides in Agriculture and Environment. Berlin: Springer-Verlag, 1998.

49. CL McKenzie, RL Byford. Continuous, alternating, and mixed insecticides affect development of resistance in the horn fly (Diptera: Muscidae). J Econ Entomol 86(4): 1040–1048, 1993.
50. D Rugg, AC Kotze, DR Thompson, HA Rose. Susceptibility of laboratory selected and field strains of the *Lucilia cuprina* (Diptera: Calliphoridae) to ivermectin. J Econ Entomol 91(3):601–607, 1998.
51. DM Dunbar, RL Beard. Present status of milky disease of Japanese and oriental beetles (*Popillia japonica*, *Anomala orientalis*) in Connecticut. J Econ Entomol 68(4):453–457, 1975.
52. WH McGaughey, RW Beeman. Resistance to *Bacillus thuringiensis* in colonies of Indianmeal moth and almond moth (Lepidoptera: Pyralidae). J Econ Entomol 81(1): 28–33, 1988.
53. CJ Perez, A Shelton. Resistance of *Plutella xylostella* (Lepidoptera: Plutellidae) to *Bacillus thuringiensis* Berliner in Central America. J Econ Entomol 90(1):87–93, 1997.
54. G Sinegre, M Babinot, JM Quermel, B Gaven. First field occurrence of *Culex pipiens* resistance to *Bacillus sphaericus* in southern France. Barcelona, Spain: VIIIth Eur Meeting Soc Vector Ecology, 5–8 Sept 1994.
55. MH Silva-Filha, L Regis, C Nielsen-Leroux, JF Charles. Low level resistance to *Bacillus sphaericus* in a field treated population of *Culex quinquefasciatus* (Diptera: Culicidae). J Econ Entomol 88(3):525–530, 1995.
56. DR Rao, TR Mani, R Rajendran, AS Joseph, A Gajanana, R Reuben. Development of a high level of resistance to *Bacillus sphaericus* in a field population of *Culex quinquefasciatus* from Kochi, India. J Am Mosq Control Assoc 11(1):1–5, 1995.
57. RG Luttrell, L Wan, K Knighten. Variation in susceptibility of noctuid (Lepidoptera) larvae attacking cotton and soybean to purified endotoxin proteins and commercial formulations of *Bacillus thuringiensis*. J Econ Entomol 92(1):21–32, 1999.
58. AR Abot, F Moscardi, JR Fuxa, DR Sosa-Gomez, AR Richter. Development of resistance by *Anticarsia gemmatialis* from Brazil and the United States to a nuclear polyhedrosis virus under laboratory selection pressure. Biol Control 7(1):126–130, 1996.
59. M Cahill, K Gorman, S Day, I Denholm. Baseline determination and detection of resistance to imidacloprid in *Bemisia tabaci* (Homoptera: Aleyrodidae). Bull Entomol Res 86:343–349, 1996.
60. N Prabhaker, NC Toscano, SJ Castle, TJ Henneberry. Selection for imidacloprid resistance in silverleaf whiteflies from the Imperial Valley and development of a hydroponic bioassay for resistance monitoring. Pestic Sci 51:419–428, 1997.
61. J Zhao, BA Bishop, EJ Grafius. Inheritance and synergism of resistance to imidacloprid in the Colorado potato beetle (Coleoptera: Chrysomelidae). J Econ Entomol 93(5):1508–1519, 2000.
62. D Mota-Sanchez, ME Whalon. Unpublished data.
63. R Nauen, A Elbert. Apparent tolerance of a field-collected strain of *Myzus nicotianae* to imidacloprid due to strong antifeeding responses. Pestic Sci 49:252–258, 1997.
64. Z Wen, JG Scott. Cross resistance to imidacloprid in strains of German cockroach (*Blatella Germanica*) and house fly (*Musca Domestica*). Pestic Sci 49:367–371, 1997.

65. DA Dame, GJ Wichterman, JA Hornby. Mosquito (*Aedes taeniorhynchus*) resistance to methoprene in an isolated habitat. J Am Mosq Control Assoc 14(2):200–203, 1998.
66. AR Horowitz, I Ishaaya. Managing resistance to insect growth regulators in the sweetpotato whitefly (Homoptera: Aleyrodidae). J Econ Entomol 87(4):866–871, 1998.
67. FJ Devin, I Ishaaya, AR Horowitz, I Denholm. The response of pyriproxyfen-resistant and susceptible *Bemisia tabaci* Genn (Homoptera: Aleyrodidae) to pyriproxyfen and fenoxycarb alone and in combination with piperonyl butoxide. Pestic Sci 55(4):405–411, 1999.
68. L Zhang, S Kasai, T Shono. In vitro metabolism of pyriproxyfen by microsomes from susceptible and resistant housefly larvae. Arch Insect Biochem Phys 37:215–224, 1998.
69. A De Cock, I Ishaaya, M Van De Veire, D Degheele. Response of buprofexin-susceptible and -resistant strains of *Trialeurodes vaporariorum* (Homoptera: Aleyrodidae) to pyriproxyfen and diafenthiuron. J Econ Entomol 88(4):763–767, 1995.
70. VW Waldner. Vorschlage zur Apfelwicklerbekamfung. Obstbau Weinvau 71–73, 1993.
71. H Riedl, R Zelger. Erste Ergebnisse der Untersuchungen zur Resistenz des Apfelwickers gegenuber Diflubenzuron. Obstbau Weinbau 31:107–109, 1994.
72. B Sauphanor, JC Bouvier. Cross-resistance between benzoylureas and benzoylhydrazines in codling moth, *Cydia pomonella* (L.). Pestic Sci 45:369–375, 1995.
73. E Sauphanor, JC Bouvier, V Brosse. Spectrum of insecticide resistance in *Cydia pomonella* (Lepidoptera: Tortricidae) in southeastern France. J Econ Entomol 91(6): 1225–1231, 1998.
74. CH Wearing. Cross-resistance between azinphos-methyl and tebufenozide in the greenheaded leafroller, *Planotortrix octo.* Pestic Sci 54(3):203–211, 1998.
75. TB Stone, SR Simms, PC Marrone. Selection of tobacco budworm for resistance to a genetically engineered *Pseudomonas fluorescens* containing the deltaendotoxin of *Bacillus thuringiensis*. J Invertebr Pathol 53:228–234, 1989.
76. F Gould, A Martinez-Ramirez, A Anderson, J Ferre, FJ Silva, WJ Moar. Broad-spectrum resistance to *Bacillus thuringiensis* toxins in *Heliothis virescens*. Proc Natl Acad Sci USA 89:7986–7990, 1992.
77. F Gould, A Anderson, A Reynolds, L Baumgarner, WJ Moar. Selection and genetic analysis of a *Heliothis virescens* (Lepidoptera: Noctuidae) strain with high levels of resistance to some *Bacillus thuringiensis* toxins. J Econ Entomol 88:1545–1559, 1995.
78. ME Whalon, WH McGaughey. Insect resistance to *Bacillus thuringiensis*. In: L Kim, ed. Advanced Engineered Pesticides. New York: Marcel Dekker, 1993, pp 215–231.
79. F Huang, AR Higgins, LL Buschman. Baseline susceptibility and changes in susceptibility to *Bacillus thuringiensis* subsp. *kurstaki* under selection pressure in European corn borer (Lepidoptera: Pyralidae). J Econ Entomol 90(5):1137–1143, 1997.
80. C Nielsen-Leroux, F Pasquier, FJF Charles, G Sinegre, B Gaven, N Pasteur. Resistance to *Bacillus sphaericus* involves different mechanisms in *Culex pipiens* (Diptera: Culicidae) larvae. J Med Entomol 34(3):321–327, 1997.

81. J Rodcharoen, MS Mulla. Resistance development in *Culex quinquefasciatus* (Diptera: Culicidae) to *Bacillus sphaericus*. J Econ Entomol 87(5):1133–1140, 1994.
82. GP Georghiou, JI Malik, M Wirth, K Sainato. Characterization of resistance of *Culex quinquefasciatus* to the insecticidal toxins of *Bacillus sphaericus* (strain 2362). Annu Rep Mosquito Control Res. Univ California, 1992, pp 34–35.
83. IF Goldman, J Arnold, BC Carlton. Selection for resistance to *Bacillus thuringiensis* subspecies *israelensis* in field and laboratory populations of the mosquito *Aedes aegypti*. J Invertebr Pathol 47:317–324, 1986.
84. U Estada, J Ferre. Binding of insecticidal crystal proteins of *Bacillus thuringiensis* to the midgut brush border of the cabbage looper, *Trichoplusia ni* (Hubner) (Lepidoptera: Noctuidae), and selection for resistance to one of the crystal proteins. Appl Environ Microbiol 60:3840–3846, 1994.
85. J Muller-Cohn, J Chaufaux, C Buisson, N Gilois, V Sanchis, D Lereclus. *Spodoptera littoralis* (Lepidoptera: Noctuidae) resistance to Cry IC and cross-resistance to other *Bacillus thuringiensis* crystal toxins. J Econ Entomol 89:791–797, 1996.
86. WJ Moar, M Pusztai-Carey, M Nan Faassen, D Bosch, R Frutos, C Rang, K Luo, MJ Adang. Development of *Bacillus thuringiensis* Cry IC resistance by *Spodoptera exigua* (Hubner) (Lepidoptera: Noctuidae). Appl Environ Microbiol 61:2086–2092, 1995.
87. WH McGaughey. Insect resistance to the biological insecticide *Bacillus thuringiensis*. Science 229:193–195, 1985.
88. WH McGaughey, DE Johnson. Indianmeal moth (Lepidoptera: Pyralidae) resistance to different strains and mixtures of *Bacillus thuringiensis*. J Econ Entomol 85(5): 1594–1600, 1992.
89. LS Bauer, CN Koller, DL Miller, RM Hollingworth. Laboratory selection of the cottonwood leafbeetle. J Invertebr Pathol 60:15–25, 1994.
90. LS Bauer. Resistance: A threat to the insecticidal crystal proteins of *Bacillus thuringiensis*. Fla Entomol 78:414–443, 1995.
91. J Ferre, MD Real, J van Rie, S Jansens, M Peferoen. Resistance to the *Bacillus thuringiensis* bioinsecticide in a field population of *Plutella xylostella* is due to a change in a midgut membrane receptor. Proc Natl Acad Sci USA 88:5119–5123, 1991.

9

New Technologies for the Delivery of Pesticides in Agriculture

Robert E. Wolf
Kansas State University
Manhattan, Kansas, U.S.A.

1 INTRODUCTION

The need to protect our environment from the hazards of using crop protection products has sparked several technological improvements in application equipment. Many rules and regulations have been upgraded and new ones established in recent years to put increased emphasis on the safety issues that relate to our food supply and the application industry. The Worker Protection Standard was put in place to specifically protect agricultural workers and pesticide handlers from exposures while working with pesticides. A more recent regulation, the Food Quality Protection Act, has changed the way the U.S. Environmental Protection Agency regulates pesticides. This law has resulted in label changes that reduce the amount of pesticide used and lower the potential for exposure. This can be accomplished in various ways, such as reduced rates, alternative application methods, increased worker re-entry intervals, and reduced number of pesticide applications.

As a result of the various regulations, efforts to increase operator safety and improve application efficiency and effectiveness, and consideration of ways to reduce the amounts of pesticides applied are influencing equipment development. Researchers are evaluating ways to reduce the drift of crop protection products from treated areas. Also, reduced exposure to those who mix, load, and handle pesticides is being mandated. Containment structures and mixing–loading pads are being constructed to protect the groundwater.

All users of pesticides are confronted with several potential hazards. Those who mix, load, apply, and handle pesticides have a risk of exposure, but they also can cause environmental harm. Misapplication, spills, and unsafe application techniques are all major sources of contamination for humans, wildlife, and water. Because pesticides are likely to be a part of the pest management system for the foreseeable future, ways to reduce risks in the use of pesticides must be practiced.

Because it is essential to protect our environment during the use of pesticides, marked improvements in application technologies have been developed. Variable rate applications, prescription rates of crop protection products, direct injection, closed handling systems, onboard dry and liquid application systems, control systems, spot sprayers, shielded sprayers, air assist systems, new nozzle designs, and tank-rinsing devices are examples of technological changes that have affected the pesticide application industry. There has also been a major effort to reduce the amount of chemicals used. Chemical companies are developing new products that are effective at very low rates and designed for targeted applications with equipment that can apply precisely the correct amount when and where it is needed.

Efficient use of inputs has always been the goal of agriculture. Chemical registrants, farmers, and chemical dealers are becoming more sophisticated and have concern for the environment. Public scrutiny of chemical use and regulations limiting the use of agricultural chemicals make it essential that technological developments be forthcoming to address environmental concerns. Most dealers and growers are ready to evaluate any new developments or practices. In addition to a general discussion of application equipment, this chapter examines the new technology available for pesticide application that will protect the environment from pesticide contamination.

2 BASIC APPLICATION SYSTEMS

Better application equipment and new techniques that allow for smaller dosages of pesticides and reduced drift have become increasingly important in minimizing harmful effects of pesticides on applicators and the environment. Changes in the application equipment places increased responsibility on those who apply pesticides to be knowledgeable about the equipment being used. It is not essential to know about all types of application equipment, but a very good understanding of application equipment in general will be beneficial to the readers of this chapter. The following sections are devoted to helping readers understand the basic application systems.

Liquid and granular formulations are the most common forms of agricultural pesticides. Application devices are available in various types and sizes, each designed for a specific application, ranging from aerosol cans to airplanes. Each of these devices has its distinct uses and features.

The types of sprayers used to apply pesticide products include hand-operated sprayers, low-pressure powered sprayers, high-capacity powered sprayers, airplane sprayers, and special sprayers for selective application of pesticides. Devices for granular application are also used for a variety of pesticides, either by broadcast application or by row or band application for covering wide swaths or narrow strips over the crop row.

2.1 Manual Sprayers

Hand-operated sprayers, such as compressed air and knapsack sprayers, are designed for spot treatment and for areas unsuitable for larger units. They are relatively inexpensive, simple to operate, maneuverable, and easy to clean and store. Compressed air or carbon dioxide is used in most manual sprayers to apply pressure to the supply tank and force the spray liquid through a nozzle.

2.1.1 Compressed Air Sprayers

Pressure for most compressed air sprayers is provided by a manually operated air pump that fits into the top of the tank and supplies compressed air to force the liquid out of the tank and through a hose. A valve at the end of the hose controls the flow of liquid. Shaking the tank provides agitation for this system. Because the pressure varies so much, manual sprayers can result in a nonuniform application. A recent enhancement is the addition of a pressure control valve to maintain a constant pressure. The sprayer could also be fitted with a pressure gauge to monitor the tank pressure.

In some compressed air sprayer units, a precharged cylinder of air or carbon dioxide is used to provide pressure. These units include a pressure-regulating valve to maintain uniform spray pressure.

2.1.2 Knapsack Sprayers

As the name indicates, a knapsack sprayer is carried on the operator's back. Pressure is maintained by a piston or diaphragm pump that is operated either by hand or by a small engine. An air chamber helps "smooth out" pump pulsation. Spray material in the tank is agitated by a mechanical agitator or by bypassing part of the pumped solution back into the tank.

2.2 Hand-Held Spray Guns

Spray guns range from those that can produce a low flow rate with a wide-cone spray pattern or a flooding or showerhead nozzle pattern to those that can produce a high flow rate with a solid narrow-stream spray pattern. Spray guns with showerhead nozzles are commonly used to make commercial lawn applications. Four factors are critical for delivering the correct rate uniformly over the application area when using a showerhead type of nozzle: (1) The exact pressure must be monitored; (2) a proper spraying speed must be maintained; (3) a uniform motion technique must be used; and (4) a constant nozzle height and angle with reference to the ground must be maintained. When the spray gun is used, one should be aware of the difficulty in obtaining a uniform spray.

2.3 Low-Pressure Field Sprayers with Booms

Low-pressure sprayers equipped with spray booms are more commonly used than any other kind of application equipment. Tractor-mounted, pull-type, and self-propelled sprayers are available in many models, sizes, and prices. Application volumes can vary from 5 to over 100 gallons per acre (gpa).

3 SPRAYER COMPONENTS

All low-pressure sprayers have several basic components, including a pump, a tank, agitation devices, flow-control assemblies, strainers, hoses and fittings, booms, nozzles, and, typically, electronic or computerized components to help improve the accuracy of the application process. A brief description of each of these components follows.

3.1 Pumps

The pump is the "heart" of the sprayer. Sprayer pumps are used to create the hydraulic pressure required to deliver the spray solution to the nozzles and then atomize it into droplets. The most common types of pumps available for applying pesticides are roller, centrifugal, diaphragm, and piston pumps. For low-pressure sprayers the centrifugal and roller pumps are the most common, but the diaphragm pump is becoming more popular. Either a diaphragm or piston pump is commonly used where higher pressures are needed to move spray product through long lengths of hose such as in turf or roadside applications.

Regardless of the type of pump, it must provide the necessary flow rate at the desired pressure. It should pump enough spray liquid to supply the gallons per minute (gpm) required by the nozzles and the tank agitator, with a reserve capacity of 10–20% to allow for some flow loss as the pump becomes worn.

Table 1 lists the characteristics of the four types of sprayer pumps discussed here.

3.2 Tanks

The spray tank should have adequate capacity for the job. Tanks should also be clean, corrosion-resistant, easy to fill, and suitably shaped for mounting and effec-

TABLE 1 Common Pump Types and Characteristics for Sprayers

Characteristic	Roller	Centrifugal	Diaphragm	Piston
Cost	Low	High	Medium	High
Displacement	Positive; self-priming; requires relief valve	Nonpositive; needs priming; relief valve not required	Positive; self-priming; requires relief valve	Positive; self-priming; requires relief valve
Drive mechanism	PTO; gas engine drives; electric motors	PTO; hydraulic; gas engines; electric motors	PTO; hydraulic; gas engines	PTO; gas engines; electric motors
Adaptability	Compact and versatile	Good for abrasive materials; handles suspensions and slurries well, needs higher rpm	Compact for amount of flow and pressure developed	Wide range of spraying applications; dependable
Durability	Parts to wear, replace	Very durable; not much wear	No corrosion of internal parts	Parts to wear, replace
Serviceability	Easy to work on and repair	Simple maintenance extends life	Low maintenance	Potential for high maintenance
Pressure range	Up to 300 psi	Up to 180 psi	Up to 725 psi	Up to 400 psi
Output volume	2–74 gpm; high volumes for size; proportional to pump speed	Up to 190 gpm; high volumes for size and weight; proportional to pump speed	3.5–6 gpm; proportional to pump speed	Low, up to 10 gpm; proportional to pump speed, independent of pressure
Speed, rpm	540, 1000	Up to 6000; requires speed-up mechanism; very efficient at higher speeds	540	540
Comments	Best choice for farmers	If hydraulically driven, then no PTO required, popular in commercial agricultural applications; running pump dry is a problem	Good for higher pressure requirements; popular for horticultural applications; pump can run dry	Similar to an engine; low capacity

gpm, gallons per minute; psi, pounds per square inch; PTO, power take off; rpm, revolutions per minute.

tive agitation. The openings on the tank should be suitable for pump and agitator connections. Tanks that are not transparent should have a sight gauge or other external means of determining the fluid level. Sight gauges should have shutoff valves to permit closing in case of failure. The primary opening of the tank should be filled with a cover that can be secured to avoid spills and splashes. It also should be large enough to facilitate cleaning of the tank. A drain should be located at the bottom so that the tank can be completely emptied.

Tanks are commonly constructed of stainless steel, polyethylene, and fiberglass. The materials used will influence the cost of the tank, its durability, and its resistance to corrosion.

3.3 Agitation Devices

Agitation requirements depend largely on the formulation of the chemical being applied. Soluble liquids and powders do not require special agitation once they are in solution, but emulsions, wettable powders, and liquid and dry flowable formulations will usually separate if they are not agitated continuously. Separation causes the concentration of the pesticide spray to vary greatly as the tank empties. Improper agitation may also result in plugging of the parts of the spray distribution system. For these and other reasons, thorough agitation is essential.

Hydraulic jet agitation is the most common method used with low-pressure sprayers. Jet agitation is simple and effective. A small portion of the spray solution is circulated from the pump output back to the tank, discharging it under pressure through holes in a pipe or through special agitator nozzles.

The amount of flow needed for agitation depends on the chemical used as well as on the size and shape of the tank. Foaming can occur if the agitation flow rate is too high or remains constant as the tank empties. Using a control valve to gradually reduce the amount of agitator flow can prevent foaming.

3.4 Flow Control Assemblies

Roller pumps, diaphragm pumps, and piston pumps usually have a flow control assembly consisting of a bypass-type pressure regulator or relief valve, a control valve, a pressure gauge, and a boom shutoff valve. Bypass pressure relief valves usually have a spring-loaded ball, disk, or diaphragm that opens with increasing pressure so that excess flow is bypassed back to the tank, thus preventing damage to the pump and other components when the boom is shut off. When the control valve in the agitation line and the bypass relief valve in the bypass line are adjusted properly, the spraying pressure will be regulated.

Because the output of a centrifugal pump can be reduced to zero without damaging the pump, a pressure relief valve and separate bypass line are not needed. The spray pressure can be controlled with simple gate or globe valves. It is preferable, however, to use special throttling valves designed to accurately control the spraying pressure. Electrically controlled throttling valves are becoming popular for remote pressure control.

Because nozzles are designed to operate within certain pressure limits, a pressure gauge must be included in every sprayer system. The pressure gauge must be used for calibrating and while operating in the field. Select a gauge that is suitable for the pressure range that you will be using.

A quick-acting boom cutoff or control valve allows the sprayer boom to be shut off while the pump and the agitation system continue to operate. Electric solenoid valves, which eliminate inconvenient hoses and plumbing, are also available.

3.5 Strainers

Three types of strainers are commonly used on low-pressure sprayers: tank filler strainers, line strainers, and nozzle strainers. The strainer size numbers (20 mesh, 50 mesh, etc.) indicate the number of openings per inch. Strainers with high mesh numbers have smaller openings than strainers with low mesh numbers.

Coarse-basket strainers are placed in the tank filler opening to prevent twigs, leaves, and other debris from entering the tank as it is being filled. A 16 or 20 mesh tank filler strainer will retain lumps of wettable powder until they are broken up, helping to provide uniform tank mixing.

A suction line strainer is used between the tank and a roller pump to prevent rust, scale, or other material from damaging the pump. A 40 or 50 mesh strainer is recommended. A suction line strainer is not usually needed to protect a centrifugal pump, except against large pieces of foreign material.

The inlet of a centrifugal pump must not be restricted. If a strainer is used, it should have an effective straining area several times larger than the area of the suction line. It should also be no smaller than 20 mesh and should be cleaned frequently. A line strainer (usually 50 mesh) should be located on the pressure side of the pump to protect the spray nozzles and agitation nozzles.

Small-capacity nozzles must have a strainer of the proper size to stop any particle that might plug the nozzle orifice. Nozzle strainers vary in size depending on the size of the nozzle tip used, but they are commonly 50 or 100 mesh.

3.6 Hoses and Fittings

All hoses and fittings should be of a suitable quality and strength to handle the chemicals at the selected operating pressure. A good hose is flexible and durable and resistant to sunlight, oil, and chemicals. It should also be able to hold up under the rigors of normal use, such as twisting and vibration. Two widely used materials that are chemically resistant are ethylene vinyl acetate (EVA) and ethylene propylene dione monomer (EPDM). A special reinforced hose must be used for suction lines to prevent their collapse.

Sometimes the pressure greatly exceeds the average operating pressures. These peak pressures usually occur as the spray boom is shut off. For this reason, the sprayer hoses and fittings must always be in good condition to prevent a possible rupture that could cause spills or cause the operator to be sprayed with the chemical.

As liquid is forced through the spray system, the pressure drops due to the friction between the liquid and the inside surface of the hoses, pipes, valves, and fittings. The pressure drop is especially high when a large volume of liquid is forced through a small-diameter hose or pipe. It is not uncommon to have a drop in pressure of 10–15 psi between the outlet of the pump and the end of the spray boom.

To minimize pressure drop, spray lines and suction hoses must be the proper size for the system. The suction hoses should be airtight, noncollapsible, as short as possible, and as large as the opening on the intake side of the pump. A collapsed hose can restrict flow and "starve" a pump, decreasing the flow as well as causing damage to the pump or the pump seals.

Other lines, especially those between the pressure gauge and the nozzles, should be as straight as possible with a minimum of restrictions and fittings. The proper size for these lines varies with the size and capacity of the sprayer. A high fluid velocity should be maintained throughout the system. If the lines are too large, the velocity will be low and the pesticide may settle out from the suspension and clog the system. If the lines are too small, an excessive drop in pressure will occur.

3.7 Booms

The boom on the sprayer provides a place to attach the nozzles in order to obtain a uniform distribution of the pesticide across the application target. Boom length and height will vary depending on the type of application. Boom stability is important in achieving uniform spray application. The boom should be relatively rigid in all directions. It should not swing back and forth or up and down. The boom should be constructed to permit folding for transport. The boom height should be adjustable.

3.8 Nozzles

The spray nozzle is the final part of the distribution system. The selection of the correct type and size is essential for each application. The nozzle determines the amount of spray applied to an area, the uniformity of the application, the coverage of the sprayed surface, and the amount of drift. One can minimize the drift problem by selecting nozzles that give the largest droplet size while providing adequate coverage at the intended application volume and pressure. Although nozzles have been developed for practically every kind of spray application, only a few types are commonly used in pesticide applications. An emphasis on nozzle design over the past few years has resulted in a vast improvement in spray quality. A few of the commonly used nozzle types for boom sprayer applications are described below.

3.8.1 Extended Range Flat-Fan Nozzles

Extended range flat-fan nozzles are frequently used for soil and foliar applications when better coverage is required than can be obtained from the flooding flat-fan, Turbo®flood (Spraying Systems Co., Wheaton, IL), or RA Raindrop® nozzles (Delavan Spray Technologies, Bamberg, SC). Extended range flat-fan nozzles are available in both 80° and 110° fan angles. The pattern from this type of nozzle has a tapered edge distribution. Because the outer edges of the spray pattern have reduced volumes, it is necessary to overlap adjacent patterns along a boom to obtain uniform coverage. Regardless of the spacing and height, for maximum uniformity in the spray distribution, the spray patterns should overlap about 40–50% of the nozzle spacing. Foam markers are commonly used to help operators keep track of swath width overlap requirements on multiple passes.

For soil applications, the recommended pressure range is 10–30 psi. For foliar application when smaller drops are required to increase the coverage, higher pressures, 30–60 psi, may be required. However, the likelihood of drift increases when higher pressures are used.

3.8.2 Even Flat-Fan Nozzles

Even flat-fan nozzles are different from the extended range flat-fan nozzle. They are designed to apply uniform coverage across the entire width of the spray pattern, thus overlap is not required. They should be used only for banding pesticides over the row. The nozzle height and spray fan angle determine the bandwidth.

3.8.3 Flooding Flat-Fan Nozzles

Flooding flat-fan nozzles produce a wide-angle, flat-fan pattern and are used for applying herbicides and mixtures of herbicides and liquid fertilizers. The nozzle spacing should be 40 in. or less. These nozzles are most effective in reducing drift when they are operated within a pressure range of 8–25 psi. Pressure changes affect the width of the spray pattern more with the flooding flat-fan nozzle than with the extended range flat-fan nozzle. In addition, the distribution pattern is usually not as uniform as that of the extended range flat-fan tip. The best distribution is achieved when the nozzle is mounted at a height and angle that allow 100% overlap. Uniformity of application depends on the pressure, height, spacing, and orientation of the nozzles. Pressure directly affects droplet size, nozzle flow rate, spray angle, and pattern uniformity. At low pressures, flooding nozzles produce large spray drops; at high pressures, these nozzles produce smaller drops than flat-fan nozzles at an equivalent flow rate.

The spray distribution of flooding nozzles varies greatly with changes in pressure. At low pressures, flooding nozzles produce a fairly uniform pattern across the swath, but at high pressures the pattern becomes heavier in the center and tapers off toward the edges. The width of the spray pattern is also affected by pressure. To obtain an acceptable distribution pattern and overlap, one should operate flooding nozzles within a pressure range of 8–25 psi.

Nozzle height is critical in obtaining uniform application when using flooding nozzles. Flooding nozzles can be mounted vertically to spray backward, horizontally to spray downward, or at any angle between vertical and horizontal. When the nozzle is mounted horizontally to spray downward, heavy concentrations of spray tend to occur at the edges of the spray pattern. Rotating the nozzles 30–45° from the horizontal will usually increase the pattern uniformity over the recommended pressure range of 8–25 psi.

3.8.4 Turbulation Chamber Nozzles

The most recent nozzle design improvements incorporate the preorifice concept with an internal turbulation chamber. This not only creates larger droplets but also improves the uniformity of the spray pattern. Turbulation chamber nozzles are available in a Turbo flood tip and in a Turbo flat-fan design.

Turbo Flood Nozzles. Turbo® flood nozzles combine the precision and uniformity of extended range flat spray tips with the clog resistance and wide-angle pattern of flooding nozzles. The design of the Turbo flood nozzle increases droplet size and distribution uniformly. The increased turbulence in the spray tip causes an improvement in pattern uniformity over that of existing flooding nozzles. At operating pressures of 10–40 psi, Turbo flood nozzles produce larger droplets than standard flooding nozzles. Having larger droplets reduces the number of drops of driftable size in the spray pattern; thus, Turbo flood nozzles work well in drift-sensitive applications. Turbo flood nozzles, because of their improved pattern uniformity, need 50% overlap to obtain properly uniform application.

Turbo Flat-Fan Nozzles. The Turbo flat-fan design shows great improvement in pattern uniformity compared to the extended range flat-fan and other drift reduction flat-fan designs. Turbo flat-fan nozzles are wide-angle preorifice nozzles that create larger spray droplets across a wider pressure range (15–90 psi) than comparable low-drift tips, reducing the amount of driftable particles. The unique design of the nozzles allows them to be mounted in a flat-fan nozzle body configuration. The wide spray angle will allow for 30 in. nozzle spacing and 50% overlap to achieve uniform application across the boom width.

3.8.5 Raindrop Nozzles

RA Raindrop® nozzles are used when spray drift is a major concern. When operated within a pressure range of 20–50 psi, these nozzles deliver a wide-angle, hollow-cone spray pattern and produce fewer small drops than flooding nozzles. For a uniform spray pattern, space the nozzles no more than 30 in. apart and rotate them 30° from the vertical axis. The RA Raindrop® nozzles are best used with soil-applied herbicides and can replace traditional flood nozzles for greater control of drift. Although the large droplets produced aid in drift control, they may result in less coverage than is required for some foliar pesticides. Heavier application rates can improve coverage. RA Raindrop® nozzles should be set to give 100% overlap.

3.8.6 Wide-Angle Full-Cone Nozzles

Wide-angle full-cone nozzles produce large droplets over a wide range of pressures in applications of pesticides. The in-line, or straight-through, design of the nozzles uses a counter-rotating internal vane to create controlled turbulence. The design allows the formation of a 120° spray angle over a pressure range of 15–40 psi. This nozzle provides a solid pattern with a uniform spray distribution and requires only about 25% overlap.

3.8.7 Drift Reduction Preorifice Nozzles

"Low-drift" nozzles are now available that will effectively reduce the development of driftable fines in the spray pattern. One design uses a preorifice located on the entrance side of the nozzle to effectively create a flow restriction, resulting in lower exit spray pressures and larger spray droplets. The term associated with this nozzle design is "drift reduction flat-fan nozzle." Drift reduction flat-fan nozzles produce a pattern similar to an extended range flat-fan pattern while effectively lowering the exit pressure of the nozzle. The lowered exit pressure creates a larger droplet spectrum with fewer driftable fines, minimizing the off-target movement of the spray.

Several styles of drift reduction flat-spray nozzles are currently available. All are very similar in design. With a larger droplet size, drift reduction preorifice nozzles can replace conventional flat-fan 80° and 110° tips in broadcast applications where spray drift is a problem. The recommended pressure for drift reduction preorifice nozzles is 30–60 psi. They require the same 50% overlap as the extended range flat-spray tips. An alternative to the preorifice nozzle is a larger extended range flat-fan nozzle operated at a lower pressure.

3.8.8 Air Assist Nozzles

Air assist nozzle technology involves the use of air incorporated into the spray nozzle to form an air–fluid mix. Several designs are currently being marketed and are commonly referred to as air induction or venturi nozzles. Basically, with the venturi design the air is entrapped in the spray solution at some point within the nozzle. To accomplish the mixing, some type of inlet port and venturi are typically used to draw the air into the tip under a reduced pressure. The air helps to atomize the solution and provides energy to help transport the droplets to the target. By increasing the size of the spray droplets, venturi nozzles reduce the spray drift by minimizing the smaller driftable fines created in a spray tip. The air induction or venturi nozzles are more expensive than conventional flat-fan and other drift reduction nozzle designs.

4 NOZZLE MATERIALS

Spray nozzle assemblies consist of a body, cap, check valve, and nozzle tip. Various types of bodies and caps (including color-coded versions) and multiple-nozzle bodies are available with threads as well as quick-attaching adapters. Nozzle tips are interchangeable or molded into the nozzle cap and are available in a wide variety of materials, including hardened stainless steel, stainless steel, brass, ceramic, and various types of plastic. Hardened stainless steel and ceramic are the most wear-resistant materials, but they are also the most expensive. Stainless steel tips have excellent wear resistance with either corrosive or abrasive materials. Plastic tips are resistant to corrosion and abrasion and are proving to be very economical tips for applying pesticides. Brass tips have been very common, but they wear rapidly when used to apply abrasive materials such as wettable powders and are corroded by some liquid fertilizers.

5 APPLICATIONS FOR GRANULAR PRODUCTS

Drop (gravity) and rotary (centrifugal) spreaders are available for applying granular pest control products. Drop spreaders are usually more precise and deliver a more uniform pattern than rotary spreaders. Because the granules drop straight down, there is also less chemical drift. Some drop spreaders will not handle larger granules, however, and ground clearance can be a problem. Moreover, because the edges of a drop-spreader pattern are well defined, any steering error will cause missed or doubled strips. Drop spreaders also usually require more effort to push than rotary spreaders.

Every drop or rotary spreader should be calibrated for proper delivery rate with each product and operator because of variability in the product, the operator's walking speed, and environmental conditions. The easiest method for checking the delivery rate of a spreader is to spread a weighed amount of product on a measured area (at least 1000 ft^2 for a drop spreader and 5000 ft^2 for a rotary spreader) and then weigh the product remaining in the speader to determine the rate actually delivered.

6 APPLICATION EQUIPMENT AND TECHNIQUES FOR MINIMIZING PARTICLE DRIFT

The misapplication of crop protectant products is a major concern in the application industry. One form of misapplication is spray drift. Although drift cannot be completely eliminated, the use of proper equipment and application techniques will maintain drift deposits within acceptable limits. The initial recommendation for drift control is to read the pesticide label. Instructions are given to ensure the safe and effective use of pesticides with minimal risk to the environment. Chemical company surveys indicate that a large percentage of drift complaints involve application procedures not specified on the label.

There are two ways that chemicals move downwind to cause damage: vapor drift and particle drift. Vapor drift is associated with the volatilization of pesticide molecules and then movement off-target. Particle drift is the off-target movement of spray particles formed during or after the application. The amount of particle drift depends mainly on the number of small "driftable" particles produced by the nozzle. Although excellent coverage can be achieved with extremely small droplets, decreased deposition and increased drift potential limit the minimum size that will provide effective pest control.

6.1 Factors Affecting Spray Drift

Several equipment and application factors greatly affect the amount of spray drift that occurs: the type of nozzle and orientation, pressure, boom height, and spray volume. The ability to reduce drift is no better than the weakest component in the spraying procedure. See the summary of recommended procedures for reducing particle drift injury provided by Table 2 in Section 6.2.

As previously mentioned, the potential for drift must be considered when selecting a nozzle type. Of the many types of nozzles available for applying pesticides, a few, especially those using the newer technology, are specifically designed for reducing drift by reducing the amount of small driftable spray particles in the spray pattern. Higher pressures and nozzles with lower flow rates will also lead to more drift by producing finer spray droplets. Changing pressure alone will also change the flow rate per nozzle and the overall application rate.

Spray height is also an important factor in reducing drift losses. Mounting the boom closer to the ground (without sacrificing pattern uniformity) can reduce drift. Nozzle spacing and spray angle determine the correct spray height for each nozzle type. Wide-angle nozzles can be placed closer to the ground than nozzles producing narrow spray angles. On the other hand, older style wide-angle nozzles also produce smaller droplets. When this occurs, the advantages of lower boom height are negated to some extent. However, the newer technology wide-angle drift reduction nozzles have actually been designed to reduce the number of small droplets and will assist in the reduction of drift at lower heights.

The use of larger nozzles is another means of minimizing drift. Increasing the spray volume by using higher capacity spray tips (usually at lower pressures to maintain constant flow rates) results in larger droplets that are less likely to move off-target. The only effective means of reducing drift by increasing spray volume is to increase the nozzle size.

TABLE 2 Summary of Recommended Procedures for Reducing Particle Drift Injury

Recommended procedure	Example	Explanation
Select nozzle type that produces coarse droplets.	Raindrop, wide-angle full-cone, Turbo flood, Turbo flat-fan, air induction/venturi.	Use droplets as large as practical to provide the necessary coverage.
Use lower end of pressure range.	Use 20–40 psi for Raindrop, less than 25 psi for other types. Air-assist will require above 40 psi.	Higher pressures generate many more small droplets with greater drift potential (less than 150 μm).
Lower boom height.	Use a boom height as low as possible to maintain uniform distribution. Use nozzle drops for systemic herbicides in corn.	Wind speed increases with height. A boom height a few inches lower can reduce off-target drift.
Increase nozzle size.	If normal application volume(s) is/are 15–20 gpa, increase to 25–30 gpa.	Larger capacity nozzles will reduce spray deposition off-target.
Spray when wind speeds are less than 10 mph and moving away from sensitive plants.	Leave a buffer zone if sensitive plants are downwind. Spray buffer zone when wind changes.	More of the spray volume will move off-target as wind increases.
Do not spray when the air is completely calm.	Absolutely calm air generally occurs in early morning or late evening, usually associated with a temperature inversion.	Calm air reduces air mixing and leaves a spray cloud that may move slowly downwind at a later time.
Use a drift control additive when needed.	Several conventional polyacrylamides and the newer biodegradable polymers are available.	Drift control additives increase the average droplet size produced by the nozzles.

Although not directly an equipment factor, one of the best tools available for minimizing drift damage is the use of drift control additives in the spray solution to increase the spray droplet size. Tests indicate that in some cases downwind drift deposits are reduced by 50–80% with the use of drift control additives. Drift control additives make up a specific class of chemical adjuvants and should not be confused with products such as surfactants, wetting agents, spreaders, and stickers. Drift control additives are formulated to produce a droplet size spectrum with fewer small droplets.

A number of drift control additives are commercially available, but they must be mixed and applied according to label directions in order to be effective. Some products are recommended for use at a rate of 2–8 oz per 100 gal of spray solution. Increased rates may further reduce drift but may also cause nozzle distribution patterns to be nonuniform. Drift control additives will vary in cost depending on the rate and formulation but are comparatively inexpensive for the amount of control provided. It is wise to test these products in each spray system to ensure that they are working properly before adapting this practice. Not all products work equally well for all systems. They do not eliminate drift, however, and common sense must still remain the primary factor in reducing drift damage.

6.2 Strategies to Reduce Spray Drift

Table 2 provides a summary of strategies that when used in combination will result in the best chance of minimizing drift. One strategy used alone will not necessarily prevent drift. A combination of strategies will provide the best insurance against the off-target movement of the crop protectant product used.

7 ELECTRONICS FOR PRECISE APPLICATION

Whether it is simply a monitor, a spray-rate controller, or a more sophisticated computer system, more and more operators are using spray apparatus equipped with electronic hardware and specially designed software to improve their application accuracy. Whatever the application requirements, electronic systems provide the versatility and intelligence to improve the efficiency and make the application process more precise and automatic.

The basic principle of operation for electronic control systems is the use of one or more sensors to measure a condition and a central processing unit (CPU) to translate the signal for display and for activating a process. Sensors are the keys to electronic control systems that monitor speed, flow, flow rate, pressure, clogged nozzles, and boom height. Monitors simply use the variables that determine application volume (speed, flow and/or pressure, and spray width) to calculate and display the resulting volume in gallons per acre. It is up to the operator to make adjustments as necessary to apply the desired number of gpa.

A combination of the above electronic components constitutes a rate-controlling system that will automatically adjust application rates on-the-go. Rate controllers input the desired gallons per acre and control the flow rate in gallons per minute by activating a servovalve (a regulating valve in the system) to maintain the required rate of flow. As the speed sensor detects an increase or decrease in ground speed, the electronic control system will calculate a new flow rate and automatically command the servovalve to adjust the application rate back to the original desired application rate. The new variable-rate systems use computers to determine the proper rate and control the amount of chemical applied. It is important to know that the limiting factor for precise application is the spray nozzle rather than the rate controller. With these units, changing nozzle pressure influences application volume (gpa) and spray droplet size (coverage and drift); it is critical to maintain the pressure within the recommended pressure range (for example, 10–50 psi for extended range flat-fan nozzles, 20–40 psi for RA Raindrop nozzles). Pressure must increase fourfold to double the nozzle flow rate. Therefore, even with a rate controller, one must keep ground speed within a narrow range in order to maintain the spray quality desired.

To regulate the flow in proportion to travel speed, the rate of increase in nozzle pressure must vary with the square of the rate of increase in speed. For example, the pattern width and distribution pattern may also be affected. For uniform application, the travel speed must be held as nearly constant as possible, even when controlled metering systems are being used. Another advantage to the new spray nozzle technology is that there is a greater margin for variation in travel speed. These new nozzles are designed to maintain a uniform quality pattern over a wider range of pressure; thus as field speeds change and the electronic controller increases or decreases system pressure, there will be less variation in spray droplet size. The potential for drift is lessened with today's high-speed application machines, which can have dramatic speed changes as they pass through the field.

These same electronic components provide the operator the ability to detect any application malfunctions. Sensors located at critical points on the application system will alert the operator to any problems that may occur. The console will either provide an audible warning or display an error message. The system may also be capable of providing a percent application error by calculating the differences between the target rate and the actual application rate.

Improvements in electronic or computerized application systems will lead to much more technological advancement in the application of crop protection materials.

8 ON-THE-GO/ONBOARD APPLICATION SYSTEMS

Another technology that has gained widespread acceptance is onboard, on-the-go impregnation of fertilizer and herbicide products. Impregnation, the combination of liquid herbicides and fertilizer for one-pass application, originally accomplished in the fertilizer plant, can now be done with airflow applicator units that are designed to place herbicide on the fertilizer carrier at the time the fertilizer is applied in the field. Introduction of airflow applicators paved the way for this technique. On-the-go impregnation provides benefits to both the environment and the equipment operators.

A major environmental improvement with onboard impregnation is moving the impregnation process from the fertilizer facility to the field where the application takes place. Elimination of herbicide residues in the mixing equipment, odors, and contaminated dusts at the plant and reduced operator exposure are all positive factors for on-the-go impregnation. Another consideration is that it avoids having unused impregnated fertilizer left over from the mixing of excess material.

On-the-go technology is also an advantage to commercial application businesses because it results in better and more efficient use of employee time and in less employee exposure to the pesticides being used. Farmers also benefit from the reduction in field compaction due to fewer trips having to be made across the field.

With the availability of new granular herbicide formulations, application equipment is being designed to apply dry fertilizer and dry granular herbicides simultaneously. This coapplication has become a popular alternative to the original liquid impregnation process. There are now several granular herbicide products on the market that are capable of being bulk handled in closed systems and can be applied either separately or together. Closed handling systems also protect the operator from unnecessary exposure to the chemical. The coapplication process offers many of the same advantages as impregnation while at the same time limiting the need to handle liquid chemicals.

9 SITE-SPECIFIC CROP MANAGEMENT (PRECISION AGRICULTURE)

The most recent development with on-the-go application technology is the concept of prescription application. "Prescription farming," "prescription agriculture," "site-specific farming," and "site-specific crop management" are terms often used to describe this practice.

Further developments of geographical information systems (GISs) and global positioning systems (GPSs) will expand the use of site-specific farming practices and will guide the development of new sprayer technology that will be able to confine crop protectant application to specific regions of a field. This technology could lead to smaller amounts of pesticides being applied to fields that are not uniformly covered with pests. The use of site-specific application systems for crop protectants will require accurate information about the spatial distribution of pest populations and a computer-controlled applicator interfaced with a navigation system.

Typically, pesticides are broadcast on an entire field without regard to the spatial variability of the pest population in the field. This practice results in areas where no or few pests exist receiving just as much product as areas with high pest populations. Information about the distribution of pests in a field may be gathered by using any of several different approaches. One method suitable for postemergence herbicides is to map the weed distributions as close to the time of application as possible. Geographical information systems and geographical positioning systems are used to develop application maps for this purpose. Crop scouts, aerial photography, and automated sensing devices could also be used in combination with the GIS/GPS technology to develop the application maps for all types of crop pests.

Remote sensing systems are designed to provide growers with timely information about pest infestations. Remote systems typically use cameras mounted on satellites or airplanes to record accurate pest information on a total field and farm basis. Early detection of pest problems can improve the farmers' ability to remedy pest infestations. Many will contend that remote sensing may change the way we use precision agriculture in the future.

Obviously, if a sophisticated application delivery system is developed that applies pesticides where pests exist and shuts off where there are no pests, then pesticide use can be reduced and the pesticide can be more effectively placed. This practice would result in a lower environmental burden and an increase in agricultural profitability. Selective spraying, spot spraying, and intermittent spraying are different names that are attached to this application method. Technology is becoming available that makes selective spraying a possibility. This technology uses machine vision sensing and digital video cameras. At the same time, computer-processing capabilities continue to increase. Computer technology is also able to control a new solenoid-activated valve that fits into standard nozzle fittings and can be pulsed on and off at a rapid rate. The flow rate of the nozzle can be varied continuously and independently of variations in pressure and droplet size.

Many options exist for the recognition of crop pests for mapping and pesticide application. Currently, insufficient data on spatial distributions of crop pests are available to determine which method may be best. Even less information exists on the economic and environmental benefits to be derived from the adaptation of this technology. The equipment is in place to make site-specific pest control applications. However, more specific pest information is needed to make sound application management decisions.

10 VARIABLE RATE SPRAY SYSTEM (PULSE NOZZLES)

A recent development in the application process is the commercialization of an electronically controlled adjustable rate spray system. This system uses conventional nozzles that are independently controlled along the boom by a computer. Flow rate is controlled at each nozzle by means of a solenoid valve that opens and closes 10–15 times per second. Each nozzle along the boom pulses in an alternating cycle and maintains a blended uniform deposition on the target. A computer connected to a flowmeter-based rate controller controls the pulses. The pulse system replaces the fluid pressure that is typically used to control the flow rate. With this system the flow rate can change in an 8:1 ratio independently of pressure change. Pressure can vary from 10 to 100 psi without any change in flow rate. This system gives the operator flexibility in the ability to control drift because it is designed to adjust flow and pressure without changing droplet size. The system is currently being marketed as the synchro nozzle. The synchro nozzle exhibits good application possibilities in combination with site-specific and variable rate application techniques.

11 ELECTROSTATIC SPRAY

An electrostatic charge is now being used commercially to aid in the transfer and attachment of the spray particle to the target. Electrostatic spray systems are commercially available for both aerial and ground applications.

With the ground application system, the process uses the principle of contact charging the liquid solution before it reaches the nozzles. The electric charge produced by the Energized Spray Process™ (ESP) system creates a high intensity electrostatic field that helps propel the spray droplets toward the target at a high velocity. Contact charging differs from earlier electrostatic systems that used induction charging of the spray solution at the nozzle. Contact charging adds 40,000 V to the liquid spray solution in a charging chamber and then distributes the solution in the charged state to the boom and nozzles. The electrostatic spray process shows promise of increasing coverage to both the upper and lower sides of the target leaves. This is a decided benefit with fungicide and insecticide applications. However, it is not clear whether the electrostatic process will provide drift reduction benefits and prove useful in the application of herbicides.

12 HOODS AND SPRAY SHIELDS

The use of mechanical shielded booms on sprayers offers applicators and growers another potential method to reduce drift. Several design options exist with this technology. Shielded booms are designed to protect the spray from the wind as it leaves the nozzle and travels to the target. A very important concern with shielded booms is the design. Improperly designed shields can result in more drift because negative pressures may build up inside the hood and force pesticide sprays out of the hood and into the environment, resulting in drift. Research has shown the potential for reduced drift when hoods are used rather than unshielded booms. Most of the studies reported that the drift potential is very closely related to the droplet size spectrum. The smaller the droplets being sprayed, the less potential there is for a dramatic reduction in drift with the hoods. Research has also shown that hoods do not perform as well in higher wind speeds as in weaker winds. It is a common belief that full boom shields provide little potential for drift reduction in row crops although their use for cereal grains or on fallow ground may result in reduced drift. Shielded boom sprayers have not been universally adopted throughout the spray industry. However, because of the uniformity of the target area, shielded booms are becoming popular in turf applications.

The use of individual row hoods, another variation of hooded spraying, in row crop settings is gaining popularity. Hoods of this type are designed to shield certain plants while spraying nonselective herbicides between the rows. Research shows that such systems may allow growers greater flexibility with their weed control program while reducing chemical costs, improving chemical efficiency, and reducing drift.

A hood spray system that incorporates optical sensors inside the individual row hoods is being developed commercially. This system uses a beam of light to detect weeds under the hoods and between the crop rows. When the sensor detects the weed, a spray nozzle is activated to spray the detected weed. With this system, the hood can protect sensitive plants in the rows from the nonselective spray materials. It is difficult for these sensors to distinguish weeds from the growing crop; thus the hood performs two critical functions in this system: It provides a protected area in which to sense the weeds and then shields the sensitive crop from the emitted spray. An additional benefit with this technology is the increased potential for using reduced amounts of herbicides. This translates to reduced crop protection costs and could also result in less drift. This technology also provides an excellent opportunity for use in site-specific crop protectant applications in the future.

13 INJECTION SYSTEMS

Efficient and safe use of inputs has always been the goal of applicators. Direct injection is an important technological development that can be used to help the application industry reduce the problems associated with chemical application.

Direct injection is a technology that may possibly have the greatest effect on the method of applying pesticides. With direct injection, the spray tank contains only water or the carrier. Prior to exiting the nozzle, chemical formulations (liquid or dry) or specially blended materials are injected directly into the spray lines that are applying the carrier as the sprayer travels through the field. The type of mixing that occurs depends on whether the injection occurs before or after the carrier spray pump. The type of metering pump used distinguishes the types of injection systems. The systems currently on the market use either piston or cam metering pumps to inject the chemical into the carrier. Either the chemical is injected into an in-line mixer prior to spraying or a series of peristaltic pumps meter the chemical and inject it on the inlet side of the carrier spray pump.

The early direct injection systems had several limitations. These included a lag time for the chemical to reach the nozzles, improper mixing of the chemical before spraying, and inability of the units to distribute wettable powder formulations. Many of the early problems with this technology have been resolved. Improved metering pump systems have reduced chemical lag time. The use of in-line mixers has resulted in more uniform mixing. The addition of agitation to mix wettable powders allows the use of a wide variety of formulations. Systems also exist that allow for the injection of dry formulations. Direct injection technology is becoming more prominent in the agricultural application industry. Control of injection with computers makes this technology well suited to adjusting rates on-the-go and for prescription applications. Rates can be accurately controlled to take advantage of site-specific needs that require precise application. On-line printers are available to produce a permanent record of chemical use and job location. Either the injection systems are included in the electronic controlling device or they can be added on as a module to existing control devices.

Another driving force behind much of the newly developed application technology is the development of sensors and the application of controllers. Spray controllers are being integrated into spray monitor systems. Electronic devices to control application rates have been widely used for years. Controllers are designed to automatically compensate for changes in speed and application rates on-the-go. Some are computer-based and work well with new application techniques such as direct injection and variable rate application. Computers and controllers work together to place pesticide in the precise desired position at the prescribed amount. The applicator's ability to precisely place pesticides is an important environmental factor.

The acceptance of direct injection technology has been spurred by environmental concerns, concern for operator safety, regulations, and the development of new products that are effective at very low rates of application. Direct injection eliminates the need to tank-mix chemicals; thus pesticide compatibility problems are eliminated. Cleanup of equipment is minimal, and with no leftover solutions, disposal of rinsates is not a major concern. If the chemicals are in returnable containers and are handled in a closed system, the potential for operator exposure is greatly reduced. Because of the added precision and the ability to spot-spray only where the pesticides are needed with the direct injection process, a substantial savings to the producer is realized and the environmental impact is reduced. Success or failure in the pesticide application industry rests on how well we manage and reduce the negative impacts on the environment.

14 HANDLING SYSTEMS

A major emphasis for chemical companies and equipment manufacturers has been to develop new and innovative ways to make the handling of chemicals more convenient and to reduce exposure for the people who use pesticide products. Bulk-handling and mini-bulk-handling systems are available to store, transport, and handle liquid and granular pesticides. The closed systems associated with bulk tanks reduce operator contact with the chemicals and eliminate potential spillage, and, with the returnable 250–300 gal containers, container disposal is eliminated.

Closed handling systems are also being developed to store, transport, and transfer dry granular pesticides. For example, pneumatic handling systems are used to transfer granular herbicides from bulk storage at the fertilizer plant into tendering vehicles that will deliver the product by air to the applicator units in the field.

Application practices using direct injection equipment can benefit from crop protection products packaged in smaller dedicated containers as described above.

15 FUTURE DEVELOPMENTS AND CONCERNS

Pesticides will continue to play a significant role in helping farmers provide an abundant and safe food supply for people throughout the world. The application industry will continue to change to make the use of pesticides as safe as possible. Technological improvements in the application industry have occurred at a very rapid rate in recent years. As scientists continue to focus on the precision farming of tomorrow, the equipment industry will work to improve and develop the new equipment needed to achieve the goal of more effective application. Major developments in field mapping and computer application controls are being refined.

10

Evolution of the Crop Protection Industry

Robert E. Holm and Jerry J. Baron
IR-4 Project
Technology Centre of New Jersey
Rutgers University
North Brunswick, New Jersey, U.S.A.

1 A CENTURY OF PROGRESS

As we enter the twenty-first century, we have a great opportunity to look back over the tremendous progress that agriculture made in the twentieth century. Modern agriculture came into being, and the crop protection industry combined with the fertilizer, hybrid seed, and equipment industries to provide an abundance in agricultural production that few would have believed possible when the century began. Table 1 indicates the dramatic yield increases for corn, soybeans, wheat, and cotton from 1920 to 1990, which ranged from 2.5-fold for soybeans to sixfold for wheat. The increased productivity was even more striking when measured by output per farmer, which increased 13-fold from an average of 9.8 people being fed per farmer in 1930 to 129 people per farmer in 1990.

When the significant cultural practices are overlaid on the yield and productivity increases, it is clear that the most dramatic improvements were made in the last half of the century and coincided with the maximized use of hybrid seed, better and more efficient mechanization of equipment, more available and cheaper fertilizer inputs (especially nitrogen), and the development of modern crop protection tools such as the ethylenebisdithiocarbamate (EBDC) fungicides, herbicides like the phenoxies and triazines, and effective insecticides like the organo-

TABLE 1 Impact of Crop Protection Developments on Row Crop Yield and Productivity

Row crop (yield parameter)	1920	1930	1940	1950	1960	1970	1980	1990
Corn (bu/acre)	29.9	20.5	28.9	38.2	54.7	72.4	91.0	118.5
Soybean (bu/acre)	N/A[b]	13.0	16.2	21.7	23.5	26.7	26.5	34.2
Wheat (bu/acre)	13.5	14.2	15.3	16.5	26.1	31.0	33.5	39.5
Cotton (lb/acre)	187	157	253	269	446	438	404	634
Output per farmer[a]	—	9.8	10.7	15.5	25.8	75.8	115	129
Significant changes in cultural practices	Increased mechanization →					Boll weevil eradication program →		
		Hybrid corn →			95% of corn acreage			
						Use of NH_3 as cheap nitrogen source →		
				Modern crop protection chemicals →				
				EBDC fungicides	Phenoxy herbicides Triazines Organophosphates	Pyrethroids	Glyphosate	Plant biotech

[a] Number of people fed.
[b] N/A, not available.
Source: Farm Chemicals WOW 2000 America.

phosphates, carbamates, and pyrethroids. In crops like cotton, where yields had plateaued for 20 years, area-wide government programs such as boll weevil eradication had an impact on a serious pest and reopened large areas of the Southeast and South to economical cotton production.

The last decade of the century brought about the advent of plant biotechnology, which was one of the most rapidly adapted new technologies ever utilized by farmers. It has been estimated that it took 7 years for 50% of U.S. corn farmers to accept hybrid corn but only 4 years for 50% of U.S. soybean growers to accept Roundup Ready soybeans [1]. We delve into the current and projected status of the plant biotechnology revolution in a later section but begin by focusing on the crop protection industry, recognizing again that it is just one important component in an overall production management system that includes fertilizers, mechanization, and improved seeds.

2 THE CROP PROTECTION INDUSTRY—HAS IT COME FULL CIRCLE?

The modern crop protection chemical industry evolved from European and U.S. chemical companies that were formed from the 1700s to the early part of the 1900s (Fig. 1). These companies dedicated resources to separate agricultural chemical operations during the first half of the last century as stand-alone units or as part of fertilizer operations. Several of these companies—for example, Eli Lilly and Bayer—also had growing pharmaceutical businesses that were not integrated with agrichemicals but often shared compounds synthesized by their chemists between biological evaluation groups. At one time, many of the major oil companies had agrichemical operations. However, one by one (Esso in 1969, Gulf in 1975, Mobil in 1981, Shell in 1988 and 1994, and Chevron in 1991 and 1993), the oil companies sold their agrichemical businesses and products to focus on their core businesses with their shorter term investment returns. Many corporate oil company boards of directors found it difficult to reconcile the 8–10 year period needed to develop and market a new agrichemical product at a cost of $30–50 million or more with the short-term investment turnaround of drilling new wells for oil or natural gas. It was this attitude that drove companies to develop life science business units comprising pharmaceuticals, agricultural chemicals, and sometimes animal health products. From an investment viewpoint, all of these high technology, heavily research-driven enterprises have similar long product development lead times with high investment commitments. With individual agrichemical product profit margin potentials of 50% or more, the corporate investment strategies and support to link these technology-driven business units together made good business logic. In addition, many of the discovery tools discussed later could be linked or shared, resulting in potential synergies and cost savings.

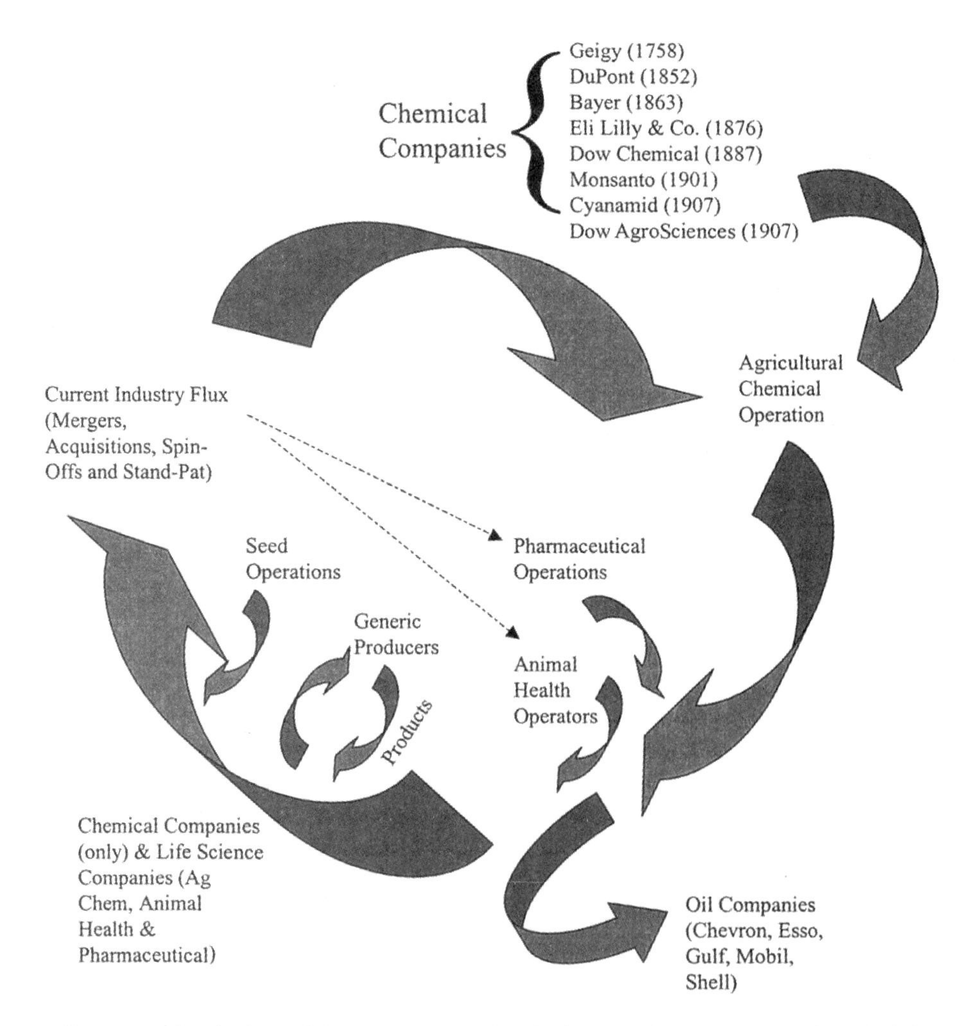

FIGURE 1 Evolution of the crop protection industry. Have we come full circle?

In the 1990s some other complicating factors came into play that had an impact on the industry. Many products patented in the 1950s–1980s came off patent and were the source of market opportunities for generic producers who had developed cost-effective manufacturing processes for many of the high-volume products. The seed industry also became a focus of the crop protection industry as a component of the plant biotechnology revolution when it became apparent that seed would be the carrier for the new technology and be a critical part in certain management systems such as herbicide-tolerant crops, as discussed in greater detail in Section 9.

What seemed to be a perfect strategy (i.e., life sciences) started to unravel in the latter part of the 1990s owing to a number of factors, mostly economically driven. After reaching record years of farm income in the early to mid-1990s, farm receipts dropped precipitously in the late 1990s and the early part of the twenty-first century owing to general worldwide feed grain surpluses and tumbling commodity prices. This situation, coupled with lower overall gross profit margins on off-patent products due to generic competition, led several companies to reconsider their life science strategies. Corporate boards were under increasing pressure from shareholders who saw lower profit margins (20% for generic agrichemicals versus more than 50% for pharmaceuticals) and public concerns with respect to plant biotechnology. They started to distance their drug and animal health operations from the agrichemical and plant biotechnology business units. A new company called Syngenta was formed in 2000 from the crop protection businesses of Novartis and Astra/Zeneca. Several of the major pharmaceutical companies (Merck in 1997, American Home Products in 2000, and Abbott in 2000) divested their crop protection business units completely, while other companies, e.g., Pharmacia (which purchased Monsanto in 2000), set up their agricultural business as a separate operating company and sold public stock in it. Many analysts saw this as a first step in total divestiture of the agricultural business unit.

3 MERGERS AND ACQUISITIONS

The divesture of the agricultural chemical businesses by the oil companies was only a small part of the turnover in the overall industry, as noted in Table 2. Although there were a few changes in the 1960s and 1970s, the trend rapidly accelerated in the 1980s and 1990s as the dynamics of life science strategies, generic producers, seed businesses, and plant biotechnology drove companies to evaluate and re-evaluate the role of agrichemicals in their operations. It is clear that the trend will continue well into the twenty-first century. Many analysts have predicted that as few as five and as many as 10 companies will emerge as the ultimate survivors, as the trend line in Figure 2 verifies. The impact has been felt not only by the major companies directly involved in the industry but also by companies manufacturing and formulating agrichemical products as support industries. Several smaller companies, including FMC, Rohm & Haas (agrichemical business acquired by Dow AgroSciences in 2001), and Uniroyal Chemical (division of Crompton Corporation), continued to survive and compete by focusing on market niches such as minor crops and the home and garden market. FMC actually made the top 10 global crop protection companies in 2000, not so much from increased sales as from consolidation in companies above them in the rankings (Table 3). In 2000 alone, Novartis and Zeneca became Syngenta, and BASF

TABLE 2 Mergers and Acquisitions in the Crop Protection Industry

Year 2000 survivor	Merged or acquired companies (year)
Aventis	AgrEvo (1999), Rhône-Poulenc (1999), Stefes (1997), Plant Genetic Systems (1996), Hoechst/Schering/Nor-Am (1994), Union Carbide (1987), Mobil (1981), Am-Chem (1970s), Nor-Am (1963), ICC/American Hoechst (1961), Hoechst (1953), Roussel/UCLAF (1946). Others include Boots, Hercules, Fisons, Boots Fisons Hercules (BFC), Morton Norwich, May and Baker, Rhodia, Chipman, American Paint, and Amchem-Rhor.
BASF	American Cyanamid (AG business of American Home Products) (2000), Micro-Flo (1998), Sandoz (part of product line, 1996), American Home Products/American Cyanamid (1994), Shell International (1994), Celamerk (1986), Cela plus Merck (Darmstadt) (1972). Others include BASF Wyandotte, Wyandotte, and BASF Colors and Chemicals.
Bayer	Gustafson (50% with C. K. Witco in 1998), Bayer Corporation (AG divisions consolidated in 1995), Bayer/Miles (1978), Chemagro (1967). Others include Mobay, Baychem, and Geary Chemical.
Dow AgroSciences	Rohm & Haas (agrichemical division in 2001), Mycogen (1996, 1998), Sentrachem (1997), Dow Elanco (Dow Chemical plus Eli Lilly) (1989). Others include Murphy Chemical and Walker Chemie.
DuPont	Pioneer Hybrid (1997, 1999), Griffin Corporation (50% in 1997), Protein Technologies (1997), Shell Chemical (U.S. business in 1988).
Monsanto	An operating company of Pharmacia Corporation (2000), Asgrow (1998), DeKalb (1998), Holden Foundation Seeds (1998), Plant Breeding International Cambridge (1998), Cargill (joint venture, 1998), Calgene (1996/97), Agracetus (1996), Chevron (home products business in 1993).
Sumitomo Chemical	Abbott (AG business in 2000), Chevron (ag business in 1991). Others include California Chemical, California Spray, and some PPG Industries products.
Syngenta	Novartis (2000), Zeneca/Astra (2000), ISK Biotech (1997), Merck (AG business in 1997), Mogen International (1997), Ciba/Sandoz (1996), Northrup King (1996), ICI Americas (1993), Stauffer Chemical (1987), Garst Seed (1985), Ciba (1970), Geigy (1970). Others include Velsicol, Zoecon, International Minerals and Chemicals (AG products), MAAG, Michigan Company, Atlas, Cannet Corp, Chipman of Canada, Fermenta ACS/Plant Protection/AB, SDS Biotech, Diamond Shamrock, Showa Denko, Ansul AG Products, Diamond Alkali, Kolher Chemical.

Source: *Farm Chemicals* WOW 2000 Special Millennium Issue, herbicide company genealogy by AP Appleby and information by RE Holm and JJ Baron.

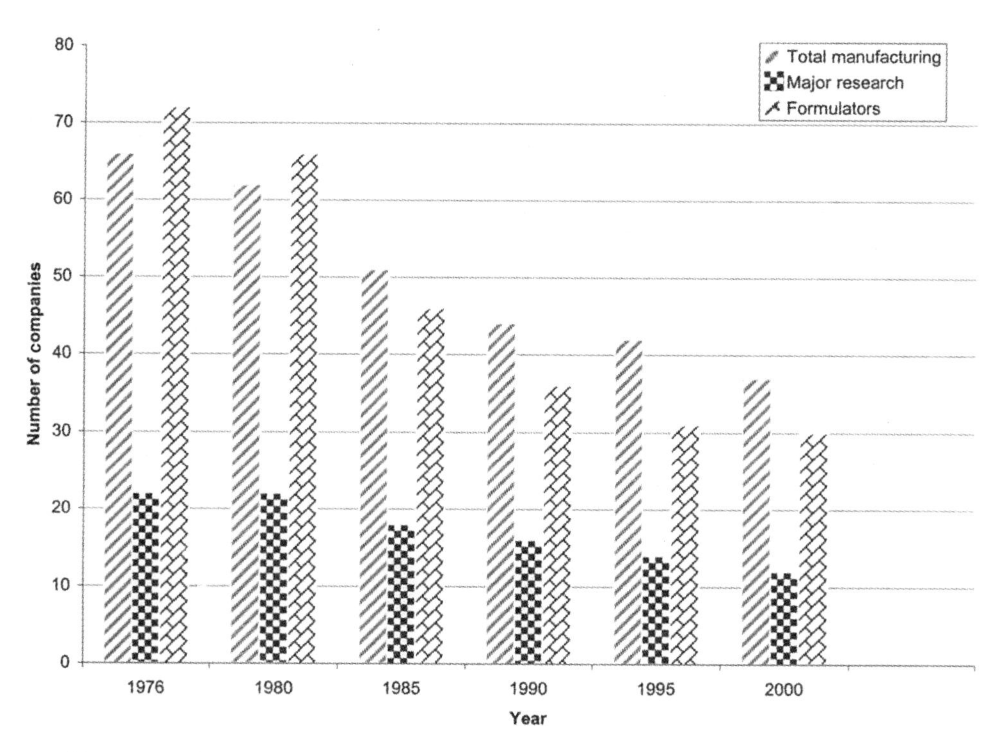

FIGURE 2 Consolidation trends in the agricultural chemical industry. (Adapted from Ref. 11.)

purchased the American Cyanamid agricultural business unit from American Home Products. The shuffling in top 10 rankings over this 5 year period is a true reflection of the constant turmoil in the industry. Also of interest in Table 3 is the flat to declining worldwide value of crop protection chemical sales over the latter part of the 1990s after increases of 2–3% per year due partially to increased use of generic products (i.e., lower cost) and the impact of herbicide-tolerant and insect-resistant crops in the United States.

Of major future interest in the merger and acquisition area will be the strategy of the Japanese agrichemical industry. While their U.S. and European counterparts have been very active in various strategies, of the major Japanese companies only Sumitomo Chemical Company has made a direct entry into the European (through Philagro) and U.S. (through Valent and Valent BioSciences) markets. The impact Ishihara Sangyo Kaisha (ISK) had in the U.S. market was greatly diminished by its product and business sale to Zeneca/Syngenta, although it has retained development rights to new products. Other companies such as

Table 3 Changes in Global Sales Leadership in Crop Protection (1996–2000)

Global rank	1996	1997	1998	1999	2000[a]
1	Novartis	Novartis	Novartis	Monsanto	Syngenta
2	Zeneca	Monsanto	Monsanto	Aventis	Aventis
3	Monsanto	Zeneca	DuPont	Novartis	Monsanto
4	AgrEvo	DuPont	Zeneca	DuPont	BASF
5	DuPont	AgrEvo	AgrEvo	Zeneca	DuPont
6	Bayer	Bayer	Bayer	Bayer	Bayer
7	RP[b]	RP[b]	RP[b]	Dow[c]	Dow[c]
8	Dow[c]	Dow[c]	Cyanamid[d]	BASF	Makhteshim-Agan
9	Cyanamid[d]	Cyanamid[d]	Dow[c]	Cyanamid[d]	Sumitomo
10	BASF	BASF	BASF	Makhteshim-Agan	FMC
Global sales[e]	29.4	29.8	31.0	29.8	29.7

[a] Estimates based on 1999 sales.
[b] Rhône-Poulenc.
[c] Dow AgroSciences.
[d] American Cyanamid/American Home Products.
[e] In billions of dollars.
Source: *Agrow* (various issues).

Kumiai Chemical, Nihon-Nohyaku, Sankyo, Hokko Chemical, Taketa Chemical, Nissan Chemical, and Nippon Soda appear to be content to license their new molecule discoveries to European and U.S.-based companies for development and marketing in countries outside Japan and Asia where they lack a major presence. With the increasing market globalization pressures and the costs of discovering new molecules, it is doubtful whether this independence can continue for long. Whether the Japanese agrichemical industry will consolidate internally within Japan, with the major global companies, or both remains to be seen. As will be noted in Section 8, many of the new chemistries being developed worldwide have their origin in Japanese laboratories.

4 IMPACT OF GENERIC PRODUCERS

From the perspective of the crop protection industry, the generic products industry has changed dramatically over the past 30 years. Generic producers used to be viewed as business opportunists by the basic manufacturers, who had invested large amounts of research dollars and business capital to discover, develop, and market new products only to see the generic producers quickly gain market entry once the products went off patent. Early off-patent product market launches by generic companies met with acrimonious lawsuits on data compensation allowed under the Federal Insecticide, Fungicide and Rodenticide Act (FIFRA). After several precedent-setting lawsuits were settled, the two camps settled into an uneasy truce. Gross profit margins of 20–30% for off-patent products versus 50–60% for patented products dictated different marketing strategies. Generic products continued to gain a foothold in markets where low-cost production was important and where farmers could not afford high production input costs such as in third world countries. It has been estimated that the market share for generic products exceeds 70% in China, 60% in India, 50% in Korea, and 40% in Taiwan.

In the 1990s, the attitude of the agricultural chemical companies toward their generic competitors started to change. In 1996, it was estimated that off-patent products accounted for over $18 billion in sales, or 58% of the global market [2]. This impact and growth could not be ignored. The Israeli company Makhteshim-Agan was formed from Makhteshim Chemical Works, Ltd. and Agan Chemical Manufacturers, Ltd. in 1996 and cracked the top 10 global sales list in 1999 with over $800 million in sales (Table 3). Makhteshim-Agan is clearly the global generic leader and remains an independent operating company (Table 4). Fernz/NuFarm (over $450 million sales in 1998) and Cheminova (over $300 million in 1998 sales) also remain independent. However, Griffin ($300 million in 1998 sales) became a 50% joint venture with DuPont and increased its 1998 sales to over $450 million with the addition of DuPont's off-patent products. DuPont gave as reasons for the joint venture (1) Griffin's knowledge of the generic business infrastructure and their proven record of managing off-patent prod-

Table 4 Generic Producers in the Crop Protection Industry

Companies acquired by or entering into joint ventures with basic manufacturers
Griffin: 50% joint venture with DuPont
Micro-Flo: BASF
Sentrachem: Dow AgroSciences
Stefes: Aventis
Top independent generic producers
Makhteshim-Agan (Israel)
Fernz/NuFarm (Australia)
Griffin (United States)
Cheminova (Denmark)
United Phosphorus (India)
Gharda (India)
CFPI (France)

Source: Generic producer information from *Farm Chemicals*, Spring 1998 issue.

ucts and (2) DuPont's strategy of focusing on basic research and patented products. Dow AgroSciences purchased Sentrachem, a $450 million (1998) generics manufacturer of glyphosate, triazines, mancozeb, carbofuran, and phenoxies in 1998. Dow indicated that the acquisition was part of their strategy to gain leadership in an industry being driven by biotechnology, consolidation, and generic competition. Along these same lines, Aventis (AgrEvo) purchased the German generics company Stefes, and BASF purchased the United States–based generic producer Micro-Flo. In a variation on the theme, Aventis (Rhône-Poulenc) created a separate operating division named Sedagri to market the company's generic product line.

Companies like Monsanto/Pharmacia who have not gone the generic route in partnerships and acquisitions have developed their own strategies for generic products. Monsanto's glyphosate became the crop protection industry's first proprietary product to exceed $1 billion in annual sales in the 1990s. Although glyphosate became a generic herbicide in much of the global market in the late 1980s and early 1990s, it did not go off-patent in the United States until 2000. Prior to that, Monsanto developed the strategy of becoming the lowest cost producer and expanding into new markets by lowering the price. According to Beer [3], the average global end user price for glyphosate technical dropped from $34/kg in 1991 to $20/kg in 1997—an 8%/yr reduction. However, agricultural uses of glyphosate increased from 42,000 tons in 1994 to just over 74,000 tons in 1997 (a 20%/yr increase) and were expected to exceed 112,000 tons in 1998. Monsanto increased production capabilities at a similar rate and was forecast to exceed 100,000 tons of glyphosate per year in 2000. Monsanto coupled an aggressive licensing program, lowered prices, and increased production with lower production costs. In the last few years, Monsanto licensed Syngenta (Novartis), Cheminova, Fernz/NuFarm, Dow AgroSciences, and BASF (Micro-Flo and Cyanamid) to sell glyphosate in stand-alone and premixed products. In addition, Monsanto used formulation technology to differentiate its Roundup glyphosate brand from the glyphosate of generic producers. It developed a formulation system termed UltraMAX that contained 25% more glyphosate than the older Roundup Ultra formulation along with formulation technology called Transorb that enhanced glyphosate uptake, translocation, and rain-fastness. They also introduced in 2000 a new formulation of glyphosate plus atrazine called ReadyMaster ATZ for the Roundup Ready corn market. With these innovations and Monsanto's dominant position in the herbicide-tolerant soybean market with its Roundup Ready program, they are in a position to maintain their glyphosate leadership position for years to come.

5 THE ROLE OF THE SEED COMPANIES

Until recent years, the association of the crop protection industry with the seed business was as remote as their relationship with the generic producers. That business approach did not change until the 1990s, when the heavy investment in plant biotechnology led to the realization that seed was the delivery system for newly discovered input traits like herbicide tolerance and insect resistance. The decade of the 1990s saw a wild scramble for seed businesses by the crop protection industry (Table 5). DuPont paid nearly $10 billion for Pioneer Hybrid in

TABLE 5 Seed Companies in the Crop Protection Industry

Seed company	Estimated sales ($ million)	Industry ownership[a] (year purchased)
Pioneer Hybrid	1900	DuPont (1997, 1999)
Novartis	1000	Syngenta
Limagrain	600	15% by Aventis
Savia	600	Independent
DeKalb	400	Monsanto (1998)
Asgrow	300	Monsanto (1998)
Advanta/Garst	200	Syngenta
Delta and Pine Land	200	Independent
Mycogen	150	Dow AgroSciences
Cargill	100	Independent

[a] Company totals (in millions): DuPont, $1900; Syngenta, $1200; Monsanto, $700.

Source: *Farm Chemicals* (various issues) and RE Holm and JJ Baron.

two steps (20% in 1996 and the remaining 80% in 1999) to gain control of the leading hybrid seed corn producer, which had over 40% of the U.S. market. Monsanto followed with acquisitions of DeKalb, the number two U.S. hybrid seed corn producer, with over 12% market share, and Asgrow, a leading soybean producer, to become the third largest seed producer. Syngenta, thanks to the seed business acquisitions of its parent companies (Zeneca with Advanta/Garst, Ciba with Ciba Seeds, and Sandoz with Northrup King), became the world's second largest seed company behind Pioneer/DuPont. Monsanto's stake in the global seed market could have been larger if it had held on to the cottonseed operations it purchased (Stoneville Pedigreed Seeds, acquired as part of the Calgene purchase) or had agreed to buy (Delta and Pine Land). Delta and Pine Land had over 70% of the U.S. cottonseed market, whereas Stoneville Pedigreed Seed held about 15% of the domestic cottonseed market. Because of antitrust concerns over control of such a large share of the cottonseed market, Monsanto sold Stoneville to Emergent Genetics in 1999. They later decided not to follow through on the Delta and Pine Land purchase and ended up paying an $81 million termination fee. Dow AgroSciences is another major player in the seed business with its

TABLE 6 Seed as a Multicomponent Delivery System

System	Technology	Benefits
Germplasm	Proprietary germplasm Hybridization systems Marker-assisted breeding tools	Higher yields Stress tolerance (drought, salt, cold, etc.) Insect and disease resistance Qualitative traits Nutritional value
Transgenes	Gene discovery, expression, and delivery	Insect, virus, and disease resistance Herbicide tolerance Quality traits
Seed protectants	Seed treatment chemicals and technology Safener technology	Soil and plant systemic insect and disease protection Increased crop tolerance to selected herbicides
Product variations	Processing, coating Pelleting, priming	Easier planting Improved germination Uniformity of emergence Soil temperature activation

purchase of Mycogen. Only Cargill; Delta and Pine Land; Limagrain, a French-based seed company that is 15% owned by Aventis (Rhône-Poulenc); and Savia, which is the largest vegetable seed producer with around 25% of the global market and 40% market share in the United States and Europe, remain independent from the crop protection industry.

The agrichemical industry views seed as a multicomponent delivery system not only for input and output traits but also for chemicals to be placed on the seed for protection against plant disease and insects (Table 6). Many of the new insecticides and fungicides being developed are highly active at low rates, are taken up by germinating seedlings, and are translocated to emerging and new foliage for 6–8 weeks or longer to protect systemically against insects and plant pathogens. Bayer, Syngenta, Aventis, Uniroyal Chemical, and others have developed strategic business plans to focus on the seed treatment market. Bayer purchased 50% share of Crompton Corporation's Gustafson seed treatment business, which is the leader in the United States. We expect the seed treatment chemical business focus to continue because of its environmental and worker exposure benefits versus foliar applications after plant emergence to control plant pests.

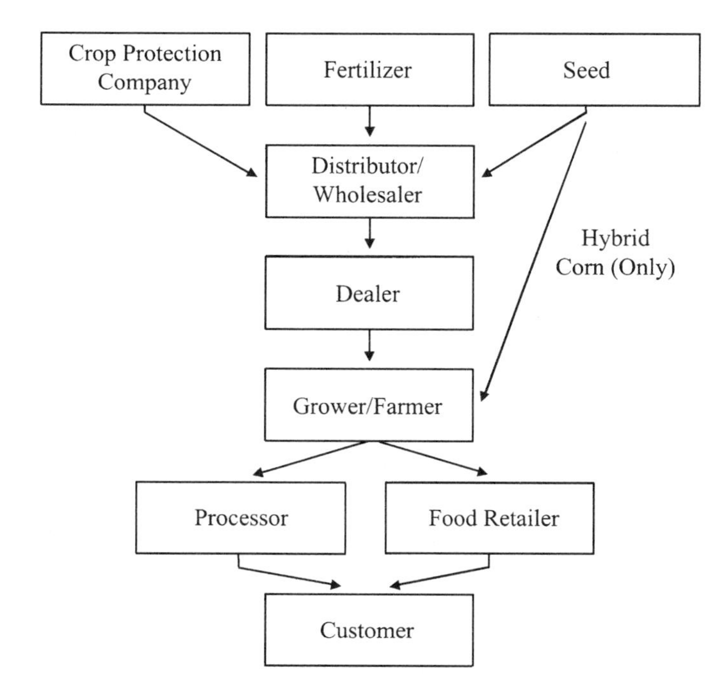

FIGURE 3 How crop protection tools reached the farmer in the past.

6 SUPPLYING CROP PROTECTION TOOLS TO THE FARMER

For many decades, from the 1920s until the early 1990s, farmers received their chemical, seed, and fertilizer inputs through the traditional chain diagramed in Figure 3. They then sold their produce to processors or food retailers for consumer purchase. The only exception was hybrid seed corn, which was sold to dealers who were usually local farmers and in turn sold to their neighbors. Now only fertilizer is sold through the traditional route. Crop protection chemicals may be sold directly to some large farms and farmer cooperatives. With the agrichemical company's big investment in the seed industry (see previous section), their involvement in seed sales has increased dramatically (see Fig. 4). Because input traits such as herbicide tolerance and insect resistance require a much greater

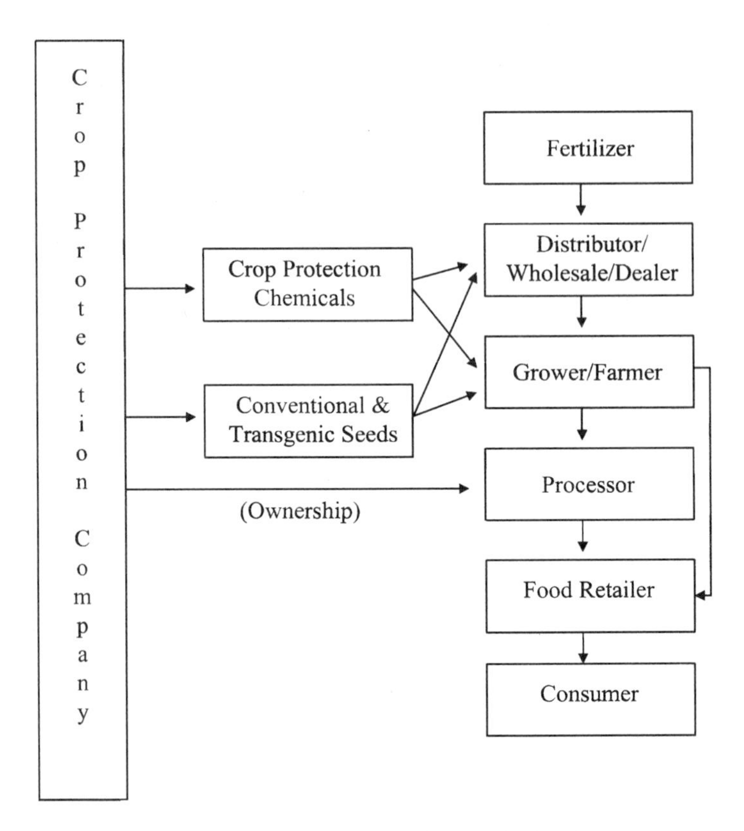

FIGURE 4 How crop protection tools reach the farmer—current and future.

level of management and technical skills to grow the treated seed, dealers and distributors are becoming much more involved in seed sales. The coupling of chemical treatments for traits such as herbicide tolerance with seed sales has been a way of maintaining grower contacts for distributors and dealers who supply both products (i.e., seed and chemicals) and the technical service to use them. Some companies have gone a step further by becoming involved in food processing. Novartis, Syngenta's parent, owns Gerber, the babyfood producer. DuPont purchased Protein Technologies, which is a global leader in the use of soybean proteins for the food industry. This has led to the term "dirt to dinner table" for the integration of DuPont's agricultural businesses ranging from supplying farmers with seed and chemicals through processing the crop for consumer use. It is likely that this trend will continue as companies look at ways to extract value from their technologies throughout the food chain.

7 THE DISCOVERY PROCESS

As with many other aspects of the industry discussed previously, such as the seed technology business integration, little changed in the processes used to synthesize and biologically evaluate new crop protection chemical candidates for 50 years from the 1940s until the early 1990s. An organic chemist could synthesize 50–100 unique new structures or 100–200 analogs of an active series annually. Biological evaluations were conducted on entire plants under greenhouse or growth chamber conditions. The limiting factor became the number of chemists a company could afford along with the biologists to conduct the empirical screening approach. By the 1970s, 100 new molecules per week or 5000 per year was the goal of many companies, and this was raised to 10,000 molecules per year in the 1980s and early 1990s. That goal changed dramatically by the mid-1990s when the dual technologies of combinatorial chemistry and high-throughput screening (HTS) pioneered by the pharmaceutical industry were adapted, especially in the life science companies that had drug and agrichemical business and research units [4]. Those discoveries led to new targets of 100,000 compounds per year by the late 1990s. However, by 1999 Aventis announced it was screening 600,000 compounds per year and would increase that number to 1 million annually. Bayer also announced at the end of the decade their intent to screen 1 million substances a year in their new $26.5 million research facility at their Monheim (Germany) Agricultural Research Center.

One of the consequences of combinational chemistry approaches for compound synthesis is the lack of pure products—i.e., the target substance is accompanied by other materials from the reaction plus starting materials. Because purification was not economically feasible before evaluation because of the high numbers involved (up to 5000 per day in some cases), the major concern switched

Table 7 New Approaches to Discovery

Company	Partner	Technology
Aventis	Cerep	Molecular modeling and combinatorial chemistry using virtual screening software for compound optimization
	3-Dimensional Pharmaceuticals	Compound libraries and technologies to optimize active compounds
	Molecular Simulations, Inc.	Molecular modeling software for lead discovery, optimization, and combinatorial chemistry approaches
	Bio Discovery	Screening and inspection of library extracts for crop production potential
Bayer	Exelixis	Use of gene-based technology to discover new insecticides and nematicides
	Paradigm Genetics	Use of gene function, bioinformatics, and new assays to develop novel screening targets for new herbicides
	ArQule	Use of combinatorial chemistry for screening to obtain several hundred thousand compounds
Dow AgroSciences	Biotica Technology	New Spinosad analogs obtained by targeting changes in biosynthetic pathway
	Integrated Genomics	DNA sequence of *Saccharopolyspora spinosa* used to improve Spinosad production through fermentation

DuPont	The Automation Partner	High-throughput system using bar-coded vials accessed by high-speed robots to prepare hundreds of thousands of molecules per day for screening
	3-Dimensional Pharmaceuticals	Use of computational and combinatorial chemical techniques to discover new molecules
	Combi Chem (acquisition)	Use of combinatorial chemistry for new leads
	Affymax Research Institute	Use of Affymax's chemical libraries for screening new crop protection targets using combinatorial chemistry and high-throughput screening approaches
	Curagen	Use of proprietary technologies to characterize the genetic component and metabolic pathways of new products
FMC	DevGen	New molecule mechanisms and target sites; high-throughput chemical library screening
Monsanto	ArQule	Use of combinatorial chemistry approach to discover new molecules
Dow AgroSciences	Cambridge Discovery Chemistry	Use of combinatorial chemistry for new leads
Syngenta	Cambridge Discovery Chemistry	R&D collaboration at Zeneca's Richmond, CA, R&D facility to provide new leads through combinatorial chemical approaches
	Rosetta Inpharmatics	Use of bioinformatic approaches to optimize the development of new lead molecules
	Novalon Pharmaceutical	Use of BioKey library of >20 billion unique biopolymers to develop high-throughput screening assays

from product quantity to quality. If biological activity was observed, the source of the activity had to be sorted out from the mixture to target the active molecule.

The biological evaluation process has always given the agricultural chemical industry a great advantage over the pharmaceutical industry in that it was possible to screen directly on the target organisms from the first evaluation. Thus, in vivo screening has always been preferred over in vitro or target-based screening. However, HTS forced assays to be miniaturized so they could be run in 96- or 384-well microtiter plate formats. More recently, the mapping of genomes and the discovery of genes coding for control of key metabolic pathways in fungi, plants, insects, and nematodes has stimulated a high level of interest in in vitro screening. The ability to use such assays within cells or whole organisms creates the opportunity to test for a specific mode of action while retaining many of the in vivo screening benefits. Of course, neither combinational chemistry nor HTS would be possible without the introduction of highly automated robotic systems. The other driving force is management of information. Information technology systems are required at all stages of the process, from chemical library design, molecular synthesis, sample management, and HTS to data capture, storage, and analysis. By combining biological screening data with genomes, scientists are gaining additional insights into an understanding of key genes that enables a more rational means to approach selected screening targets. For promising in vivo leads, the determination of unknown modes of action can now be facilitated by profiling gene expression changes brought about by the applied chemical.

The new technological approaches being explored by the crop protection companies and their technology partners are outlined in Table 7. This is not meant to be an exhaustive or all-inclusive review of the company approaches, but it is an indication of the tremendous explosion of new ideas and efforts to more effectively discover new lead molecules. By increasing the screening rate 100-fold (from 10,000 to 1 million per year) over a period of 10 years, the industry has set into motion a new discovery approach that when coupled with the emerging knowledge of genomics will continue to revolutionize the industry for years to come.

8 NEW CHEMISTRIES

Although the use of chemicals to control plant pests can be traced back many centuries, the modern era of crop protection tools can be said to have begun in the 1940s or 1950s [5]. It is a difficult task to look specifically at agrichemicals and their market introduction dates without spending a lot of effort and duplicating many previous articles. One way to evaluate the impact of the discovery and market introduction of new crop protection tools is to look at the U.S. corn and soybean herbicides over the past five decades (Table 8). The U.S. herbicide market usually accounts for over 60% of all agrichemical sales and is a prime discovery target for the industry. The 1990s gave us almost 60% of all the corn herbi-

Table 8 U.S. Introductions of Corn and Soybean Herbicides

Decade	Corn	Soybean
1950s	1	0
1960s	5	5
1970s	4	4
1980s	3	13
1990s	18[a]	16[a]
Total	31	38

[a] Includes transgenic crop herbicide tolerance.

cides marketed over the five decades 1950–2000, while the soybean herbicides introduced in the 1990s accounted for almost 40% of all the soybean herbicides introduced since 1950. This demonstrates the tremendous impact of the discovery process described in the previous section on the development and marketing of new crop protection tools. Similar results were seen in the insecticide/miticide and fungicide areas comparing the 1990s to the 1950s. The major discoveries of new molecules for crop protection in the 1990s are listed in Table 9 for herbicides and plant growth regulators, Table 10 for insecticides, miticides, and nematicides, and Table 11 for fungicides. One new term that has entered the vocabulary of the industry more recently is reduced risk or safer materials. One overlying factor in the discovery and development efforts of registrants has been the increasing influence of regulatory pressures. Dr. Reed (see Chap. 4) gives the background and details of these regulatory issues. Briefly, FIFRA '88 made the industry take a hard look at reregistering products registered prior to 1984. Companies compared the costs of defending older products against the investment required to discover and develop new ones. Many companies decided to reduce their product defense expenditures, especially on minor crops, as noted by Guest and Schwartz in Chapter 7 of this volume. The new focus on molecules that could be used at lower use rates and with safer mammalian and environmental safety profiles was validated by the Food Quality Protection Act (FQPA) in 1996. As noted by Reed in Chapter 4, FQPA brought about new standards for food and environmental safety including an additional 10× safety factor for children along with aggregate (food plus water plus environmental) exposure and cumulative risk from products with the same modes of action. The Reduced Risk status classification enacted by the U.S. Environmental Protection Agency (USEPA) (PR Notice 97-3, 1997) spelled out a specific set of standards for new chemistries concerning mammalian, environmental, aquatic, wildlife, and avian toxicity and other parameters. Molecules passing this strict set of standards received a Reduced Risk classification and preferred regulatory treatment at the USEPA. Many companies made Re-

Table 9 New or Recently Registered Crop Protection Chemicals—Herbicides and Plant Growth Regulators

Compound[a]	Company	Trade name
Azafenidin (H)	DuPont	Milestone
Bispyribac-sodium[c] (H)	Valent	Regiment
Carfentrazone-ethyl[b,c] (H)	FMC	Aim, Shark, Affinity
Cloransulam[c] (H)	Dow AgroSciences	Firstrate
Diclosulam[c] (H)	Dow AgroSciences	Strongarm
Diflufenzopyr[b,c]	BASF	Distinct
Dimethenamid-p[b,c] (H)	BASF	Frontier
Flufenacet[c] (H)	Bayer	Axiom
Flumetsulam[c] (H)	Dow AgroSciences	Broadstrike
Flumiclorac[b,c] (H)	Valent	Resource
Flumioxazin[c] (H)	Valent	Valor
Fluthiacet-methyl[b] (H)	Syngenta	Action
Halosulfuron[c] (H)	Monsanto/Gowan	Permit, Sempra
Imazamox[b,c]	BASF	Raptor
Imazapic[b,c] (H)	BASF	Cadre
Isoxaflutole[c] (H)	Aventis	Balance
Mesotrione[b,c] (H)	Syngenta	Callisto
Oxadiargyl (H)	Aventis	Topstar
Oxasulfuron (H)	Syngenta	Expert, Dynam
Prohexadione-Ca[b,c] (PGR)	BASF	Apogee
Prosulfuron[c] (H)	Syngenta	Peak
Pyrithiobac sodium[c] (H)	DuPont	Staple
Quinclorac[c] (H)	BASF	Facet, Paramount
Rimsulfuron[b,c] (H)	DuPont	Matrix
Sulfentrazone[c] (H)	FMC	Authority
Sulfosulfuron[c] (H)	Monsanto	Maverick
Tepraloxydim[c] (H)	BASF	Equinox, Aramo
Thiazopyr[b,c] (H)	Dow AgroSciences	Visor
Triflusulfuron[c] (H)	DuPont	Upbeet
Tralkoxydim[b,c] (H)	Syngenta	Achieve
Trinexapec-ethyl[c] (PGR)	Syngenta	Palisade

[a] H = herbicide; PGR = plant growth regulator.
[b] In the USEPA Reduced Risk classification.
[c] Registered by the USEPA.
Source: Ref. 12.

TABLE 10 New or Recently Registered Crop Protection Chemicals—Insecticides, Miticides, and Nematicides

Compound[a]	Company	Trade name
Acetamiprid[b,c] (I)	Aventis	Assail, Adjust, Pristine
Bifenzate[b,c] (M)	Uniroyal	Floramite, Acramite
Bifenthrin[c] (I)	FMC	Brigade, Capture
Buprofezin[b,c] (I)	Aventis	Applaud
Clofentezine[c] (M)	Aventis	Apollo
Cyfluthrin[c] (I)	Bayer	Baythroid
Emamectin benzoate[c] (I)	Syngenta	Proclaim, Strategy
Etoxazole[b,c] (M)	Valent	Secure
Fenoxycarb[c]	Syngenta	Comply
Fenpropathrin[c] (I)	Valent	Danitol
Fipronil[b,c] (I)	Aventis	Regent
Fosthiazate (N)	ISK	Nemathorin
Halofenozide/RH-0345[b,c] (I)	Dow AgroSciences	Mach 2
Hexythiazox[c] (M)	Gowan	Savey
Imidacloprid[c] (I)	Bayer	Admire, Gaucho, Provado
Indoxacarb[b,c] (I)	DuPont	Steward, Avaunt
Lambda-cyhalothrin[c] (I)	Syngenta	Karate, Warrior
Lufenuron (I)	Syngenta	Match
Methoxyfenozide[b,c] (I)	Dow AgroSciences	Intrepid
Milbemectin[b] (M)	Gowan/Sankyo	Milbeknock
Novaluron[b] (I)	Makteshim-Agan	Rimon
Pymetrozine[b,c] (I)	Syngenta	Fulfill
Pyridaben[c] (I,M)	BASF	Pyramite
Pyriproxyfen[b,c] (I)	Valent	Knack, Distance, Esteem
Spinosad[b,c] (I)	Dow AgroSciences	Success, Spintor
Tebufenozide[b,c] (I)	Dow AgroSciences	Confirm, Mimic
Tebupirimphos[c] (I)	Bayer	Aztec (w/cyfluthrin)
Tefluthrin[c] (I)	Syngenta	Force
Thiamethoxam (I)	Syngenta	Actara, Cruiser, Adage
Triazamate (I)	Dow AgroSciences	Aphistar

[a] I = insecticide; M = miticide; N = nematicide.
[b] In the USEPA Reduced Risk classification.
[c] Registered by the USEPA.
Source: Ref. 12.

TABLE 11 New or Recently Registered Crop Protection Chemicals—Fungicides

Compound[a]	Company	Trade name
Acibenzolar[a,b]	Syngenta	Actigard
Azoxystrobin[a,b]	Syngenta	Abound, Quadras, Heritage
Cymoxanil[b]	DuPont	Curzate
Cyproconazole[b]	Syngenta	Alto
Cyprodinil[a,b]	Syngenta	Vangard, Switch (w/ fludioxonil)
Difenoconazole[b]	Syngenta	Dividend
Dimethomorph[b]	BASF	Acrobat
Famoxadone	DuPont	Famoxate, Charisma, Equation
Fenamidone	Aventis	Reason
Fenbuconazole[b]	Dow AgroSciences	Indar, Enable
Fenhexamid[a,b]	Arvesta	Elevate
Fluazinam[a,b]	Syngenta	Omega
Fludioxonil[a,b]	Syngenta	Maxim, Scholar, Switch (w/ cyprodinil)
Flutolanil[b]	Gowan	Moncut
Harpin protein[a,b]	Eden Biosciences	Messenger
Kresoxim-methyl[a,b]	BASF	Sovran, Cygnus
Myclobutanil[b]	Dow AgroSciences	Rally, Nova
Picoxystrobin	Syngenta	Acanto
Propamocarb-HCl[b]	Aventis	Tattoo
Propiconazole[b]	Syngenta	Tilt, Orbit
Pyraclostrobin	BASF	Headline, Cabrio
Pyrimethanil	Aventis	Scala
Quinoxyfen	Dow AgroSciences	Arius, Quintec
Spiroxamine	Bayer	Proper, Impulse, Hogger
Tebuconazole[b]	Bayer	Folicur, Elite, Raxil
Trifloxystrobin[a,b]	Bayer	Flint, Twist
Triflumizole[b]	Uniroyal	Procure, Terraguard
Trifluzamide	Dow AgroSciences	RHO753[c]
Zoxamide[a,b]	Dow AgroSciences	Gavel

[a] In the USEPA Reduced Risk classification.
[b] Registered by the USEPA.
[c] Trade name not available.
Source: Ref. 12.

duced Risk chemistries a primary discovery and development goal, as can be noted by the number of products with this classification in each table. Discovery and development of many of these chemistries were made possible by many of the technologies noted in the previous section, especially combinational chemistry and high-throughput screening, with more targeted assay systems.

However, not all new discoveries came from the traditional synthesis route. Two notable exceptions were Spinosad from Dow AgroSciences and the strobilurin fungicides from Syngenta (Zeneca) and BASF. Spinosad is a naturally occurring mixture of two active components, 85% spinosyn A and 15% spinosyn B, produced by a bacterium originally isolated from a soil sample taken from a Jamaican rum distillery. These compounds are macrolides with a unique tetracyclic ring system with different attached sugars. Spinosad is produced by fermentation and has been widely adapted in crop protection systems because of its high efficacy on target insects and safety to beneficial species making it ideal for use in integrated pest management systems. Its unique mode of action has resulted in a lack of cross-resistance problems.

The origin of azoxystrobin can be traced back to a family of fungicidal natural products that are derivatives of β-methoxyacrylic acid—the strobilurins, audemansins, and myxothiazols [6]. The strobilurins are formed in several genera of small fungi that typically grow high on beech trees. The fungicidal activity of these natural products relies on their ability to inhibit fungal mitochondrial electron transport at a specific cytochrome b binding site. Because no commercial fungicides had that specific mode of action, the class of strobilurin chemistry made excellent synthetic targets because, unlike Spinosad, the natural products were unsatisfactory themselves for agricultural purposes due to unsuitable physical properties, insufficient activity, and fermentation scale-up costs. Several companies, noted previously, initiated analog synthesis programs around this chemistry that resulted in kresoxim-methyl, azoxystrobin, picoxystrobin, and trifloxystrobin.

These examples are meant to highlight the diversity in the discovery process that has led to the explosion of new chemistries with unique modes of action as outlined in Tables 9–11. With all of the powerful combinational chemistry, high-throughput screening, genomic, information technology, and robotic tools available to industry chemists, biologists (traditional and molecular), and biochemists, we expect this new discovery momentum to continue as we enter the twenty-first century. However, a few factors could slow this trend in new-molecule discovery. The increasing consolidation of companies noted previously is placing pressures on company management to hold research costs level or decrease them slightly. Because the biotechnology programs in many companies compete for the same research dollars as the traditional chemical discovery programs, the traditional approaches are being squeezed as the newer approaches gain momentum. Time will tell whether efficiencies in new compound synthesis and screening will offset funding decreases. However, if the past is any indicator, the future holds many exciting new developments that will certainly surprise and possibly astound us when they occur and when we take a backward look in another 25 years.

9 PLANT BIOTECHNOLOGY

Perhaps no technology in twentieth century agriculture has raised more hopes and elicited more concern than plant biotechnology. The application of emerging molecular biological discoveries to agriculture had its humble beginnings at companies like Monsanto in the 1970s. This low-key, long-range research effort intensified in the 1980s, with many skeptics believing that no practical good would result for the tens and hundreds of millions of dollars invested in this technology. However, those doubts disappeared rapidly in the 1990s as the input benefits of herbicide-tolerant and insect-resistant major row crop plants became evident. As mentioned earlier, the technology adoption rate for Roundup Ready soybeans (i.e., soybean plants tolerant to the nonselective herbicide glyphosate) was twice as rapid as the adoption rate for hybrid corn, and the new seed accounted for over half of the U.S. crop in 1999 (Table 12). By 1999, over one-fourth of the U.S. corn crop and half the U.S. cotton crop were genetically modified varieties. Stacked genes (i.e., plants with two traits such as Roundup Ready and Bt insect-tolerant) were beginning to make significant market inroads, especially in cotton.

On a global basis, the adoption rate of this technology has been quite variable (Fig. 5). The United States (72%), Argentina (17%), and Canada (10%) accounted for 99% of all the global acreage planted to genetically modified soybeans (54%), corn (28%), cotton (9%), and canola (9%). Herbicide-tolerant crops accounted for 71% of the planted acres, and insect-resistant plants with the Bt

TABLE 12 Impact of Plant Biotechnology on U.S. Crop Acreage

Technology	Year introduced	Millions of acres planted			
		1996	1997	1998	1999
BXN cotton	1995	0.15	0.17	0.8	1.1
Roundup Ready soybeans	1996	1.4	9	27	35
Bt corn	1996	1.0	6	15	21
Bt cotton	1996	1.8	2.5	2.4	2.4
Roundup Ready cotton	1997	0	0.43	2.3	3.4
Roundup Ready and Bt cotton	1997	0	0.07	0.48	1.6
Liberty Link corn	1997	0	0.7	2	5
Roundup Ready corn	1998	0	0	0.9	2.3

Source: *Agrow* (various issues) and USDA acreage reports.

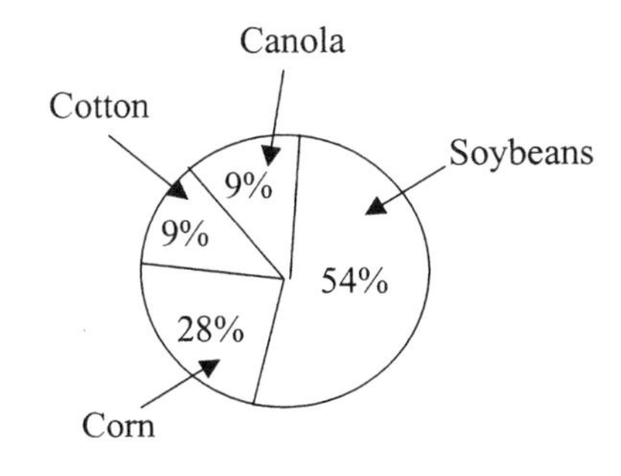

By Crop

	MM Acres
Soybeans	53.2
Corn	27.6
Cotton	8.9
Canola	8.9
Total	98.6

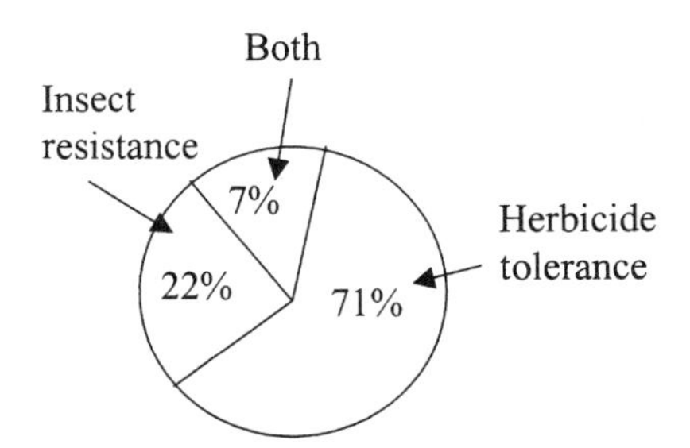

By General Trait

	MM Acres
Herbicide tolerance	70.0
Insect resistance	21.7
Both	6.9
Total	98.6

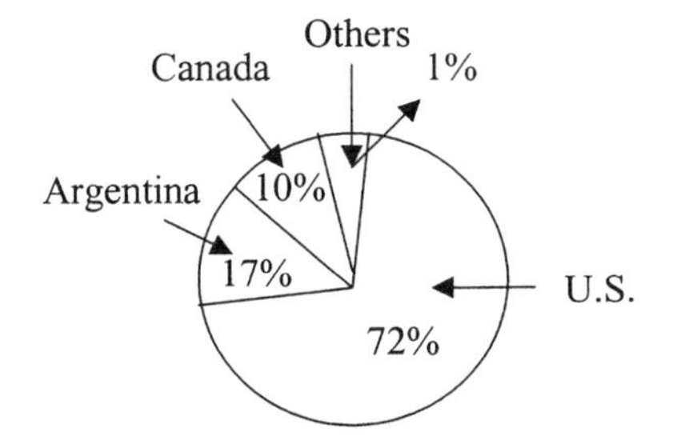

By Country

	MM Acres
U.S.	71.0
Argentina	16.8
Canada	9.8
Others*	1.0
Total	98.6

FIGURE 5 Impact of plant biotechnology on global crop acreages. (*) "Others" includes China, Australia, South Africa, Mexico, Spain, Portugal, France, Romania, and Ukraine. (Data from *Chemical Engineering News*, Nov 1, 1999, and *Agrow*, Nov 12, 1999.)

gene accounted for 22%; the remaining 6% were planted to crops with both traits. However, the technology has not been as warmly received in Europe, where only 10 plant biotechnology products had passed European regulatory review in 1999 compared to over 60 in the United States. Japan appeared to be more positive and had approved 20 plant biotechnology products.

In Europe, consumer concerns resulted in a backlash against foods derived from plant biotechnology and crops, particularly in the United Kingdom. This has led to some European countries refusing to import grain grown from genetically modified crops and reportedly cost U.S. corn growers over $200 million in lost export sales in 1999. In the United States, the Food and Drug Administration, the Department of Agriculture, and the USEPA have regulatory policies in place to approve plant biotechnology and resulting food products and have deemed them safe for animal and human consumption. The EU concerns have forced U.S. farmers to reconsider planting intentions for this new technology. Many believe that U.S. acreage, especially in corn and soybeans, planted to transgenic crops will plateau and possibly decline by 5–10% in the early years of the twenty-first century until the issues with Europe and other parts of the world are resolved. However, in a 2000 report entitled "World Agricultural Biotechnology: GMO," the Freedonia Group, a market research firm based in Cleveland, OH, projects that transgenic crops will increase in acreage by 13% annually (worldwide) and will occupy 155 million acres by 2004 [8]. The countries leading the next stage are expected to be Brazil and China. It is predicted that future growth in plant biotechnology products will come from three sources: (1) stacked seed carrying multiple traits, (2) novel feed and cereal grain crops such as wheat and hay that will become popular in emerging markets that rely on grains for food and livestock feed, and (3) development of crops containing beneficial value-added output traits that appeal directly to consumers such as vitamin-enriched rice or cholesterol-lowering corn.

Perhaps the last point is the crux of the problem with consumers in some parts of the world. The industry has focused its entry into this market on input traits that are of value to the farmer but not the consumer (Table 13). Although

TABLE 13 Impact of Biotechnology on Input Traits of Plants

Present
Herbicide tolerance: BXN cotton; Roundup Ready soybeans, corn, cotton, and canola; Liberty Link corn, rice, sugar beets, canola, and soybeans; SR corn
Insect resistance: Bt corn, cotton, potatoes, sweet corn
Other: Virus-resistant potatoes, squash, and papaya
Future
Herbicide tolerance: Acuron gene for tolerance to protoporphyrinogen oxidase (PPO) herbicides (CGA 276,854)—many crops; Isoxaflutole tolerance in row crops
Insect resistance: Corn resistant to corn rootworm, cotton resistant to boll weevil, etc.
Disease resistance: Resistance to foliar diseases, reduced aflatoxin production, etc.

a few crops have been commercialized with output traits such as altered tomato ripening (i.e., longer shelf life), they were not commercial successes or recognized by the consumers as being beneficial to them (Table 14). High oil corn was planted (over 3 million acres in 1999) by growers for its high livestock feed value, but again, consumers did not see any benefit accruing to them. However, grains enriched with vitamins or with lower saturated fat content will have health benefits for consumers and over time should help win public acceptance for these foods with new or altered nutritional properties. In the meantime, the crop protection industry is investing heavily in new plant biotechnology approaches in partnership with independent biotechnology companies, many of which also actively work with the pharmaceutical industry to develop new input and output traits (Table 15). The next big input trait due for 2003 market launch is resistance to corn rootworm, the most serious insect pest for U.S. corn, which is the second largest insecticide market (cotton being the largest). This technology could dramatically reduce the use of soil insecticides (many of which are organophosphates and carbamates) as occurred with foliar insecticides with Bt cotton. The National Cotton Council estimated in 1999 that cotton farmers used 84,000 fewer gallons of insecticides on Bt cotton than in 1998, with a net benefit to cotton producers of $92 million [9]. Similar benefits were quantified for soybean growers in a report that estimated a net annual saving of $220 million ($380 million gross

TABLE 14 Impact of Biotechnology on Output Traits of Plants

Present
Altered tomato ripening for improved shelf life
High solids tomatoes
High amylose and waxy corn
High lauric acid canola
High oil corn
Future
Corn with high lysine and methionine and low phytate
Soybeans high in lysine, methionine, and lauric and stearic acids
Rice high in vitamin A
Soybeans low in linoleic acid and saturated fats
Corn with low nitrogen requirements, improved pH tolerance, and modified starch
Soybeans with resistance to cyst nematodes, viruses, and fungi
Cotton with improved or altered fiber characteristics and colored fiber
Canola with high laurate, myristate, and oleic and euricic acid contents and low linoleic acid content
Potatoes with high solids, low sugars, and reduced browning for improved processing

Table 15 New Approaches to Crop Protectant Discovery

Company	Partner[a]	Technology
Aventis	Plant Tec (acquisition)	Carbohydrate metabolism—enhance starch content and quality
	Biogemma (JV)	CSIRO's Plex Gene technology to control DNA expression and develop new plant traits
	Lynx	Genetic mapping/DNA analysis technologies to develop new crop varieties
	Agritope (JV Agronomics)	Novel genes through ACTTAG gene approach developed by Salk Institute—both input and output traits
	NetGenics	Bioinformatics software for crop research
	Vilmorin Clause and MAH Plant Genomic Fund (JV)	Identify/discover genes for vegetable crop improvement, especially bacterial pathogens, nematode and virus resistance, and drought tolerance
BASF	Metanomics (JV)	Plant genotype analysis
	Sun Gene (JV)	Identify plant genes for stress-resistant crops and plants with improved quality traits
	S Valöf Weibull	Herbicide-tolerant canola; modified starch quality in potato; disease and insect resistance; specialty quality traits
	Freiburg University	Improved quality traits
	HYSEQ	High-throughput genomics and bioinformatics technologies to develop plant-produced pharmaceuticals and nutraceuticals
Bayer	Lion BioScience	Use of high-throughput sequencing, genome mapping, and analysis of plant–fungus interactions to control corn smut
Dow AgroSciences	Interlink	Identification of novel genes for pest and disease resistance
	Aventis (Alliance)	Development of a range of transgenic crops expressing herbicide tolerance (glyphosate, BXN, and isoxazoles), insect resistance, and agronomic quality traits—initial focus on cotton and sugar beets

DuPont	Lynx Therapeutics	Manipulation and analysis of large libraries of DNA molecules to isolate genes for new crop varieties and crop protection products
	John Innes Centre and Sainsbury Lab with Zeneca	Use of advanced genomics techniques to develop new wheat products with improved agronomic, industrial, and food uses
	Maxygen	Genomic approaches for crop protection and grain quality improvement
Monsanto	IBM Technology Alliance	Identification and mapping of genomes of major plant diseases
	Genzyme Molecular Oncology	Serial analysis gene expression for plant gene libraries
	Maxygen	Use of genomic approaches to develop new crop protection and quality traits
	Paradigm Genetics	Use of gene discovery and function approaches to develop new products
	Millennium Pharmaceuticals	Gene mapping technology for agricultural applications
	Incyte Pharmaceuticals	High-throughput gene sequencing and biotransformation technologies
	Gene Trace	Genotyping methods for determining genetic composition of individual species
Syngenta	Genzyme Molecular Oncology	Serial analysis of gene expression for plant growth and disease applications
	Incyte Pharmaceuticals	Genomic approaches with wheat, corn, and rice for new varieties
	Diversa	Identification and optimization of genes and pathogens to develop transgenic crops with improved pest resistance and quality traits

[a] JV = joint venture.

savings minus the $160 million paid for technology fees) to U.S. soybean growers who used Roundup Ready soybeans. In addition, there was a reduction of herbicides applied on 16 million acres [10].

Beyond the current input traits and projected input and output traits lies the potential to use transgenic plants as factories for the production of specialty chemicals such as pharmaceuticals. There is growing belief that high value chemicals produced by fermentation can be more efficiently produced by plants. With increased knowledge of the plant genome and more sophisticated gene transformation and expression systems in plants, the future for plant biotechnology may be bright indeed!

10 MAJOR TREND SUMMARY

The twentieth century brought about dramatic increases in food production, especially in the United States, Europe, and Asia, with a significant decrease in the number of farmers. The agrichemical industry played a key role in this technology revolution along with hybrid seeds, sophisticated mechanization, and the availability of improved fertilizer and fertility practices. The crop protection industry has undergone what may be a complete cycle from a chemical-related industry that has had extensive associations with the oil and pharmaceutical industries back to mainly a chemically oriented industry. Mergers and acquisitions have concentrated the industry into fewer and larger companies that can afford the extensive research and development costs (usually 10% of sales) to discover and bring new products to market. The discovery ante has been raised significantly (100-fold from 10,000 molecules screened per year to 1 million per year) at major companies due to the twin innovations of combinatorial chemistry and robotic high-throughput screening. Generic producers have staked out low-cost positions on off-patent products and partnered with major manufacturers in order to remain competitive in the largest markets. Perhaps the most surprising trend has been the association of the industry with the seed business, which was once thought to be as uncomplementary and nonsynergistic as the fertilizer business. Of course, advances in plant biotechnology over the 1980s and 1990s made clear that the combination of the technologies had great synergism, especially for input traits like herbicide tolerance and insect resistance. The new chemistries being brought to the market as we start the twenty-first century are much safer to humans and the environment than products introduced in the 1940s and 1950s, although they require significantly greater management sophistication due to their directed target site modes of action and integration into plant biotechnology production systems.

11 THE FUTURE

It is always risky to try to forecast developments in any area of expertise, especially one like agriculture in general and crop protection chemicals and plant biotechnology specifically. However, major trends are evident. Consolidation of the industry will continue and may end only when five or ten major companies have the majority of the global agrichemical business. This clearly appears to be the case for the United States and Europe. However, it is unclear whether this trend will spread to Japanese companies, which, with a few notable exceptions, have resisted the pressure to consolidate—both internally and with their European and U.S. competitors. The industry will continue to deal with off-patent products by merging with low-cost producers and/or spinning off separate companies to market generic products, especially in third world countries where high new product costs are not economically viable in local production situations.

It will be interesting to see whether the current disassociation of the crop protection business units from their pharmaceutical parents will continue, but the current trend is clearly in that direction. Long-term profit margins for agrichemicals will be below those for pharmaceuticals and above those for bulk and specialty chemicals, making them less attractive for drug company shareholders. The alignment of the agrichemical industry with seed companies will likely continue after a period of consolidation to digest the extensive activity that occurred during the 1990s. Seed is the delivery system for both plant biotechnology traits and new, highly active molecules that can provide systemic insect and disease protection in seedlings and young plants by the use of coated seed treatments. The current input traits will expand to disease tolerance and resistance along with tolerance to adverse environmental parameters such as drought, salt, cold, and heat. These traits will benefit mainly farmers, who will be able to increase yields and productivity to feed the growing global population on poorer soil and under more marginal environmental conditions. Consumer attitude, especially in Europe, toward genetically enhanced crops will likely change as more nutritionally enhanced food products reach the market and are appreciated by consumers for what they can do to help fight disease, extend life expectancy, and improve overall food nutritional quality. In addition, the reduced risk chemistries now in the pipeline will be used extensively to produce nutritious food with minimal impact on the environment. These new products will eventually displace the products currently in the market for managing severe pest outbreaks and will be used in integrated pest management strategies to increase plant resistance to pests.

The future of the crop protection industry, with its integration of seed and plant biotechnology integration, has never been brighter.

REFERENCES

1. AM Thayer. Ag biotech food: Risky or risk free? Chem and Eng News, Nov 1, 1999, pp 11–19.
2. B Kantz. Global move to post-patent. Farm Chem Int, Spring 1998, pp 12–13.
3. A Beer. Glyphosate—Still growing after all these years. Agrow, No. 324 (Mar 12, 1999), pp 21–23.

4. J Ormond. Changes in the approach to new chemical discovery. Agrow Supp, Autumn 1999, pp 37–38.
5. RE Holm. A quantum leap. Farm Chem Spec Millennium Issue, WOW 2000 America, pp 14–16.
6. JM Clough, CRA Godfrey, JR Godwin, RSI Joseph, C Spinks. Azoxystrobin, A novel broad spectrum systemic fungicide. Pesticide Outlook, August 1996, pp 16–20.
7. Anon. Global GM crop plantings up 44% in 1999. Agrow, No. 340 (Nov 12, 1999), p 22.
8. Anon. Global Demand for GMO's to Grow 13% Annually, According to New Report. Pestic Toxic Chem News, Apr 13, 2000, pp 1 and 12.
9. PC Burnett. An obvious cotton benefit. Cotton Farming, April 2000, p 68.
10. LP Gianessi, JE Carpenter. Agricultural Biotechnology: Benefits of Transgenic Soybeans. Washington, D.C.: National Center for Food and Agricultural Policy, April 2000, p 103.
11. D Botts. Living with FQPA—Involving stakeholders in an evolving process. Presented at workshop held on Mar 2, 2000, in Tampa, Florida by the Florida Fertilizer and Agricultural Association and the Florida Fruit and Vegetable Association.
12. JJ Baron. IR-4 New Products/Transition Solution List—March 2001. IR-4 Newsletter, Spring 2001, Vol. 32, No. 1, pp 1–22.

Index